SOTHIS PRESS

Sothis Press aims to completely reimagine the public domain book with an emphasis on universal access to captivatingly designed print editions. We believe that technology will enable every single book ever published to be perpetually accessible in beautiful print form. Sothis is beauty in ubiquity.

www.sothis.press
1 Waterhouse Square, London, ECIN 2ST

Sheet-Metal Work

A Manual of Practical Self-Instruction in the Art of Pattern Drafting and Construction Work in Light-and Heavy-Gauge Metal, Including Skylights, Roofing, Cornice Work, Etc

William Neubecker

SOTHIS PRESS

ISBN 978-1-0368-0664-4

This edition is published by
Sothis Press © 2023. The company
is registered in England and Wales.
The Company Number is 14900389.

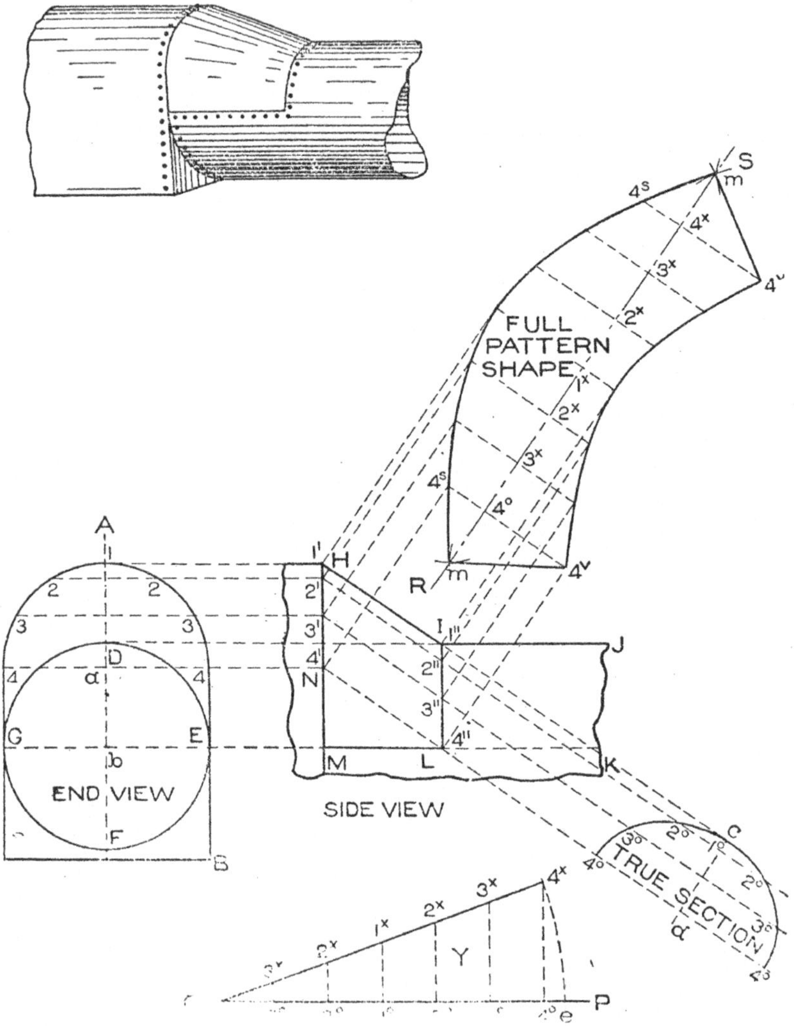

PATTERN FOR GUSSET SHEET OF BOILER, SHOWING DEVELOPMENT.

Sheet-Metal Work

A Manual of

PRACTICAL SELF-INSTRUCTION IN THE ART OF PATTERN DRAFTING AND CONSTRUCTION WORK IN LIGHT- AND HEAVY-GAUGE METAL, INCLUDING SKYLIGHTS AND ROOFING, CORNICE WORK, ETC.

By WILLIAM NEUBECKER
Instructor, Sheet-Metal Department
New York Trade School

ILLUSTRATED

CHICAGO
AMERICAN SCHOOL OF CORRESPONDENCE
1909

Foreword

IN recent years, such marvelous advances have been made in the engineering and scientific fields, and so rapid has been the evolution of mechanical and constructive processes and methods, that a distinct need has been created for a series of *practical working guides*, of convenient size and low cost, embodying the accumulated results of experience and the most approved modern practice along a great variety of lines. To fill this acknowledged need, is the special purpose of the series of handbooks to which this volume belongs.

¶ In the preparation of this series, it has been the aim of the publishers to lay special stress on the *practical* side of each subject, as distinguished from mere theoretical or academic discussion. Each volume is written by a well-known expert of acknowledged authority in his special line, and is based on a most careful study of practical needs and up-to-date methods as developed under the conditions of actual practice in the field, the shop, the mill, the power house, the drafting room, the engine room, etc.

¶ These volumes are especially adapted for purposes of self-instruction and home study. The utmost care has been used to bring the treatment of each subject within the range of the common understanding, so that the work will appeal not only to the technically trained expert, but also to the beginner and the self-taught practical man who wishes to keep abreast of modern progress. The language is simple and clear; heavy technical terms and the formulæ of the higher mathematics have been avoided, yet without sacrificing any of the requirements of practical instruction; the arrangement of matter is such as to carry the reader along by easy steps to complete mastery of each subject; frequent examples for practice are given, to enable the reader to test his knowledge and make it a permanent possession; and the illustrations are selected with the greatest care to supplement and make clear the references in the text.

¶ The method adopted in the preparation of these volumes is that which the American School of Correspondence has developed and employed so successfully for many years. It is not an experiment, but has stood the severest of all tests—that of practical use—which has demonstrated it to be the best method yet devised for the education of the busy working man.

¶ For purposes of ready reference and timely information when needed, it is believed that this series of handbooks will be found to meet every requirement.

Table of Contents

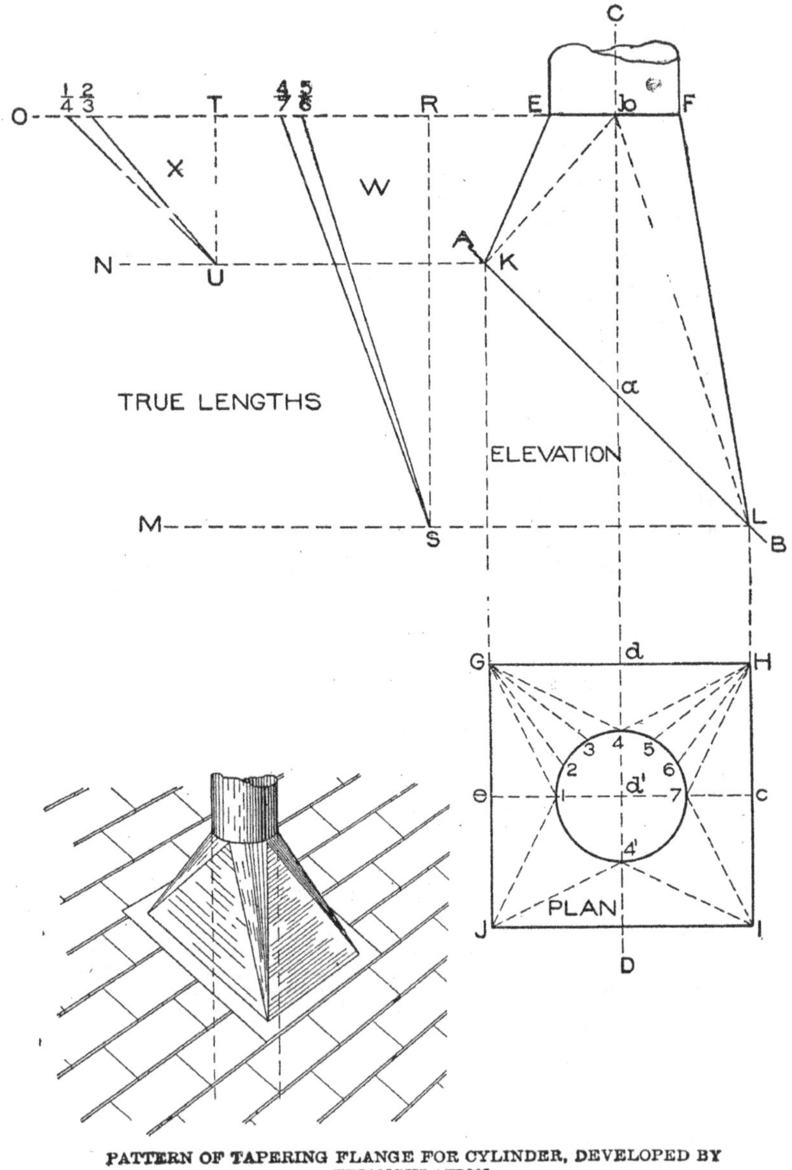

PATTERN OF TAPERING FLANGE FOR CYLINDER, DEVELOPED BY TRIANGULATION.

SHEET-METAL WORK.

PART I.

The sheet-metal worker of today who wishes to succeed must know far more than was necessary years ago. There are many good, practical sheet-metal workers in the trade who are handicapped because they are unable to lay out the patterns that arise in their daily work. Notwithstanding the introduction of labor-saving machinery, the demand for good workmen has increased. While most sheet-metal workers acquire practical knowledge in the shop, they lack the technical education necessary to enable them to become proficient as pattern cutters and draftsmen. In this course, special attention is given to the fundamental principles that underlie the art and science of pattern drafting.

Practical workshop problems will be presented, such as arise in everyday practice, thus giving the student the practical experience that usually comes only after long association with the trade.

CONSTRUCTION.

In constructing the various articles made from sheet metal, various gauges or thicknesses of metal are used. For all gauges from No. 20 to No. 30 inclusive, we assume in the development of the pattern, that we are dealing with no thickness, and we make no allowance for bending or rolling in the machine. But where the metal is of heavier gauge than No. 20, allowance must be made for shrinkage of the metal in the bending and rolling operations, which will be explained in connection with development in heavy sheet-metal work. What has been said about wiring, seaming, and transferring patterns in the Tinsmith's Course is applicable to this course also. It is sometimes the case that the capacity of a vessel or article must be determined, when the rules given in Mensuration should be followed. When figuring on sheet-metal work, the specifications sometimes call for various metals, such as galvanized sheet iron or steel, planished iron, heavy boiler plate, band iron, square or round rods for bracing, etc., zinc, copper, or brass; and the weight of the metal must often be calculated together with that of stiffening rods, braces, etc. On this account it is necessary to have tables which can be consulted for the various weights.

TABLES.

There is a wide difference between gauges in use, which is very annoying to those who use sheet metal rolled by different firms according to the various gauges adopted. It would be well to do away with gauge numbers, and use the micrometer caliper shown in Fig. 1, which determines the thickness of the metal by the decimal or fractional parts of an inch.

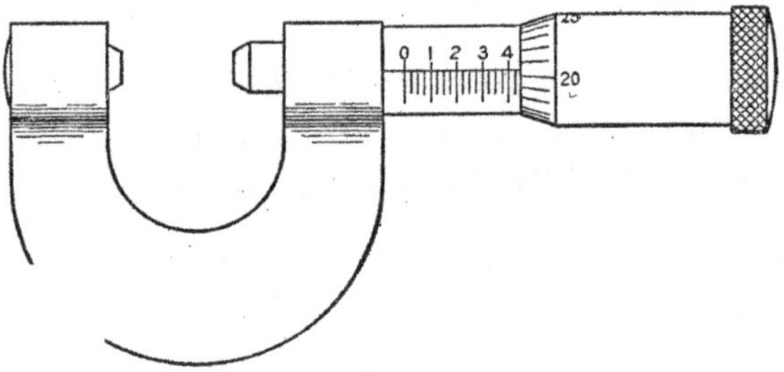

Fig. 1.

This is the most satisfactory method for the average mechanic who works sheet metal manufactured by firms using different gauges. The tables on pages 61 to 74 can be consulted when occasion arises.

SHOP TOOLS.

In allowing edges for seaming and wiring, we must bear in mind that when a seam is to be grooved by hand or machine the allowance to be made to the pattern should conform to the rolls in the machine or the hand tools in use. The edges of the pattern are usually bent on the sheet-iron folder, or brake, while the seam can be seamed or grooved with the hand groover or giant grooving machine. Where round pipe work is done in lengths up to 3 feet, the slip roll former is used, while square or rectangular pipes are bent up on the brake in 8-foot lengths. Where pipes, elbows, stove bodies, furnace shells, metal drums, etc., are made, the sheets are cut square on the large squaring shears, rolled, grooved, and stiffened, by beading both ends in the beading machine, using ogee rolls. There is also a special machine for seaming the cross seams in furnace pipes, also a set of machines for the manufacture of elbows used in sheet-metal work. As before mentioned, if these machines are at hand, it will be well to make slight modifications in the patterns so that both the machines and patterns may work to advantage.

PATTERNS OBTAINED BY VARIOUS METHODS.

In this course will be explained the four methods used in developing patterns for sheet-metal work, namely, parallel line, radial line, triangulation, and approximate developments. What was said on parallel and radial line developments in the Tinsmith's Course is applicable to this course also.

INTERSECTIONS AND DEVELOPMENTS.

The following problems on parallel line developments have been selected because they have a particular bearing on pipe work arising in the sheet-metal trade. All of the problems that will follow should be carefully studied, drawn on cheap paper, and proven by cardboard models. These models will at once show any error in the patterns which might otherwise be overlooked. As only the Examination Plates are to be sent to the School, the student should draw all the other plates given in this course.

The first problem to be drawn is shown in Fig. 2, being the intersection between a cylinder and octagonal prism. In drawing these problems for practice, make the cylinder and octagonal prism both 2 inches in diameter. The height of the cylinder from B to E should be $4\frac{1}{2}$ inches; and the length of the prism from G to H, 3 inches. Let A represent the plan of the cylinder, shown in elevation by B C D E; and F, the section of the prism, shown in plan by G H I J. Number the corners of the section F as shown, from 1 to 4 on both sides; and from these points draw horizontal lines intersecting the plan of the cylinder at $2'3'$ and $1'4'$ on both sides as shown. Establish a convenient intermediate point of intersection between the corners of the prism, as *a* and *a* in A, from which draw horizontal lines intersecting the section F at a', a', a', and a'. Take a tracing of the section F with its various intersections, and place it in its proper position as shown by F^1, in the

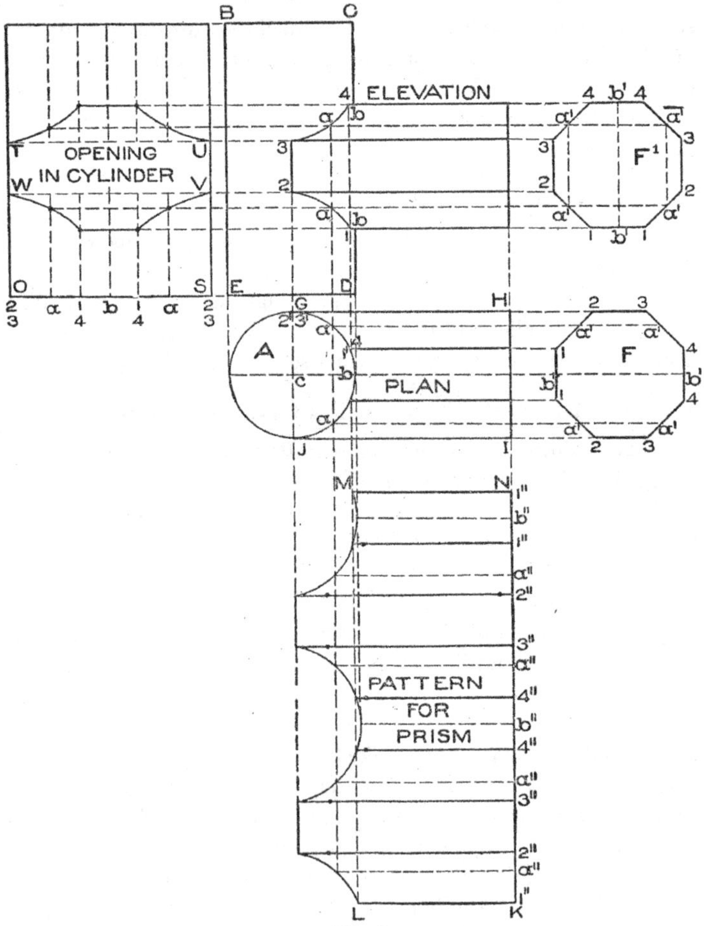

Fig. 2.

center of the cylinder B C D E, allowing the section to make a quarter turn, and bringing the points b' b' at the top and bottom on a vertical line, while in the section F, b' b' are on a horizontal line. From the various intersections in F^1, draw horizontal lines intersecting vertical lines drawn from similarly numbered intersections in the plan A, as shown in elevation. A line drawn through these points will represent the joint between the cylinder and prism.

For the development for the prism, extend the line H I in plan as N K, upon which place the stretchout of all the points contained in the section F, as shown by similar figures and letters on N K. Through these points, at right angles to N K, draw lines which intersect with lines drawn from similarly numbered points and letters in plan, at right angles to J I. Trace a line through points thus obtained, and K L M N will be the desired pattern. To obtain the development for the opening in the cylinder, extend the line D E in elevation as S O, upon which place the stretchout of all the points contained in the half-circle A, as shown by similar numbers and letters on S O. At right angles to S O and through these points, draw lines intersecting horizontal lines drawn from intersections having similar numbers and letters in elevation, thus obtaining the intersections shown by T U V W, which will be the shape of the opening to be cut into one-half of the cylinder.

In Fig. 3 is shown the intersection between a hexagonal and quadrangular prism, the hexagonal prism being placed in elevation at an angle of 45° to the base line. When drawing this problem for practice, make the height of the quadrangular prism 4½ inches, and each of its sides 2 inches. Place the hexagonal prism at an angle of 45° to the base line, placing it in the center of the quadrangular prism in elevation as shown; and inscribe the hexagonal section in a circle whose diameter is 2½ inches. Let A represent the plan of the quadrangular prism placed diagonally as shown, above which draw the elevation B C D E. In its proper position and proper angle, draw the outline of the hexagonal prism as shown by 1^v 1″ 4″ 4^v; and on 1″ 4″ draw the half section as shown by F, numbering the corners 1″ 2″ 3″ and 4″. From the corner 1′ in the plan A, draw the center line 1′ 4. Take a tracing of the half section F, and place it as shown by F^1, placing the points 1″ 4″ in F on the center line in F^1 as shown. From the corners 1, 2, 3, and 4, draw lines parallel to the center line, intersecting the two sides of A (*b* 1′ and 1′ *a*) at 2′ 3′ and 1′ 4′, as shown. From these intersections draw vertical lines, which intersect by lines drawn parallel to $4''$ 4^v from corners having similar numbers in F, thus obtaining the points of intersection 1^v 2^v 3^v and 4^v. Dropping vertical lines from the intersections on the plane $1''$ $4''$ in elevation, and intersecting similarly numbered lines in plan, will give the horizontal section of $1''$ $4''$, as shown by 1° 2° 3° and 4°.

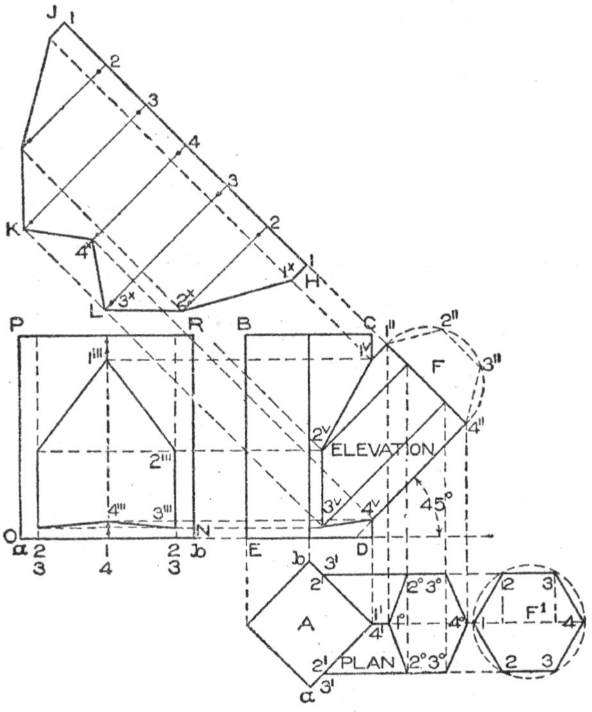

Fig. 3.

For the development of the hexagonal prism, extend the line $4''$ $1''$ as shown by H J, upon which place the stretchout of twice the number of spaces contained in the half section F, as shown by similar figures on the stretchout line H J. From these points, at right angles to H J, draw lines as shown, which intersect by lines drawn at right angles to the line of the prism from intersections 1^v to 4^v, thus obtaining the points of intersection 1^x to 4^x. Lines traced from point to point as shown by J K L H, will be the required development. The shape of the opening to be cut into the quadrangular prism, is obtained by extending the line D E in elevation as N O, upon which place the stretchout of one-half the section A, with the various points of intersection, as shown by similar figures on O N. At right angles to O N erect lines from these points, which intersect by lines drawn from similarly numbered intersections in elevation at right angles to the quadrangular prism, thus obtaining the points of intersection 1‴ to 4‴ on both sides. Then N O P R will be the half development.

Fig. 4 shows the intersection between two cylinders of equal diameters at right angles. Make the height of the vertical cylinder 3 inches, that of the horizontal cylinder 1¼ inches, and the diameters of both 2 inches. Let A represent the plan of the vertical cylinder, and B its elevation. Draw the plan of the horizontal cylinder C, shown in elevation by D placed in the center of the vertical cylinder. Draw the half section E in plan and divide it into equal parts, as shown from 1 to 3 to 1. In a similar manner draw the half section E^1 in elevation, which also divide into the same number of spaces as E, reversing the numbers as shown.

The following suggestions are given to avoid confusion in numbering the points or corners of irregular or round sections in plan and elevation. If the half section E were bent on the line 1–1 and turned upward toward the reader, and we should view this section from the front, the point 3 would be at the top, or, if bent downward, would be at the bottom; therefore the points 3 and 3 in elevation are placed at top and bottom. Now if the section E^1 in elevation were bent on the line 3-3 either toward or away from the reader, the point 1 when looking down would show on both sides as shown in plan, which proves both operations. No matter whether the form is simple, as here shown, or complicated as that which will follow, the student should use his imaginative power. Study the problem well; close your eyes and imagine you see the finished article before you, or, failing in this, make a rough model in the shop or a cardboard model at home, which will be of service. Now from the intersections in E, draw horizontal lines intersecting the circle A at 1′, 2′ and 3′ on both sides. From these points erect perpendicular lines and intersect them with horizontal lines drawn

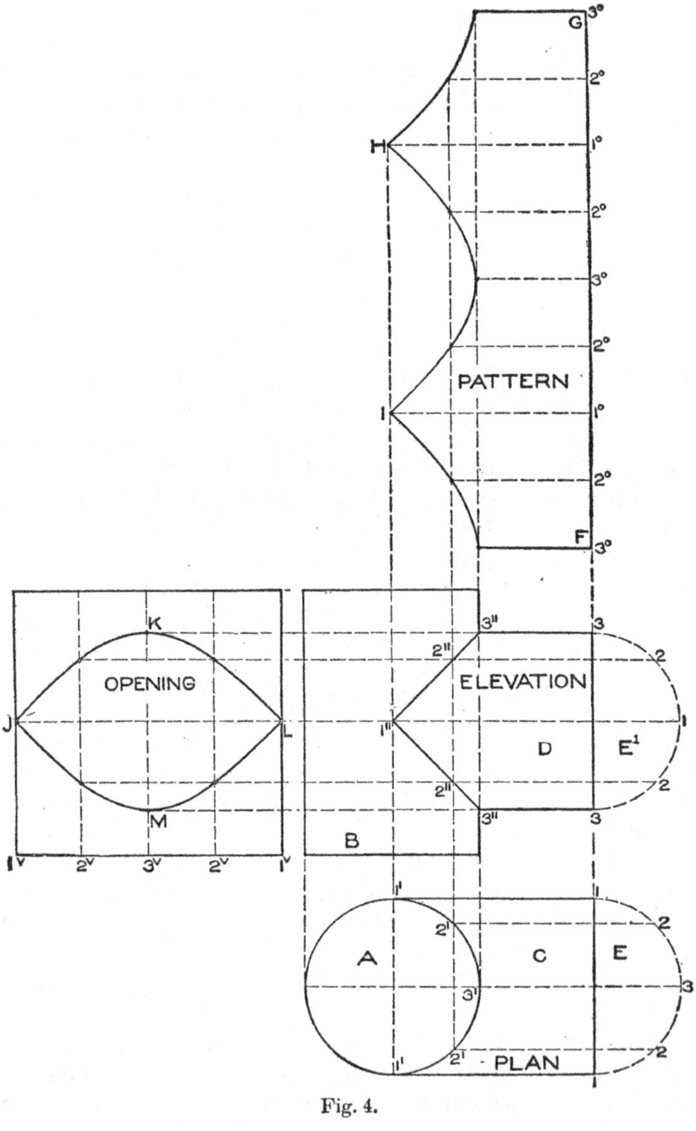

Fig. 4.

from similarly numbered intersections in E^1. Lines traced through these points 3″ 2″ 1″ and 1″ 2″ 3″ will be straight because both branches are of equal diameters.

For the development of the cylinder D in elevation, extend the line 3-3 as shown by F G, upon which place the stretchout of twice the number of spaces contained in E^1, as shown by similar numbers 3° to 1° to 3° to 1° to 3° on the stretchout line F G. From these points, at right angles to G F, draw lines, and intersect them by lines drawn parallel to the cylinder B from similar numbers in the joint line. Trace a line through these points in the development, when F G H I will be the desired shape.

For the opening to be cut into the cylinder B to receive the cylinder D, extend the base of the cylinder B as shown by 1^v 1^v, upon which place the stretchout of the half circle A in plan, as shown by similar figures on the stretchout line 1^v 1^v. From these points erect perpendiculars, which intersect by lines drawn from similarly numbered intersections in elevation at right angles to the line of the cylinder B. Trace a line through the intersections thus obtained; J K L M will be the shape of the opening.

Fig. 5 shows the intersection of two cylinders of unequal diameters at an angle of 45°. Make the diameters of the large and small cylinders 2 inches and 1¼ inches respectively; the height of the large cylinder 3 inches; and the length of the small cylinder measured from its shortest side in elevation, 1 inch, placed at an angle of 45° in the center of the cylinder B. A represents the plan of the large cylinder struck from the center *a* and shown in elevation by B. Draw the outline of the small cylinder C at its proper angle, and place the half section D in its position as shown; divide it into a number of equal spaces, as shown from points 1 to 5. Through the center *a* in plan. draw the horizontal line *a* 5; and with *b* as a center describe a duplicate of the half section D with the various points of intersection, as shown by D^1, placing the points 1 and 5 on the horizontal line *a* 5. From the intersections in D^1 draw horizontal lines intersecting the large circle A at 3′ to 3′ as shown, from which points erect perpendicular lines; intersect them by lines drawn parallel to the lines of the smaller pipe from similarly numbered intersections in D. A line traced through the points thus obtained will represent the intersection or miter joint between the two pipes.

These same principles are applicable no matter what diameters the pipes have, or at what angle they are joined, or whether the

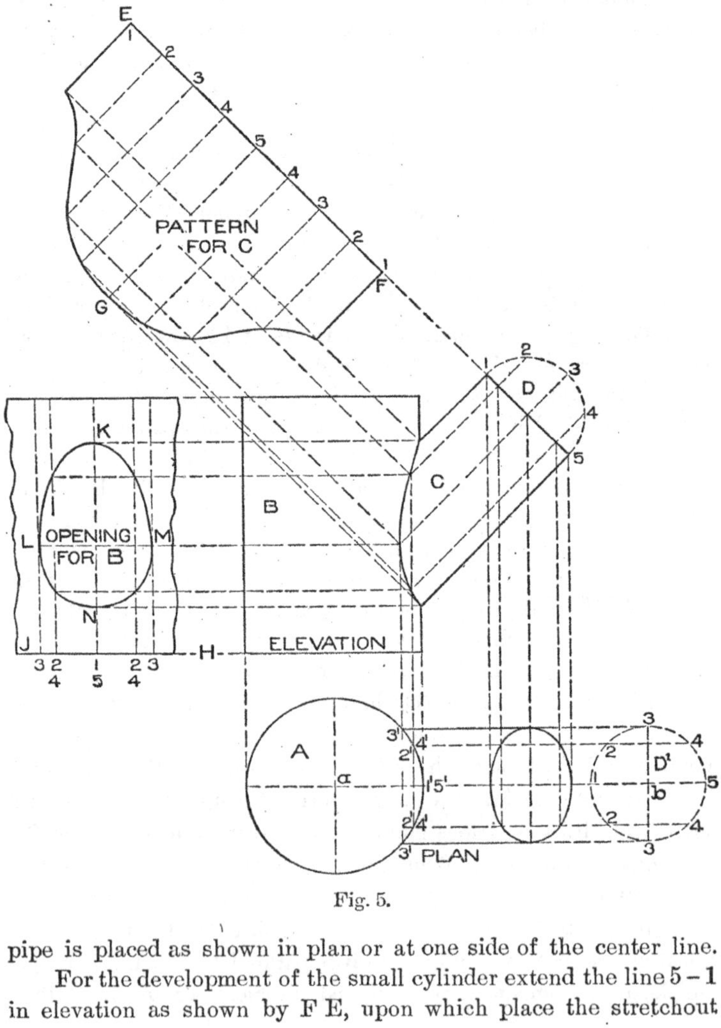

Fig. 5.

pipe is placed as shown in plan or at one side of the center line.

For the development of the small cylinder extend the line 5 – 1 in elevation as shown by F E, upon which place the stretchout of the circle D^1 in plan, or twice the amount of D in elevation, as shown by similar figures on the stretchout line F E. At right angles to F E and through these small figures, draw lines which intersect with lines drawn at right angles to the lines of the small cylinder from similarly numbered intersections in the miter line in elevation. Trace a line through the points thus obtained; E F G will be the development for the cylinder C.

To obtain the opening in the large cylinder extend the lines of the large cylinder in elevation as shown at the base by H J, upon which place the stretchout of the intersections contained in the circle A, being careful to transfer each space separately (as they are unequal) to the stretchout line H J. Through these points and at right angles to H J erect lines which intersect with horizontal lines drawn from similar points in the miter line in elevation A line traced through the points thus obtained, as shown by K L M N, will be the desired development.

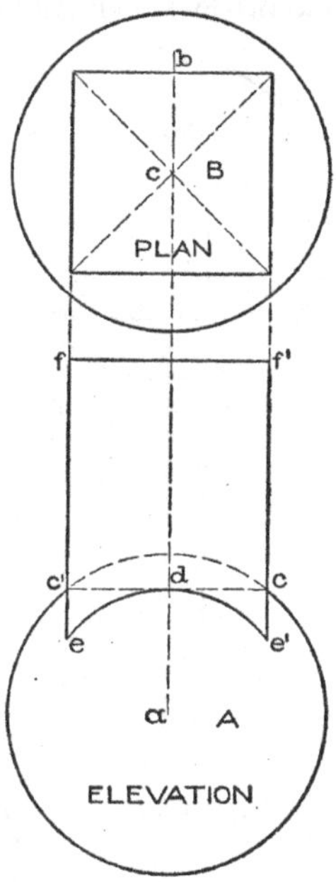

Fig. 6.

Fig. 6 shows the intersection between a quadrangular prism and sphere, the center of the prism to come directly over the center of the sphere. Make the diameter of the sphere $2\frac{1}{2}$ inches, the sides of the prism $1\frac{1}{2}$ inches, and the height from f to c' $2\frac{3}{8}$ inches. Draw the elevation of the sphere A which is struck from the center a, from which erect the perpendicular $a\ b$. With any point, as c, as a center and using the same radius as that used for A, describe the plan B. Through c draw the two diagonals at an angle of 45°, and draw the plan of the prism according to the measurements given. Now draw the elevation of the prism $f\ c'$ and $f'\ c$, the sides of the prism intersecting the sphere at c and c'. From either of these points draw a horizontal line intersecting the center line $a\ b$ at d. Then using a as a center and $a\ d$ as the radius, describe the arc $e\ e'$ intersecting the sides of the prism extended at e and e'; $f\ e\ e'\ f'$

will be the development for one of the sides of the prism. In practice the four sides are joined in one.

Fig. 7 shows the intersection of a quadrangular prism and sphere when the center of the prism is placed to one side of the center of the sphere. Make the diameter of the sphere the same as in the preceding figure; through *x* in the plan draw the 45° diagonal, and make the distance from *x* to A ½ inch, the sides of the prism 1 inch, and the height from E to *c* in elevation 1½ inches. Having drawn the elevation and plan of the sphere, construct the plan of the prism as shown by A B C D. Parallel to the center line *x y* project the prism in elevation intersecting the sphere at *a* and *c*. Now since the center of the sphere is on one of the diagonals of the prism in plan, either two of the sides meeting at one end of that diagonal, as B C and C D, will be alike, and both will be different from the other two sides A B and A D, meeting at the opposite end of the diagonal. Therefore the line F *a* in elevation will be used in obtaining the development of D C in plan, while the line E *c* will be used in obtaining the development for the two sides D A and A B in plan.

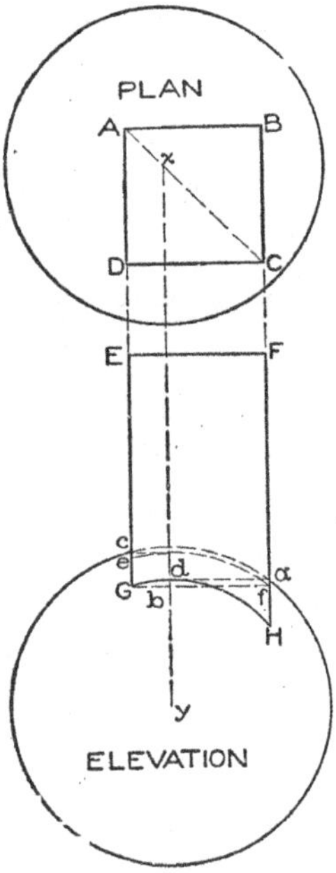

Fig. 7.

Now from *a* draw a horizontal line intersecting the center line *x y* at *b;* and using *y* as a center and *y b* as the radius, describe the arc G H intersecting the sides of the prism extended to G and H. Then E F G H is the development for each side of the prism shown in plan by D C and C B. In a similar manner, from the intersection *c* in elevation draw a horizontal line intersecting the center line *x y* at *d*. Then using *y* as center and *y d* as radius, describe an arc intersecting the sides of the prism at *e* and *f* E F *f e* will show the development for either side of the prism shown in plan by D A and A B. By connecting the points G and *f* it will be found that the line is a true horizontal line, which proves

ELBOW PATTERNS *

In all elbow work the difficulty lies in obtaining the correct rise of the miter line. By the use of a protractor this is overcome and thus the necessity of drawing a complete quadrant is avoided. Following the rule given in the illustration the rise can be easily found, when the throat and diameter of the pipe is known.

In the upper table are shown various pieced elbows, having different degrees when finished, and the various miter lines. There are six miter patterns shown, the first for a 6-pieced elbow having 90° when completed; the second for a 4-pieced 90° elbow; the third for a 3-pieced 90° elbow; the fourth for a 2-pieced 70° elbow; the fifth for a 2-pieced 90° elbow, and the sixth for a 2-pieced 105° elbow.

No matter what size of throat the elbow may have, or what diameter or number of pieces, always follow the rule given in the illustration and obtain the miter line; then place the half profile in its proper position and place the full girth of the pipe on the line shown in the pattern by similar numbers. By reversing the cut opposite the line 1–7–1 the pattern for the middle pieces is obtained, after which one cut can be placed into the other as shown on Page 48 Sheet Metal Work, Part I.

---

* The illustration referred to will be found on the back of this page.

PATTERNS FOR VARIOUS PIECED ELBOWS OF SIMILAR THROAT AND DIAMETER.

NO. OF PIECES	DIVIDE BY	DEGREE OF ELBOW	RISE OF MITER LINE
2	2	105	$52\frac{1}{2}°$
2	2	90	45°
2	2	70	35°.
3	4	90	$22\frac{1}{2}°$
4	6	90	15°
6	10	90	9°

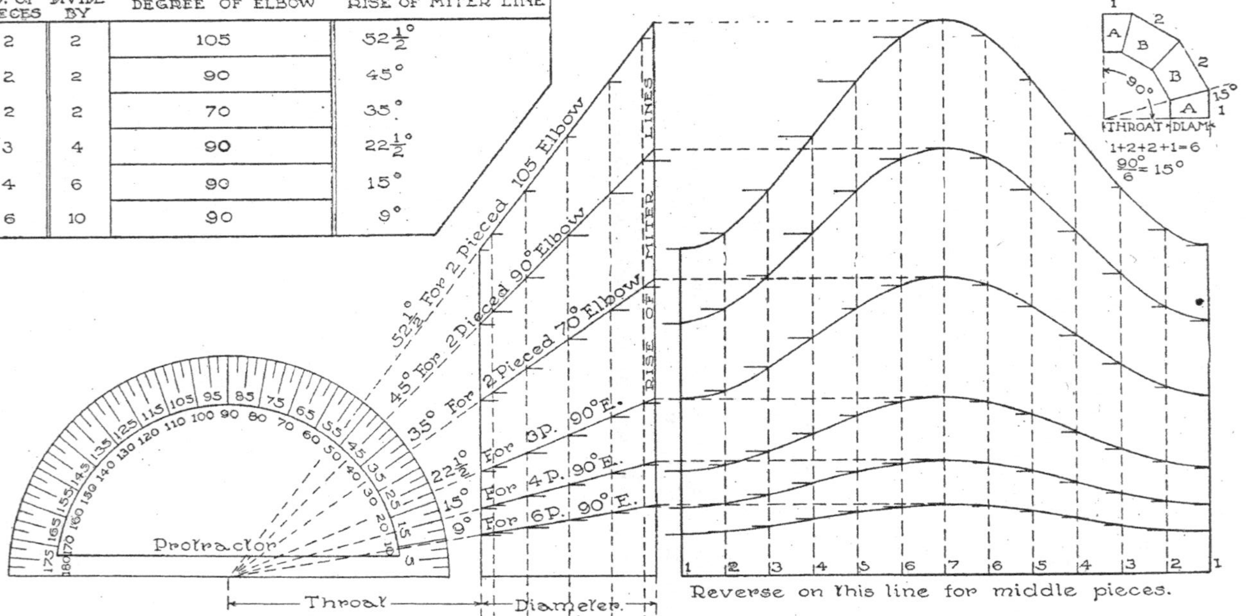

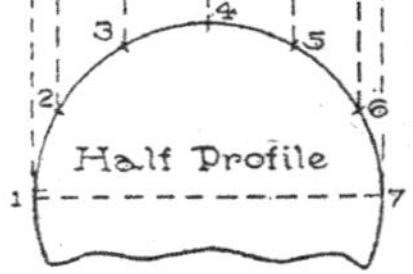

Rule: In all elbow work, no matter what degree the elbow may have, the end-piece "A" counts one, and each middle piece "B" counts two; thus in a four pieced elbow divide by six.

Note: Use above method for any pieced elbow of any given angle whose throat and diameters are similar.

the two developments. Should the plan of the prism be so placed on the sphere that all sides would be different, then two elevations would be necessary so that the intersections of all the sides could be shown.

Developments by Triangulation. In developing sheet-metal work of irregular forms, patterns are required which cannot be developed by either the parallel or radial-line methods. These irregular shapes are so formed that although straight lines can be drawn upon them the lines would not run parallel to one another, nor would they all incline to a common center. In the methods previously described, the lines in parallel developments run parallel to one another, while in radial-line developments all the lines meet at a common center. Hence in the development of any irregular article, it becomes necessary to drop all previous methods, and simply proceed to measure up the surface of the irregular form, part by part, and then add one to another until the entire surface is developed. To accomplish this, one of the simplest of all geometrical problems is made use of and shown in Part II of Mechanical Drawing, Plate V, Problem 11, entitled "To construct a triangle having given the three sides." To carry out this method it is necessary only to divide the surface of the plan or elevation of any irregular article into a number of equal parts. Use the distances in plan as the bases of the triangles, and the distances in elevation as the altitudes or heights of the triangles, or *vice versa;* and then find the hypothenuse by connecting the two given lengths.

To illustrate this simple principle Fig. 8 has been prepared. Let A B C D represent the plan of a plane surface, shown in elevation by A^1 B^1. We know that the true length of the plane is equal to A^1 B^1 and the true width is equal to A D or B C in plan. We also know that the vertical height from the bottom of the plane A^1 to the top B^1 is equal to B^1 *b* as shown. But suppose we want to obtain the true length of the diagonal line B D in plan on the developed plane. To obtain this it will be necessary only to take the length of B D, place it from *b* to D^1, and draw a line as shown from B^1 to D^1, which is the length desired.

While this may look very simple, it is all that there is to triangulation, and if the student thoroughly understands the simple principle and studies the problems which will follow, he will have no trouble in applying this principle in complicated work. To make it still clearer we will prove the length of the line B^1 D^1. Take the distance of A^1 B^1, place it in plan as shown by A B^2, and complete the rectangle A B^2 C^2 D. Draw the diagonal B^2 D, being the length sought, which will be found to equal B^1 D^1 in elevation. When drawing this problem in practice, make the plan 4 by 6 inches and the vertical height in elevation 5 inches.

In obtaining developments by triangulation, the student should use all of his conceptive powers as previously explained. Before making any drawing, he must see the article before him in his mind's eye, so to speak, before he can put it down on paper. Therefore we want to impress upon the student the necessity of drawing all the problems that will follow in this part and in the Practical Workshop Problems. It should be understood that triangulation is not given as an alternative method, but is used when no other method can be employed, and without it no true pattern could be obtained for these irregular shapes; hence the necessity of close study.

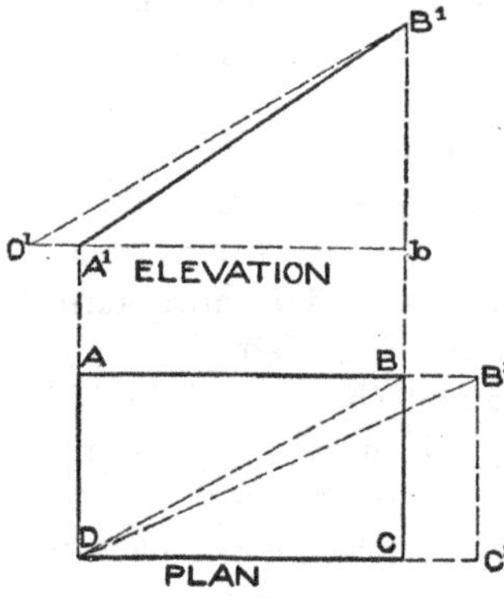

Fig. 8.

In Fig. 9 is shown an irregular solid whose base and top are triangles crossing each other, and in which the principle just explained will be put to practical test Inscribe the triangles shown in plan in a circle whose radius is equal to a 1, or $1\frac{1}{2}$ inches, and make the height of the article in elevation 2 inches. The dotted triangles 1 2 3 in plan represent the section of the article on the line 2–3 in elevation; and the solid triangle 1^1 2^1 3^1 in plan, the section on the line 2^1 3^1 in elevation. Now connect the two sections in plan by drawing lines from 1 to 2^1 and to 3^1, from 2 to 2^1 and to 1^1, and from 3 to 1^1 and to 3^1. In a similar manner connect the points in elevation as shown. It now becomes necessary to obtain a triangle giving the true length of the lines connecting the corners of the triangle in plan, and as all of these lines are equal only one triangle is necessary. Therefore take the distance from

1 to 2^1 in plan and place it on the line 3–2 extended in elevation, as shown from 2 to 1°, and draw a line from 1° to 2^1, which is the desired length.

For the pattern, proceed as is shown in Fig. 10. Take the distance of any one of the sides in the triangle, as 1–2 in Fig. 9, and place it on the horizontal line 1–2 in Fig. 10. Then using 1 and 2 as centers, with 1° 2^1 in elevation in Fig. 9 as radius, describe the arcs in Fig. 10 intersecting each other in 2^1. Then 1 2 2^1 will be the pattern for one of the sides shown in plan in Fig. 9 by 1 2 2^1. Proceed in this manner in Fig. 10 as shown by the small arcs; or a tracing may be taken of the one side 1 2 2^1, and traced as shown until six sides are obtained, which will be the full pattern and which is numbered to correspond to the numbers in plan.

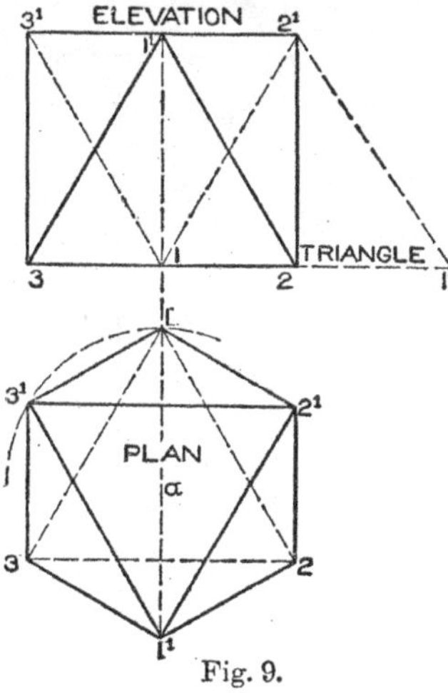

Fig. 9.

In Figs. 11, 12, and 13 are shown the methods used in developing a scalene cone. The method of obtaining the development of any scalene cone, even though its base is a perfect circle, is governed by the same principle as employed in the last problem on triangulation It is well to remember that any section of a scalene cone drawn parallel to its base will have the same shape (differing of course in size) as the base. This is equally true of articles whose

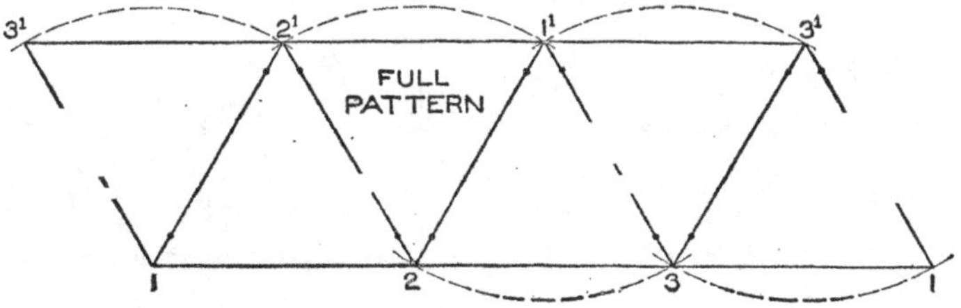

Fig. 10.

bases are in the shape of a square, rectangle, hexagon, octagon, or any other polygon. What has just been explained will be proven in connection with Fig. 11, in which A B C represents a side elevation of a scalene cone, whose plan is shown by 1 4^1 7 4 C^1. Draw any horizontal line, as A D, on which set off the distances

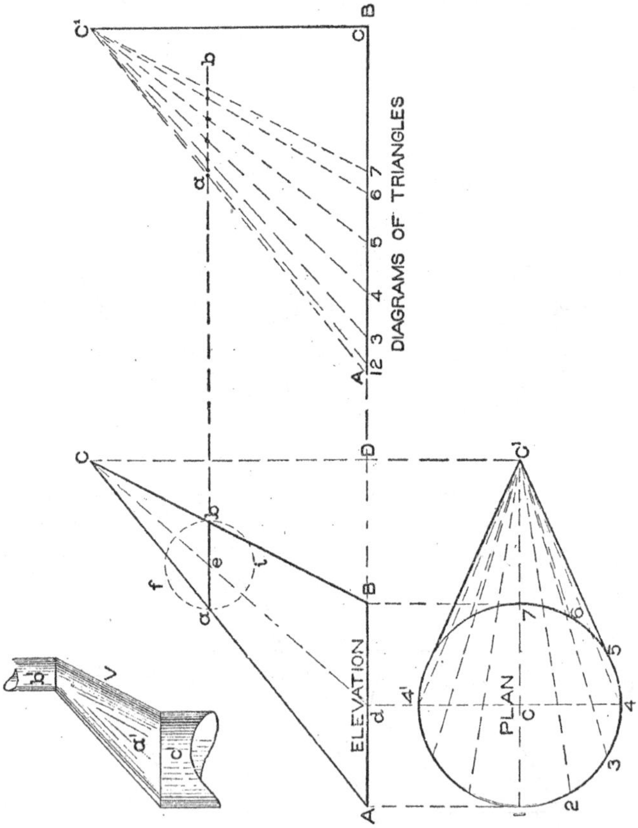

Fig. 12.

Fig. 11.

A B equal to 3 inches and B D equal to $2\frac{1}{2}$ inches, and the vertical height D C equal to $4\frac{1}{2}$ inches. Draw lines from B and A to C, which completes the elevation. In its proper position below the line A B, draw the plan of A B as 1 4 7 4^1 struck from the center C. Through C draw the horizontal line C C^1, and intersect it by a vertical line drawn from the apex C in elevation, thus obtaining the apex C^1 in plan. Draw lines from 4 and 4^1 to C^1, which completes the plan.

As both halves of the scalene cone are symmetrical, it is necessary only to divide the half plan 1 4 7 into a number of equal spaces as shown by the small figures 1 to 7, and from points thus obtained draw radial lines to the apex C^1. Then these lines in plan will represent the bases of triangles which will be constructed, whose altitudes are all equal to D C in elevation. Therefore in Fig. 12 draw any horizontal line, as A B, and from any point, as C, erect the perpendicular line C C^1 equal in height to D C in Fig. 11. Now from C^1 in plan take the various lengths of the lines 1 to 7 and place them on the line A B in Fig. 12, measuring in every instance from the point C, thus obtaining the intersections 1 to 7, from which lines are drawn to the apex C^1. Then these lines will represent the true lengths of similarly numbered lines in plan in Fig. 11.

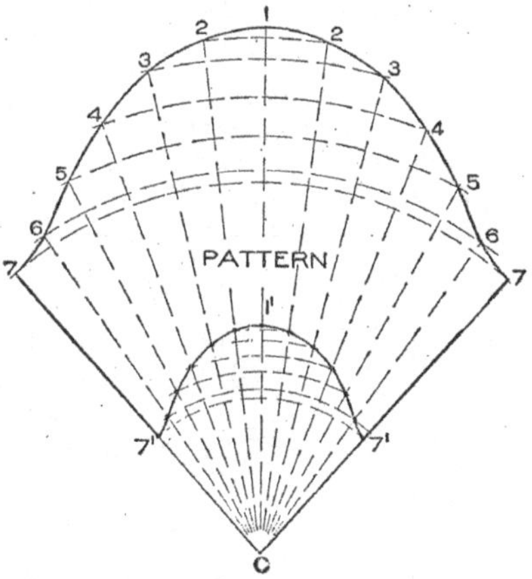

Fig. 13.

For the pattern proceed as is shown in Fig. 13. With C as center and radii equal to C^1 7, 6, 5, 4, etc., in Fig. 12, describe the arcs 7–7, 6–6, 5–5, 4–4, etc., in Fig. 13 as shown. Now assuming that the seam is to come on the short side of the cone, as C B in Fig. 11, set the dividers equal to one of the equal spaces in the plan; and starting on the arc 7–7 in Fig. 13, step from arc 7 to arc 6, to arcs 5, 4, 3, 2, and 1, and then continue to arcs 2, 3, etc., up to 7. Trace a line through these intersections as shown by 7-1-7, and draw lines from 7 and 7 to C, which completes the pattern.

Now to prove that any section of an oblique or scalene cone cut parallel to its base, has a similar shape to its base (differing in size), draw any line as a b in Fig. 11 parallel to A B. From C in plan erect a vertical line intersecting the base line A B at *d*, from which draw a line to the apex C, cutting the line *a b* at *e*. Then the distances *e a* and *e b* will be equal; and using *e* as a center and *e b* as radius, describe the circle *a f b i*, which is the true section on *a b*. Then *a b* B A will be the frustum of a scalene cone. Extend the line *a b* parallel to A D, cutting the diagram of triangles in Fig. 12 from *a* to *b*. Then with radii equal to the distances from C^1 to the various intersections on the line *a b*, and using C in Fig. 13 as center, intersect similarly numbered radial lines drawn from 7 to 1 to 7 to the apex C. A line traced as shown from 7′ to 1′ to 7′ will be the desired cut, and 7–7–7′–7′ will be the pattern for the frustum. The practical use of this method is shown in diagram V in Fig. 11; *a*′ is the frustum of the oblique cone, on the ends of which are connected round pipes *b*′ and *c*′.

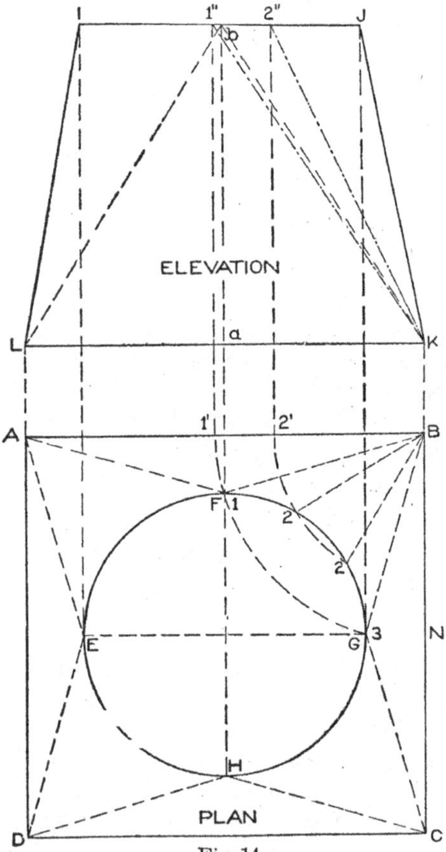

Fig. 14.

It is shown in Fig. 14 how in an irregular solid whose base is square and top is round, both top and bottom on horizontal planes are developed. The corners in plan F B G, G C H, H D E and E A F should be considered as sections of scalene cones. Proceed by drawing the plan A B C D $3\frac{1}{2}$ inches square, which represents the plan of the base of the article; and the circle E F G H $2\frac{1}{2}$ inches in diameter, which shows the plan of the top of the article; the vertical height to be 3 inches, shown from *a* to *b*. As the circle is in the center of the square, making the four corners symmetrical, it is necessary only to divide the one-quarter circle into a number of equal parts as shown by the small figures 1, 2, 2, 3, from which draw lines to the apex B. Complete the elevation as shown by I J K L. Now using B as center, and radii equal to B 1 and B 2 in plan, describe arcs intersecting A B at 1′ and 2′ as shown. From these points erect perpendiculars intersecting the top of the article I J

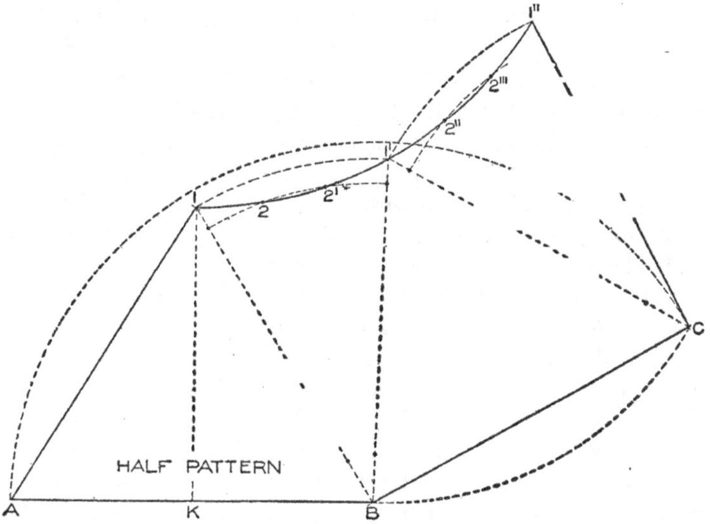

Fig. 15.

in elevation at 1″ and 2″, from which draw lines to K. Then K 1″ and K 2″ will be the true lengths of the lines shown in plan by B 1 and B 2 respectively on the finished article.

For the half pattern proceed as follows: In Fig. 15 draw any horizontal line, as A B, equal in length to A B in plan in Fig. 14. Now with K 1″ as radius and A and B in Fig. 15 as centers, describe arcs intersecting each other at 1 From 1 drop a vertical line intersecting A B at K. Then 1 K should equal J K in elevation in Fig. 14, which represents the true length through G N in plan.

Now with radii equal to K 1″ and K 2″ in elevation, and with B in Fig. 15 as center, describe the arcs 1–1′ and 2–2′. Now set the dividers equal to one of the spaces in G F in plan in Fig. 14; and starting at 1 in Fig. 15, step off arcs having similar numbers as shown by 1, 2, 2′, 1′. Now using 1 B as radius, and 1′ as center, describe the arc B C, and intersect it by an arc struck from B as center and with B A as radius, as shown at C. Take a tracing of 1 B 1′ and place it as shown by 1′ C 1″. Now connect the various intersections by drawing lines from 1 to A to B to C to 1″ to 1′ to 1, which completes the half pattern. The triangular pieces 1 A B or 1′ B C will represent the flat sides of the article shown in plan by 1 A B or 3 B C respectively in Fig. 14; and the cone patterns 1–1′ B and 1′–1″ C in Fig. 15, the sections of the scalene cones 1–3–B and H–G–C respectively in plan in Fig. 14. This same rule is applicable whether the top opening of the article is placed exactly in the center of the base or at one side or corner. Various problems of this nature will arise in Practical Workshop Problems; and if the principles of this last problem are thoroughly understood, these will be easily mastered.

Approximate Developments. In developing the blanks or patterns for sheet-metal work which requires that the metal be hammered or raised by hand, or passed between male and female dies in foot or power presses, circular rolls, or hammering machines, the blanks or patterns are developed by the approximate method, because no accurate pattern can be obtained. In all raised or pressed work in sheet metal, more depends upon the skill that the workman has with the hammer, than on the patterns, which are but approximate at their best. While this is true, it is equally true that if the workman understands the scientific rule for obtaining these approximate patterns a vast amount of time and labor can be saved in bringing the metal to its proper profile. If the true rule for averaging the various shapes and profiles in circular work is not understood, the result is that the blank has either too little or too great a flare and will not form to its proper profile and curve. Before proceeding to describe the approximate development methods, attention is called to the governing principle underlying all such operations. We have previously shown how the patterns are developed for simple flaring ware; in other words, how to develop the frustum of a cone. The patterns for curved or any other form of circular or hammered work are produced upon the same principle. The first illustration of that principle is shown in Fig. 16, in which A B C D represents a sphere 3 inches in diameter composed of six horizontal sections, struck from the center *a*.

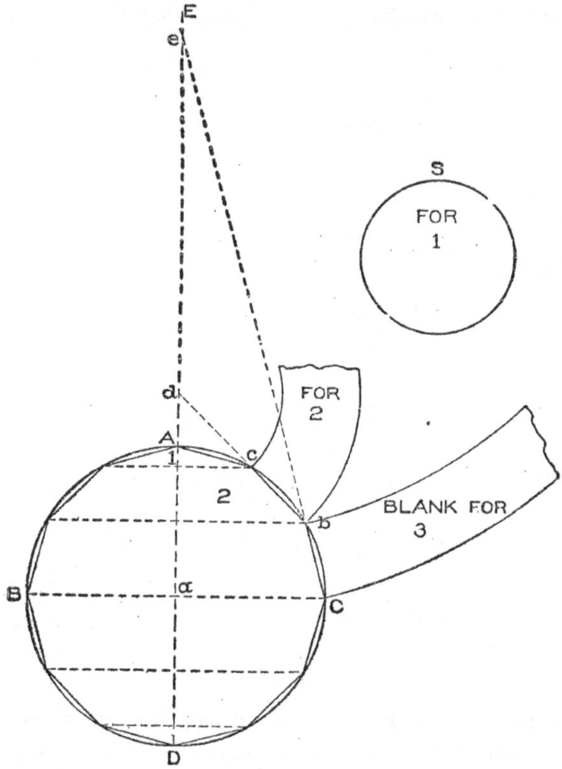

Fig. 16.

Divide the quarter circle A C into as many parts as there are sections required in the half sphere (in this case three), and draw horizontal lines through the ball as shown. The various radii for the patterns are then obtained by drawing lines through C *b*, *b c*, and *c* A. Thus C *b* extended meets the center line E D at *e*, which is the center for striking the blank for number 3, using the radii *e b* and *e* C. In similar manner draw a line from *b* to *c*, extending it until it meets E D at *d*. Then *d c* and *d b* will be the radii for blank number 2, while A *c* is the radius for blank 1 shown at S. The lengths of the pattern pieces are determined in the same manner as would be the case with an ordinary flaring pan in producing the patterns for tin ware, and will be explained

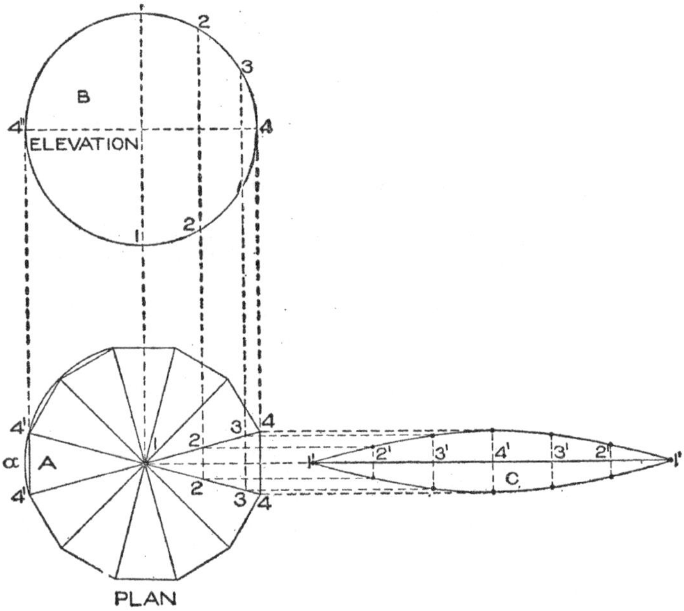

Fig. 17.

thoroughly in the Practical Workshop Problems which will shortly follow.

In Fig. 17 is shown another elevation of a sphere composed of twelve vertical sections as shown in plan view. While the method used for obtaining the pattern is by means of parallel lines, and would be strictly accurate if the sections in plan remained straight as from 4 to 4, the pattern becomes approximate as soon as we start to raise it by means of machine or hammer to conform to the profile B in elevation, because the distance along the curve *a* from 4′ to 4′

in plan is greater than a straight distance from 4 to 4. The pattern by this method is obtained as follows: Let B represent the elevation of the sphere, and A the plan of the same, which is divided into as many sides as the sphere is to have vertical sections, in this case 12, being careful that the two opposite sides 4–4 and 4′ 4′ in plan run parallel to the center line as shown. Make the diameter of the sphere 4–4″ 3 inches. Divide the half elevation into an equal number of spaces as shown from 1 to 4 to 1, and from these points drop lines at right angles to 4–4″ intersecting the miter lines 1–4 in plan as shown. Now draw any horizontal line, as 1′–1′, upon which place the stretchout of 1–4–1 in elevation as shown by 1′–4′–1′ on the line 1′–1′ in C. Through these points draw lines at right angles to 1′–1′, which intersect by lines drawn from similarly numbered intersections on the miter lines 1–4 in plan, at right angles to 4–4. A line traced through points thus obtained as shown by C will be the desired pattern.

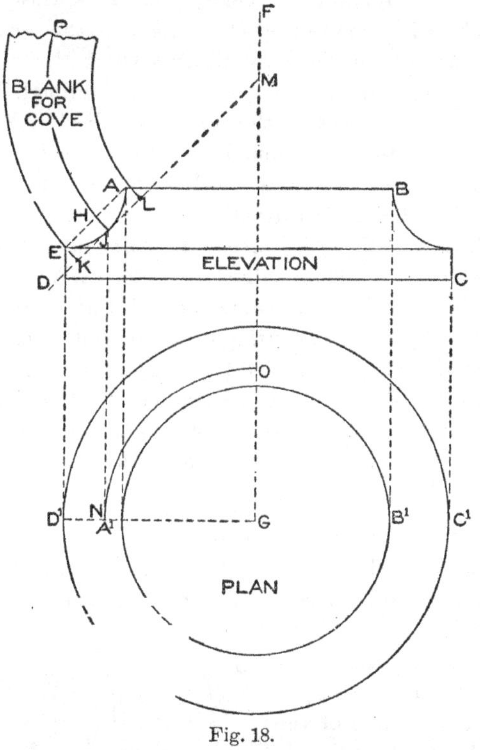

Fig. 18.

In Fig. 18 is shown the principle used in obtaining the radii with which to develop the blank for a curved or circular mould when it is to be hammered by hand. In this connection, only the principle employed will be shown, leaving the full development and also the development for patterns which are to be raised by hand

and hammered by machine, to be explained in problems which will follow in Practical Workshop Problems. Draw this problem double the size shown. First draw the elevation A B C D, and through the elevation draw the center line F G. Then using G as a center, draw the circles A^1 B^1 and C^1 D^1 representing respectively the horizontal projections of A B and C D in elevation. Now draw a line from A to E in elevation, connecting the corners of the cove as shown. Bisect A E and obtain the point H, from which at right angles to A E draw a line intersecting the cove at J. Through J parallel to A E draw a line intersecting the center line F G at M. Take the stretchout from J to A and from J to E and place it on the line J M as shown respectively from J to L and from J to K. Then will M L and M K be the radii with which to strike the pattern or blank for the cove. From J drop a vertical line intersecting the line D^1 G in plan at N. Then with G as center strike the quarter circle N O. Now using M as center and M J as radius, strike the arc J P. Then on this arc, starting from J, lay off 4 times the stretchout of N O in plan for the full pattern. It should be understood that when stretching the cove A E, the point J remains stationary and the metal from J to L and from J to K is hammered respectively toward J A and J E. For this reason is the stretchout obtained from the point J.

PRACTICAL WORKSHOP PROBLEMS.

In presenting the 32 problems which follow on sheet-metal work, practical problems have been selected such as would arise in every-day shop practice.

In this connection we wish to impress upon the student the necessity of working out each and every one of the 32 problems. Models should be made from stiff cardboard, or, if agreeable to the proprietor of the shop, the patterns can be developed at home, then cut out of scrap metal in the shop during lunch hour, and proven in this way.

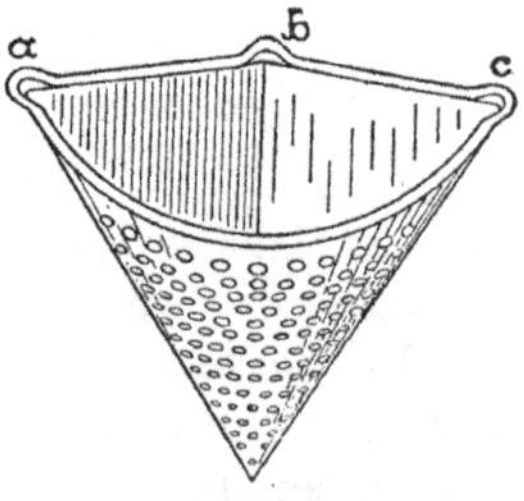

Fig. 19.

Our first problem is shown in Fig. 19, and is known as a sink drainer. It is often the case that the trap under the kitchen sink is choked or blocked, owing to a collection of refuse matter. To avoid this a sink drainer is used, and is fastened in position through the wire loops *a*, *b* and *c*. The refuse matter is poured into the drainer, from which it is easily removed after the fluid has passed through the perforations. These drainers may be made of tin or of black or galvanized iron, but where a good job is wanted 16-ounce copper should be used. To obtain the pattern for any sized drainer, proceed as follows: First draw the plan of the drainer A B C in Fig. 20, making A B and B C each two inches and forming a right angle. Then using B as center and A B as radius, draw the arc A C. In its proper position above the plan construct the side elevation, making E D 2 inches high, and draw the line F D. Then will F E D be the side elevation. Divide the arc A C into equal spaces as shown by the small figures 1 to 5. For the pattern use F D as radius, and with D in Fig. 21 as center strike the arc 1 5. From 1 draw a line to D and step off on 1–5 the same number of spaces as contained in A C in plan in Fig. 20, as shown by similar figures in Fig. 21. Draw a line from 5 to D. Then will 1–5–D be the pattern for the front of the strainer, in which perforations should be punched as shown. To join the sides of this pattern, use 1 and 5 as centers, and with either F E or A B in Fig. 20 as radius, describe the arcs E and E[1] in Fig. 21. Now using D as center and D E in Fig. 20 as radius, intersect the arcs E and E[1] as shown in Fig. 21. Draw lines from 1 to E[1] to D to E to 5, which completes the pattern, to which edges must be allowed for wiring at the top and seaming at the back.

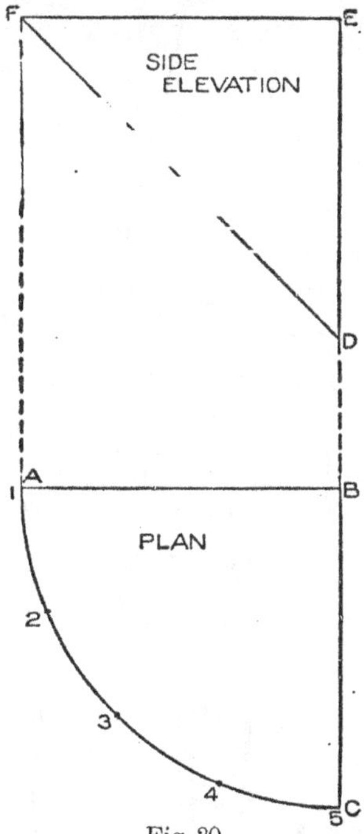

Fig. 20.

When joining a faucet or stop cock to a sheet-metal tank it is usual to strengthen the joint by means of a conical "boss," which is indicated by A in Fig. 22. In this problem the cone method is employed, using principles similar to those used in developing a frustum of a cone intersected by any line. Therefore in Fig. 23 let

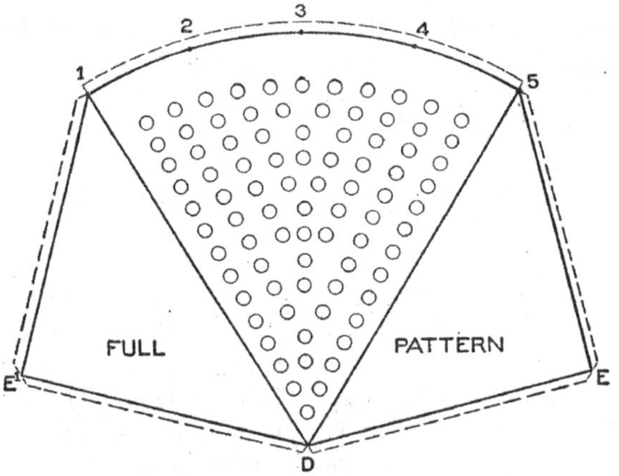

Fig. 21.

A B represent the part plan of the tank, C portion of the faucet extending back to the tank line, and F G H I the conical "boss" to fit around a faucet. When drawing this problem make the radius of the tank D A equal to $3\frac{1}{2}$ inches, and from D draw the vertical line D E. Make the distance from G to H equal to $2\frac{3}{4}$ inches, the diameter of the faucet F I $1\frac{1}{4}$ inches and the vertical height K C $1\frac{1}{2}$ inches Draw a line from G to H intersecting the center line D E at K. Then using K as center describe the half section G J H as shown. Divide J H into equal parts shown from 1 to 4, from

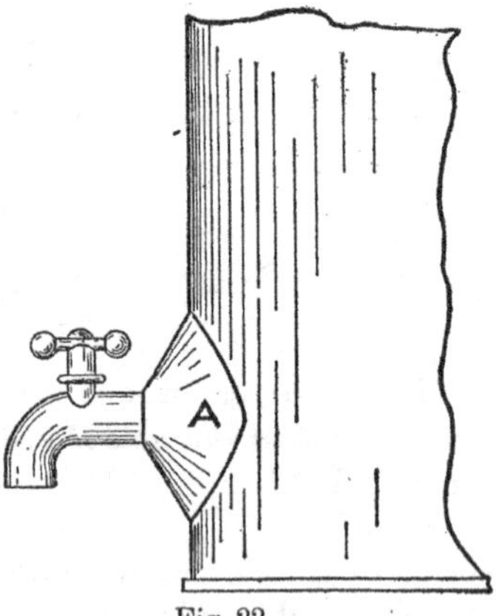

Fig. 22.

which drop vertical lines intersecting the line G H as shown, from which draw radial lines to the apex E cutting the plan line of the tank A B as shown. From these intersections draw horizontal lines intersecting the side of the cone H I at 1, 2′, 3′, and 4′. Now use E as center, and with radius equal to E 1 describe the

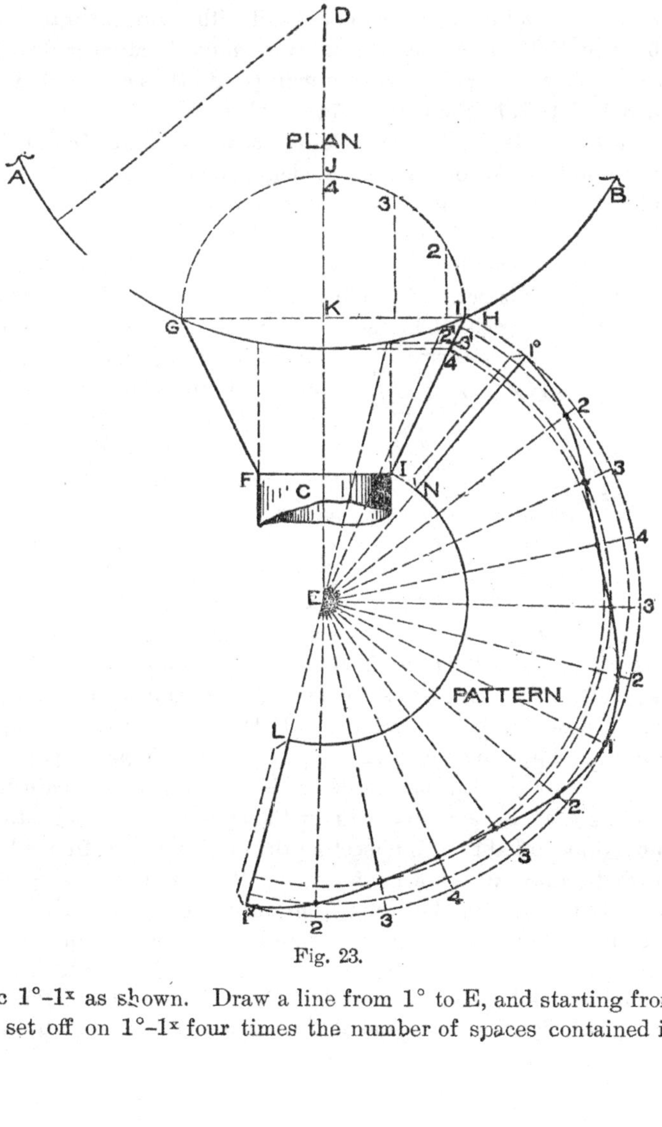

Fig. 23.

arc 1°–1ˣ as shown. Draw a line from 1° to E, and starting from 1° set off on 1°–1ˣ four times the number of spaces contained in

J H in plan, as shown by similar numbers on 1° 1ˣ. Draw a line from 1ˣ to E, and with E I as radius describe the arc N L intersecting the radial lines 1° E and 1ˣ E at N and L respectively. From the various numbers on the arc 1° 1ˣ draw radial lines to the apex E; and using E as center and with radii equal to E 4′, E 3′, and E 2′, draw arcs intersecting similarly numbered radial lines as shown. Trace a line through points thus obtained; then will N 1° 1 1ˣ L be the pattern for the "boss."

In Fig. 24 is shown what is known as a hip bath. In drawing out the problem for practice the student should remember that it is similar to the preceding one, the only difference being in the outline of the cone. Make the top of the cone I B in Fig. 25 equal to 3¼ inches, the bottom C D 1¾ inches, the vertical height from K to 5′ 2½ inches, the diameter of the foot E F 2¼ inches, and the vertical height 5′–5″ ¼-inch. Through the center of the cone draw the center line K L, and at pleasure draw the outline of the bath as shown by A J B. It is immaterial of what outline this may be, the principles that follow being applicable to any case. Thus, in the side elevation, extend the lines B C and A D until they intersect the center line at L. In similar manner extend the sides of the foot piece E D and F C until they intersect the center line at R. Now with 5′ as center and with radius equal to 5′ D or 5′ C, describe the half section C H D, which divide into equal spaces as shown by the small figures 1 to 9. From the points of division erect vertical lines meeting the base line of the bath D C at points 1, 2′, 3′, etc., to 9. From the apex L and through these points draw radial lines intersecting the outline B J A, from which horizontal lines are drawn intersecting the side of the bath B C as shown from 1 to 9. For the pattern for the body use L as center, and with L C as radius draw the arc F L¹. Now starting at any point, as 1, set off on F L¹ twice the stretchout of D H C as shown by similar numbers on the arc F L¹. From the apex L and through the small figures draw radial lines, which intersect by arcs

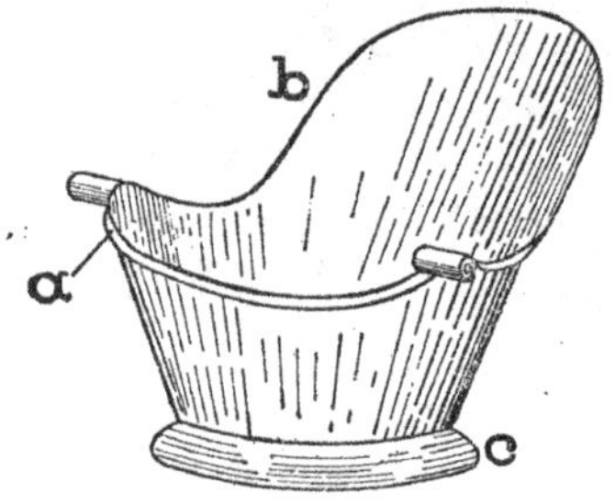

Fig. 24.

struck from L as center with radii equal to similarly numbered intersections on B C. Trace a line through points thus obtained, and L^1 M N P F will be the pattern for the body of the bath, to which laps should be added at the bottom and sides for seaming.

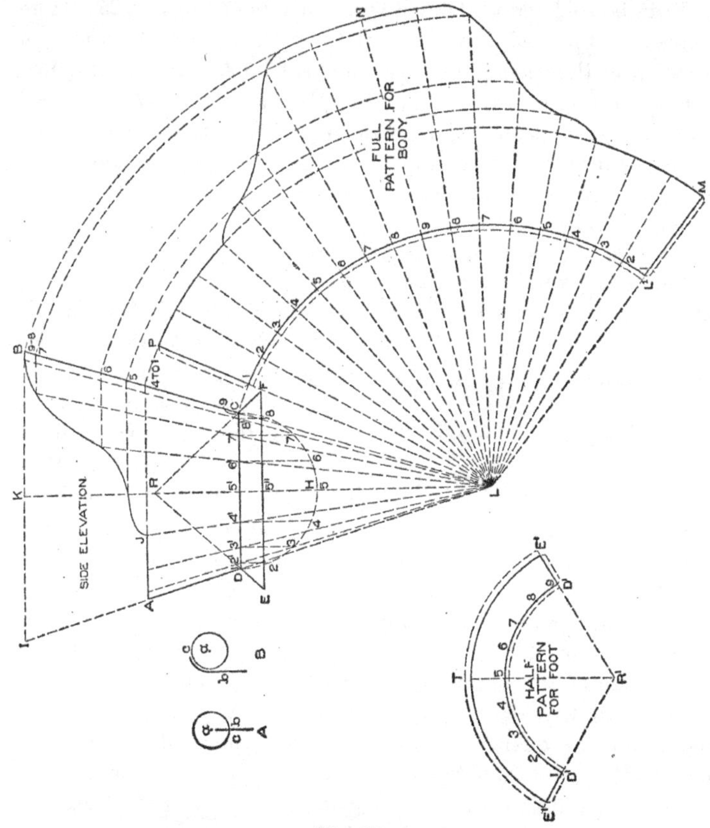

Fig. 25.

The pattern for the foot is obtained by using as radii R D and R E, and striking the pattern using R^1 as center, the half pattern being shown by E^1 T E^1 D^1 D^1, and the distance D^1 D^1 being equal to the stretchout of the half section D H C in side elevation.

It is usual to put a bead along the edges of the top of a bath as shown at *a* and *b* in Fig. 24. For this purpose tubing is sometimes used, made of brass, zinc, or copper and bent to the required shape; or zinc tubes may be rolled and soldered by hand, filled with heated white sand or hot rosin, and bent as needed. The tube or bead can be soldered to the body as shown in (A) in Fig. 25. Here *a* represents the bead, in which a slot is cut as *c*, and which is then slipped over the edge of the bath and soldered. Another method is shown in (B), in which the bath body *b* is flanged over the bead *a* and soldered clean and smooth at *c*, being then scraped and sandpapered to make a smooth joint. A wired edge is shown at *c* in Fig. 24, for which laps must be allowed as shown in Fig. 25 on the half pattern for foot.

In Fig. 26 is shown the perspective view of a bath tub; these tubs are usually made from IX tin or No. 24 galvanized iron. The bottom and side seams are locked and thoroughly soldered, while the top edge is wired with handles riveted in position as shown at A. The method used in developing these patterns will be the cone method and triangulation. In drawing this problem

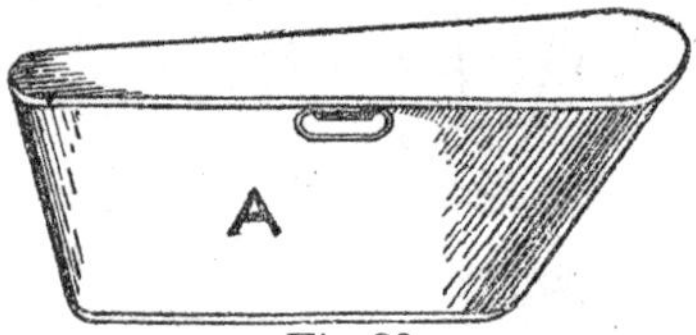

Fig. 26.

for practice (Fig. 27), first draw the center line W 8 in plan; and using *a* as center with a radius equal to 1½ inches draw the semicircle C–12 D. Now make the distance *a* to *b* 4 inches; and using *b* as center with a radius of 1⅝ inches draw the semicircle E–7–H. Draw lines from E to D and from C to H. D E 7 H C 12 D will be the plan of the bottom of the bath. In this case we assume that the flare between the top and bottom of the narrow end of the bath should be equal; therefore using *a* as center and with a radius equal to 1⅝ inches draw the semicircle A W B. At the upper end of the bath the flare will be unequal; therefore from *b* measure a distance on line W 8 of 1 inch and obtain *c*, which use as center, and with a radius equal to 2 inches describe the arc F 8 G. Draw lines from F to A and from B to G; and A F 8 G B W A will be the plan of the top of the bath. Now project the side elevation from the plan as shown by the dotted lines, making the slant height from I to R 2⅛ inches and from J to K 3⅛ inches; draw a line from K to R, and J K R I will be the side elevation of the bath tub. In constructing the bath in practice, seams are located at H G, F E,

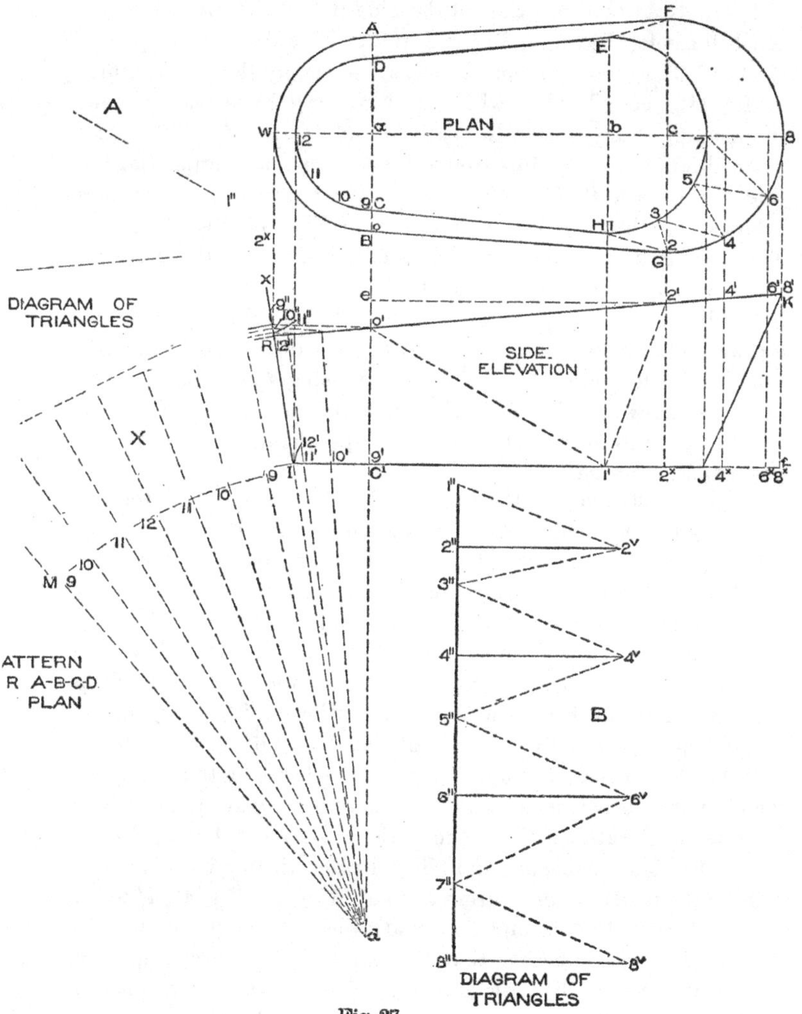

Fig. 27.

A D, and C B in plan, thus making the tub in four pieces

The lower end of the bath will be developed by the cone method as in the last two problems. From the center *a* drop a line indefinitely as shown. Extend the side R I of the side elevation until it meets the center line *a d* at *d*. Now divide the quarter circle 12–9 in plan into equal spaces as shown by the small figures 9, 10, 11, and 12, from which drop vertical lines (not shown) intersecting the bottom of the bath tub in elevation from 9′ to 12′. Then through these points from *d* draw lines intersecting the top line of the bath R K as shown, from which draw horizontal lines intersecting the side I–R extended as I X at points 9″ to 12″. Then using *d* as center and *d* I as radius, describe the arc I M, upon which place the stretchout of D 12 C in plan, as shown by similarly numbered points on L M. Through these points from *d* draw radial lines, which intersect by arcs drawn from similarly numbered intersections on I R extended, using *d* as center. Trace a line as shown, and L M N P will be the pattern for the lower end of the tub A B C D in plan. Laps should be allowed for wiring and seaming.

As the patterns for the upper end and sides will be developed by triangulation, diagrams of triangles must first be obtained, for which proceed as follows: Divide both of the quarter circles H 7 and G 8 in plan into the same number of spaces as shown respectively from 1 to 7 and from 2 to 8. Connect these numbers by dotted lines as shown from 1 to 2, 2 to 3, 3 to 4, etc. From the various points 2, 4, 6, and 8 representing the top of the bath, drop lines meeting the base line J *f* in elevation at 2^x, 4^x, 6^x, and 8^x, and cutting the top line of the bath at 2′, 4′, 6′, and 8′. Then will the dotted lines in plan represent the bases of the triangles, which will be constructed, whose altitudes are equal to the various heights in elevation. Take the various distances 1 to 2, 2 to 3, 3 to 4, 4 to 5, etc., in plan up to 8, and place them on the vertical line 1″–8″ in (B) as shown from 1″ to 2″, 2″ to 3″, 3″ to 4″, 4″ to 5″, etc., up to 8″. For example, to obtain the true length of the line 6–7 in plan, remembering that the points having even numbers represent the top line of the bath and those having uneven numbers the base line, draw at right angles to 1″–8″ in (B), from 6″, a line equal in height to 6^x–6′ in elevation, and draw a line from 6^v to 7″ in (B), which is the length desired. For the true length of 6–5 in plan it is necessary only to take this distance place it from 6″ to 5″ in (B) and draw a line from 6^v to 5″. In this way each altitude answers for two triangles. In plan draw a line from 1 to 0. Then will two more triangles be necessary, one on the line 1–0, and the other on B G or 0–2. From 2′ in elevation draw a horizontal line, as 2′ *e*, intersecting the vertical line dropped from 0 at *e*. Now take the distances 0 1 and 0 2, and place them in (A) as shown by the horizontal lines 0″–1″ and 0^x–2^x respectively. At right angles to both lines at either end draw the vertical lines 0″–0‴ and 0^x–0^v equal in height respectively to C^1 0′ and *e* 0′ in elevation. Draw in (A) lines from 2^x to 0^v and from 1″ to 0‴, which are the desired lengths. Before proceeding with the pattern, a true section must be obtained on 2′–8′ in side elevation. Take the various distances 2′ to 8′ and place them on the line 2′–8′ in Fig. 28. At right angles to 2′–8′ and through the small figures draw lines as shown. Now measuring in each and every instance from the center line in plan in Fig. 27, take the various distances to points 2, 4, and 6 and place them on similarly numbered lines in Fig. 28, measuring in each case on either side of the line 2′–8′, thus obtaining the intersections 2–4–6. A line traced through these points will be the true section on 2′–8′ in elevation in Fig. 27.

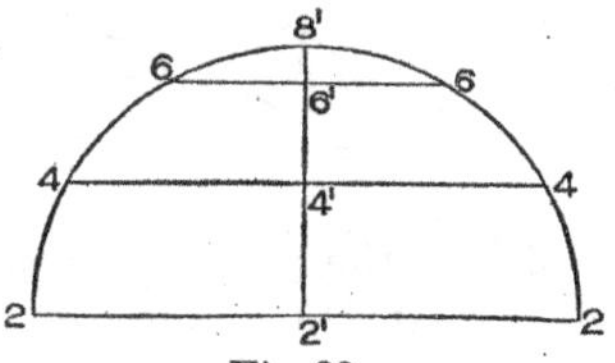

Fig. 28.

For the pattern for the upper end of the tub proceed as follows: Take the distance of 7″–8^v in (B) and place it on the vertical line 7–8 in Fig. 29. Then using 8 as center and with a radius equal to 8′–6 in Fig. 28, describe the arc 6 in Fig. 29, which intersect by an arc struck from 7 as center and with 7″–6^v in (B) in Fig. 27 as radius. Then using 7–5 in plan as radius, and 7 in Fig. 29 as center, describe the arc 5, which intersect by an arc struck from 6 as center and with 6^v–5″ in (B) in Fig. 27 as radius. Proceed in this manner, using alternately as radii first the divisions in Fig. 28, then the length of the slant lines in (B) in Fig. 27, the divisions on 7 H in plan, then again the slant lines in B, until the line 1–2 in Fig. 29 is obtained. Trace a line through points thus obtained, as shown by 2–8–7–1. Trace this opposite the line 8–7, as shown by 2′ 1′. Then will 2–8–2′–1′–7–1 be the desired pattern, to which laps must be allowed.

For the pattern for the side of the bath draw any line 9–1 in Fig. 30 equal to 9–1 in plan in Fig. 27. Now with a radius equal

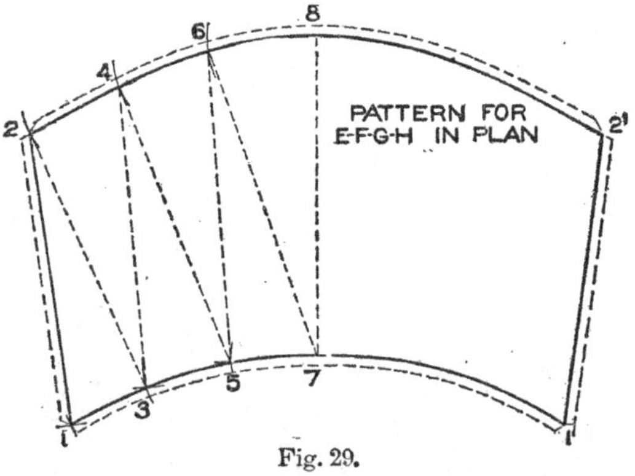

Fig. 29.

to 9–P in the pattern X and with 9 in Fig. 30 as a center, describe the arc 0, which intersect by an arc struck from 1 as center and with 1″–0‴ in (A) in Fig. 27 as radius. Now taking a radius equal to 0^{v}–2^{x} in (A) with 0 in Fig. 30 as center, describe the arc 2, which intersect by an arc struck from 1 as center, and with 1–2 in Fig. 29 as radius. Draw lines from corner to corner in Fig. 30, which gives the desired pattern, to which laps are added for seaming and wiring.

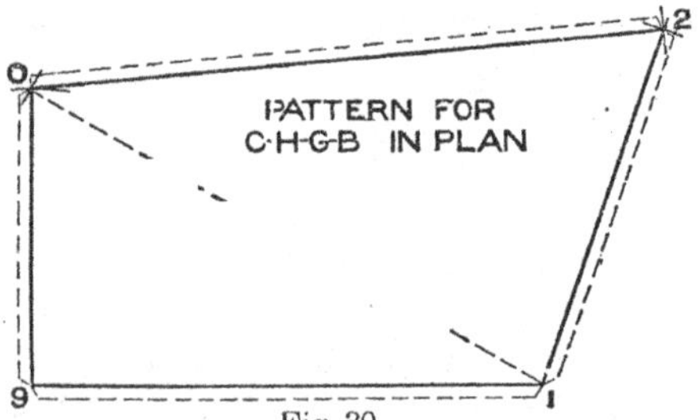

Fig. 30.

In Fig. 31 is shown a perspective view of a funnel strainer pail. These pails are usually made from IX bright tin, and the same principles as are used in the development of the pattern are applicable to similar forms, such as buckets, coal hods, chutes, etc. This problem presents an interesting study in triangulation, the principles of which have been explained in previous problems. First draw the center line C I in Fig. 32, at right angles to which draw H E and H F each equal to $1\frac{1}{4}$ inches. Make the vertical height H C $3\frac{1}{4}$ inches and C D 2 inches. Now make the vertical heights measuring from C G, to A, and to B respectively $1\frac{1}{4}$ inches, and $1\frac{1}{2}$ inches. Make the horizontal distance from C to G $2\frac{3}{4}$ inches, the diameter from G to A $1\frac{3}{8}$ inches, and from A to B $\frac{3}{4}$-inch, and draw a line from B to C. Connect points by lines; then will A B C D E F G be the side elevation of the pail. In its proper position below F E, with J as center, draw the plan K L M N. Also in its proper position draw the section on A G as O P R S Now draw the rear elevation making G^1 U and G^1 V each equal to H E, and 1″ T and 1″-1′ each equal to C D. Project a line from B in side, intersecting the center line in rear at 4′. Then through the three points 1′ 4′ T draw the curve at pleasure, which in this case is struck from the center *a*. W Y X Z represents the opening on G A in side obtained as shown by the dotted lines but having no bearing on the patterns. Pails of this kind are usually made from two pieces, with seams at the sides, as in Fig. 31. The pattern then for the back shown by C D E H in side elevation in Fig. 32 will be obtained by the cone method, struck from the center I, the stretchout on E^1 E^2 in the pattern being obtained from the half plan. The pattern for C D E H is shown with lap and wire allowances by D^1 D^2 E^2 E^1 and needs no further explanation.

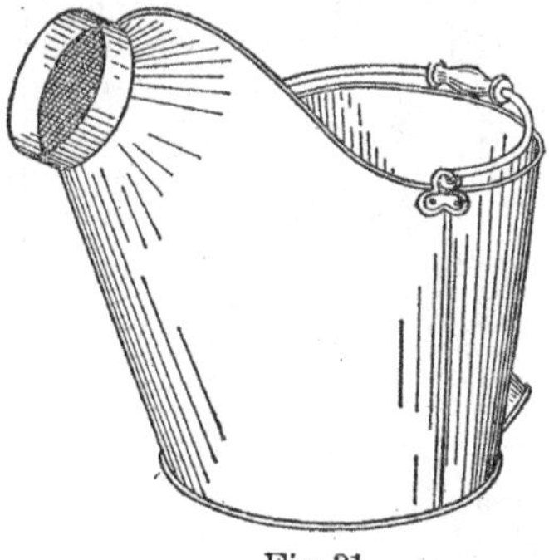

Fig. 31.

The front part of the pail shown by A B C H F G will be developed by triangulation, but before this can be done a true section must be obtained on B C, and a set of sections developed as follows: Divide one-half of 1′ 4′ T in rear elevation into equal parts as shown from 1′ to 4′, from which draw horizontal lines intersecting the line B C as shown. From these intersections lines are drawn at right angles to B C equal in length to similarly numbered lines in rear as 3′-3″, 2′-2″, and 1′-1″. Trace a line as shown, so that C 1‴ 2‴ 3‴ 4‴ will be the true half section on B C. To avoid a confusion of lines take a tracing of A B C H F G and place it as shown by similar letters in Fig. 33. Now take tracings of the half sections in Fig. 32, as H E D C, C 1‴ B, P O S, and the quarter plan N J M, and place them in Fig. 33 on similar lines on which they represent sections as shown respectively by H 9′ 8′ C, C 8 B, A 3 G, and F 9 H. Divide the half section

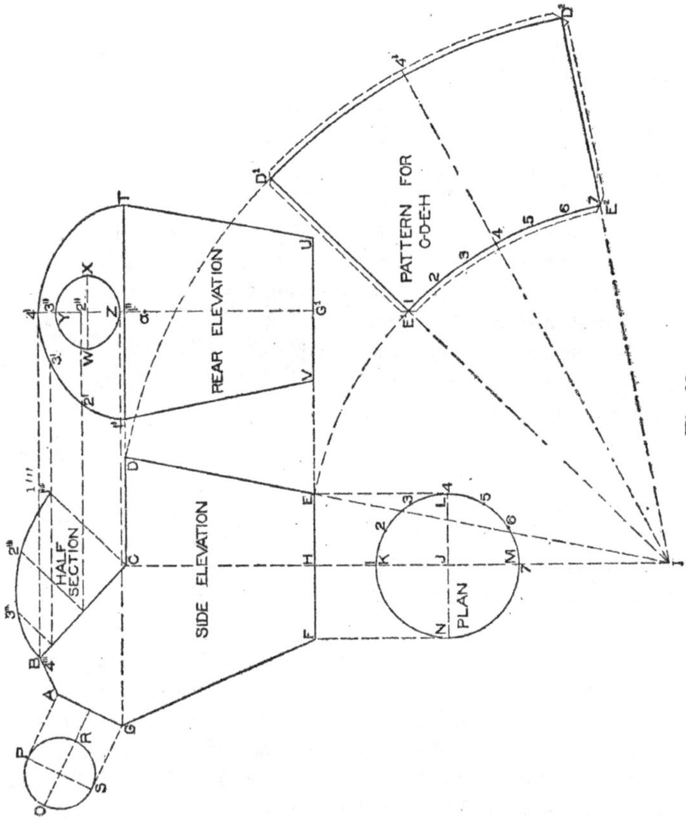

Fig. 32.

A 3 G into 6 equal parts as shown by the small figures 1 to 5. As this half section is divided into 6 parts, then must each of the sections B 8 C and F 9 H be divided into 3 parts as shown respectively from 6 to 8 and 9 to 11. As C 8′ and H 9′ are equal respectively to C 8 and H 9 they are numbered the same as shown.

Now at right angles to G A, B C, C H, and H F, and from the various intersections contained in the sections G 3 A, B 8 C, C 8′ 9′ H, and H 9 F, draw lines intersecting the base lines of the sections G A, B C, C H, and H F at points shown from 1′ to 11′. Now draw dotted lines from B to 5′ to 6′ to 4′ to 7′ to E to C, and then from H to E to 10′ to 2′, etc until all the points are

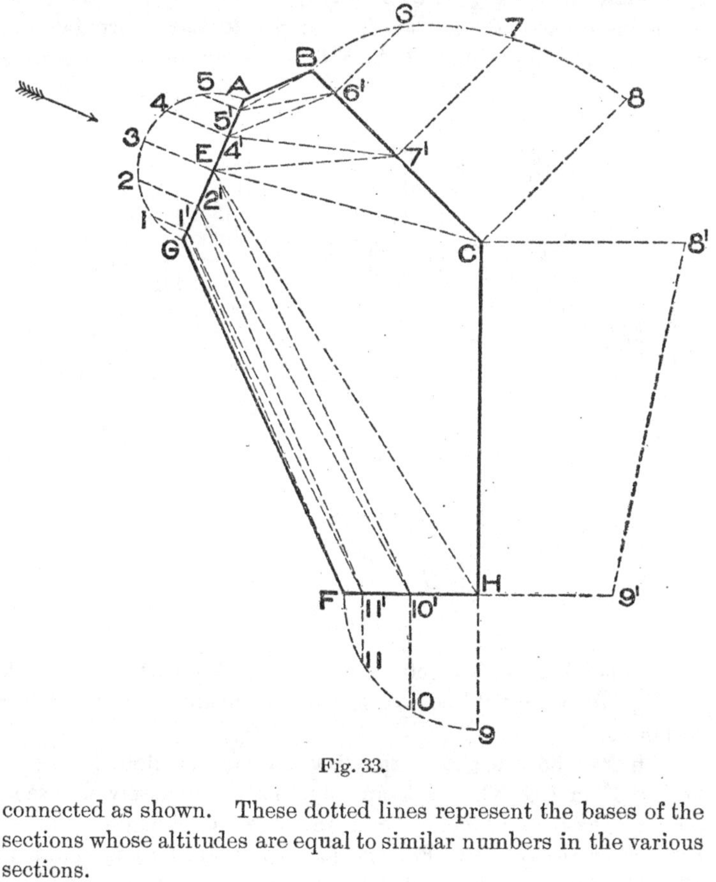

Fig. 33.

connected as shown. These dotted lines represent the bases of the sections whose altitudes are equal to similar numbers in the various sections.

In order that the student may thoroughly understand this method of triangulation as well as similar methods that will follow in other problems, the model in Fig. 34 has been prepared, which shows a perspective of Fig. 33 with the sections bent up in their proper positions. This view is taken on the arrow line in Fig. 33, the letters and figures in both views being similar. For the true sections on the dotted lines in C E A B in Fig. 33, take the lengths of the dotted lines C E, E 7′, 7′ 4′, etc., and place them on the horizontal line in Fig. 35 as shown by similar letters and figures. From these small figures, at right angles to the horizontal line, erect the vertical heights C 8, E 3, 7′ 7, etc., equal to similar

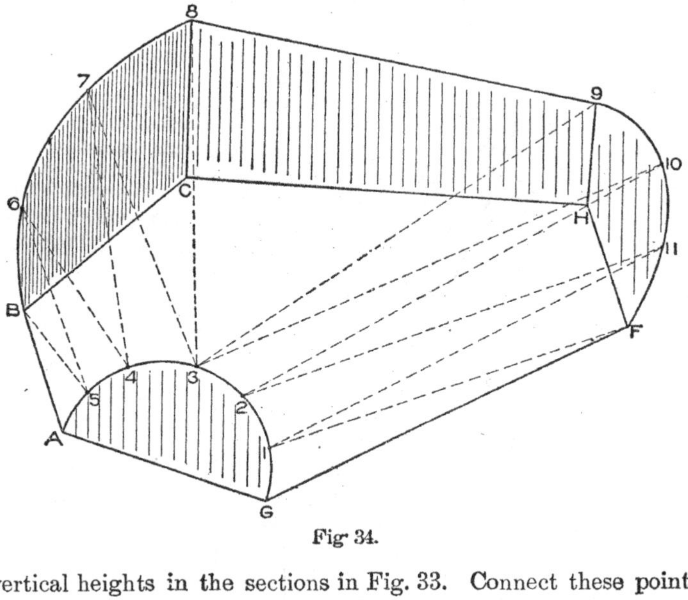

Fig. 34.

vertical heights in the sections in Fig. 33. Connect these points in Fig. 35 by dotted lines as shown, which are the desired true distances.

In Fig. 36 are shown the true sections on dotted lines in G E H F in Fig. 33, which are obtained in precisely the same manner, the only difference being that one section is placed inside of another in Fig. 36. For the pattern proceed as is shown in Fig. 37. Draw any vertical line as G F equal to G F in Fig. 33. With radius equal to G 1 and with G in Fig. 37 as center describe the arc 1, which intersect by an arc struck from F as center and with a radius equal to F 1 in Fig. 36. Now with F 11 in Fig. 33 as radius and F in Fig. 37 as center, describe the arc 11, which is intersected by an arc struck from 1 as center and with 1–11 in Fig 36 as radius. Proceed in this manner until the line 3–9 in Fig. 37 has been obtained. Then using 8′–9′ in Fig. 33 as radius and 9 in Fig. 37 as center, describe the arc 8, which is intersected by an arc struck from 3 as center and with 3–8 in Fig.

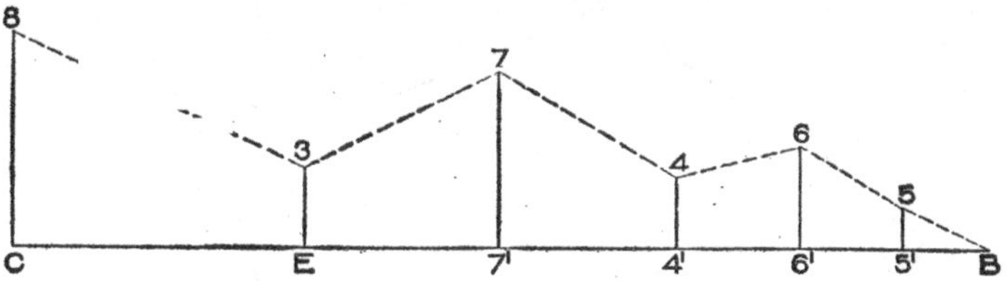

Fig. 35.

35 as radius. Now use alternately as radii, first the divisions in B 8 C in Fig. 33, then the length of the slant lines in Fig. 35, the divisions in E 3 A in Fig. 33, and again the distances in Fig. 35, until the line B A in Fig. 37 has been obtained, which is obtained from B A in Fig. 33. Trace a line through points thus obtained in Fig. 37 as shown by A B 8 9 F G A. Trace this half pattern opposite the line G F. Then will B A G A^1 B^1 8^1

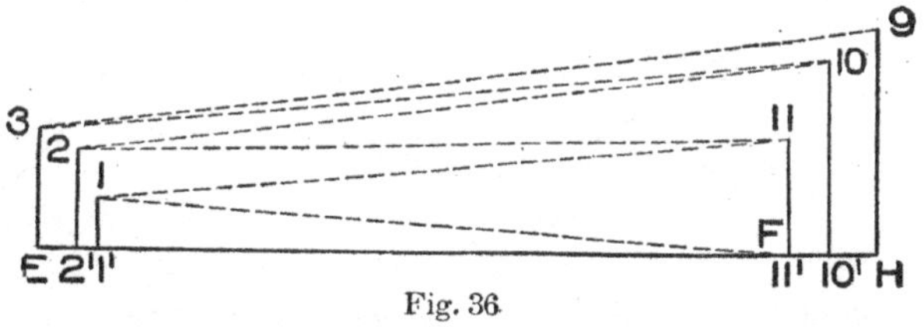

Fig. 36

9^1 F 9 8 be the pattern for the front half of the pail. If for any reason the pattern is desired in one piece, then trace one half of D^1 D^2 E^2 E^1 in Fig. 32 on either side of the pattern in Fig. 37 as shown by the dotted lines 8′ D^1 E^1 9^1 and 9 E D 8. Allow edges for wiring and seaming.

Fig 38 shows the method for obtaining the pattern for an Emerson ventilator shown in Fig. 39.

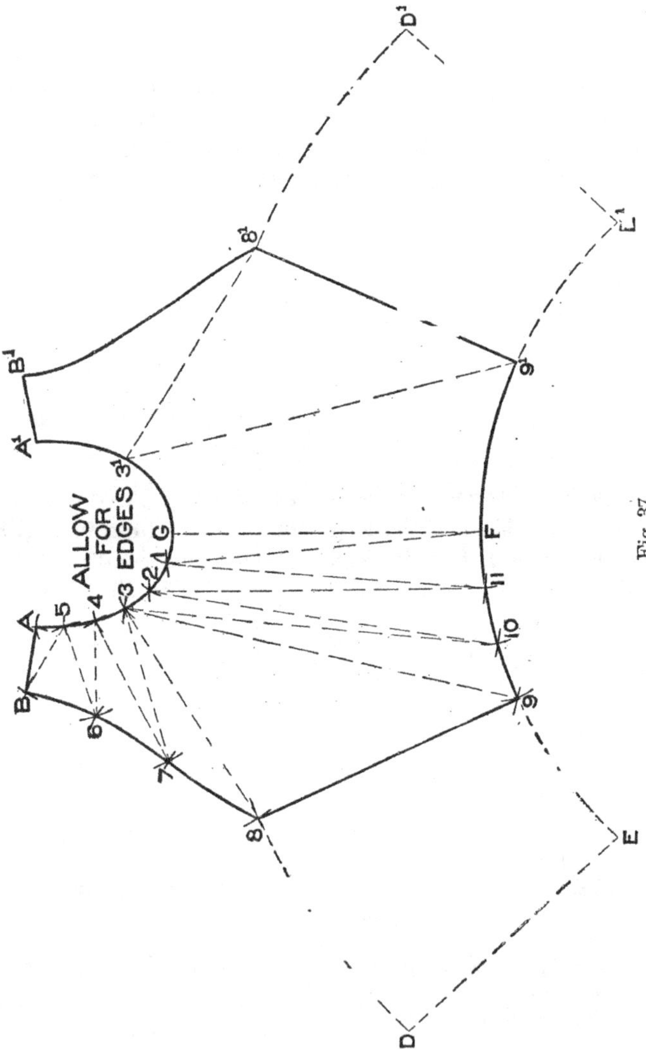

Fig. 37.

While the regular Emerson ventilator has a flat disc for a hood it is improved by placing a cone and deflector on the top as shown. To make the patterns, proceed as shown in Fig. 38. First draw the center line *a b*, on either side of which lay off

$1\frac{1}{2}$ inches, making the pipe A, 3 inches in diameter. The rule usually employed is to make the diameter of the lower flare and upper hood twice the diameter of the pipe. Therefore make the diameter of *s d* 6 inches. From *s* and *d*, draw a line at an angle of 45° to intersect the line of the pipe at *t* and *i*; this completes B. Measure 2 inches above the line *t i* and make *u m* the same diameter as *s d*. Draw the bevel of the deflector so that the apex will be $\frac{1}{2}$ inch above the line *t i* and make the apex of the hood the same distance above *u m* as the lower apex is below it. Then draw lines as shown which complete C and D.

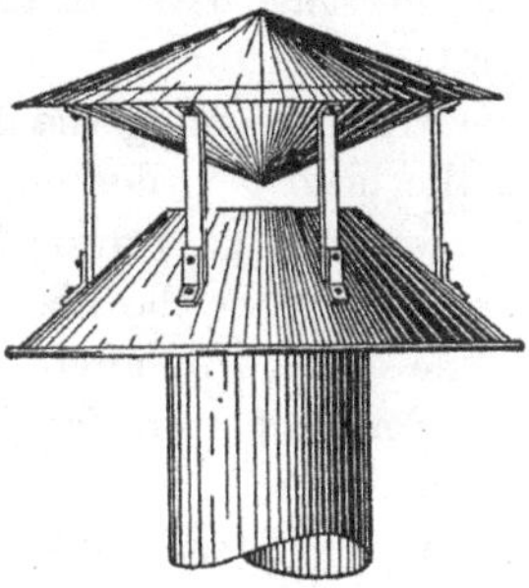

Fig. 39.

Now with *c* as a center and radii equal to *c e* and *c d* draw the quarter circles *e f* and *d h* respectively, which represent the one-

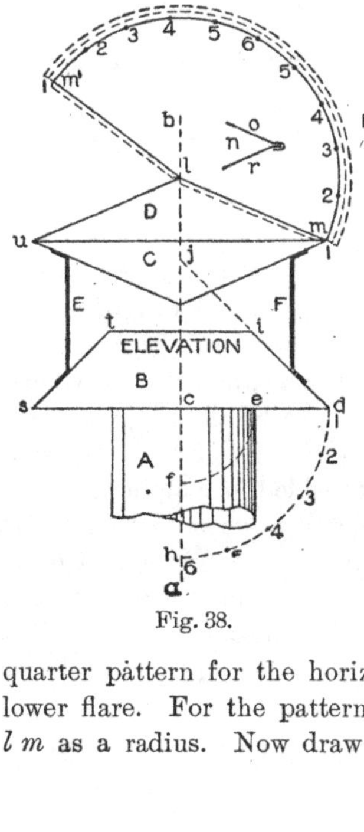

Fig. 38.

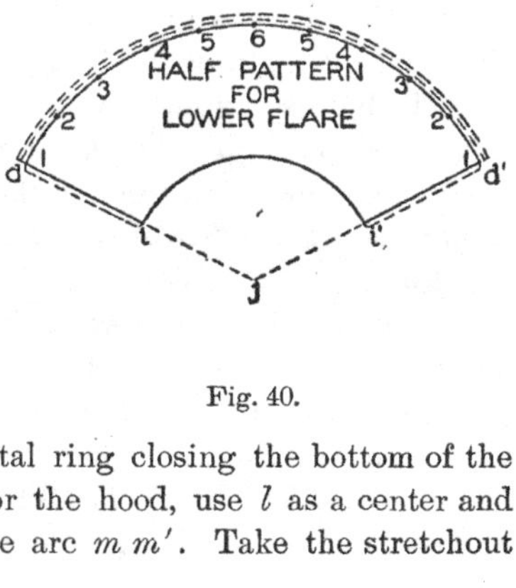

Fig. 40.

quarter pattern for the horizontal ring closing the bottom of the lower flare. For the pattern for the hood, use *l* as a center and *l m* as a radius. Now draw the arc *m m′*. Take the stretchout of the quarter circle 1 to 6 on *d h*, and place twice this amount on *m m'* as shown from 1–6–1. Draw a line from 1 to *l*. Then *m'* 6 *m l*, will be the half pattern for the hood. As the deflector has the same bevel as the hood, the hood pattern will also answer for the deflector.

When seaming the hood and deflector together as shown at *n*, the hood *o* is double-seamed to the deflector at *r*, which allows the water to pass over; for this reason allow a double edge on the pattern for the hood as shown, while on the deflector but a single edge is required. Edges should also be allowed on *e d h f*.

For the pattern for the lower flare, extend the line *d i* until it intersects the center line at *j*. Then with radii equal to *j i* and *j d* and with *j* in Fig. 40 as center describe the arcs *i i'* and *d d'*. On one side as *d* draw a line to *j*. Then set off on the arc *d d'*

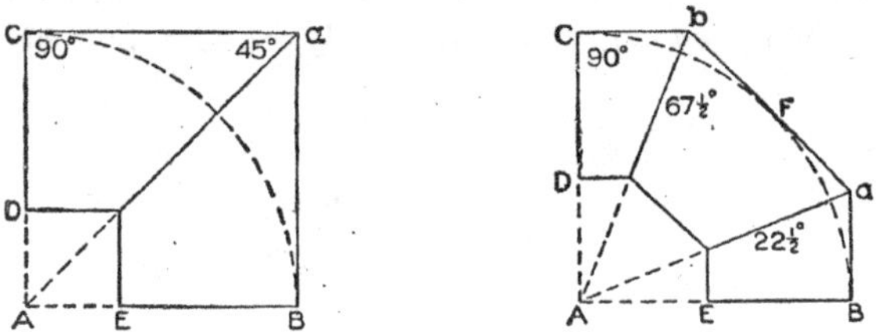

Fig. 41. Fig. 42.

twice the number of spaces contained in *d h* in Fig. 38 as shown in Fig. 40. Draw a line from *d'* to i and allow edges for seaming. Then *d d' i' i* will be the half pattern for the lower flare.

The braces or supports E and F, Fig. 38, are usually made of galvanized band iron bolted or riveted to hood and pipe. The hood D must be water tight. or the water will leak into the deflector, from which it will drip from the apex inside the building.

Elbows. There is no other article in the sheet-metal worker's line, of which there are more made in practice than elbows. On this account rules will be given for constructing the rise of the miter line in elbows of any size or diameter, also for elbows whose sections are either oval, square or round, including tapering elbows Before taking up the method of obtaining the patterns, the rule will be given for obtaining the rise of the miter line for any size or number of pieces. No matter how many pieces an elbow has, they join together and form an angle of 90°. Thus when we speak of a two-pieced, three-pieced, four, five or six-pieced elbow, we understand that the right-angled elbow is made up of that number of pieces. Thus in Fig. 41 is shown a two-pieced elbow placed in the quadrant C B, which equals 90° and makes C A B a right angle. From A draw the miter line A *a* at an angle of 45° to the base line A B. Then parallel to A B and A C and tangent to the quadrant at C and B draw lines to intersect the miter line, as shown. Knowing the diameter of the pipe as C D or E B draw lines parallel to the arms of the pipe, as shown. Then C B E D will be a two-pieced elbow, whose miter line is an angle of 45°.

In a similar manner draw the quadrant B C, Fig. 42, in which it is desired to draw a three-pieced elbow. Now follow this simple

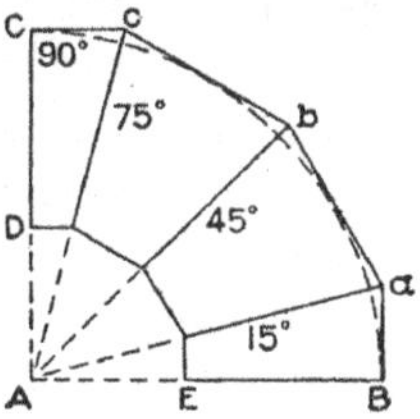

Fig. 43.

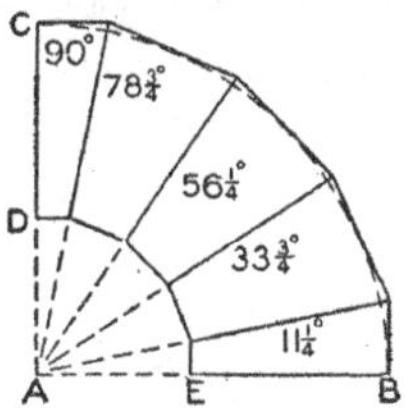

Fig. 44.

rule, which is applicable for any number of pieces: Let the top piece of the elbow represent 1, also the lower piece 1, and for every piece between the top and bottom add 2. Thus in a three-pieced elbow:

Top piece equals	1
Bottom piece equals	1
One piece between	2
Total equals	4

Now divide the quadrant of 90° by 4 which leaves 22½°. As one piece equals 22½°, draw the lower miter line A *a* at that angle to the base line A B. Then as the middle piece represents two by the above rule and equals 45°, add 45 to 22½ and draw the second miter line A *b*, at an angle of 67½° to the base line A B. Now tangent to the quadrant at C and B draw the vertical and horizontal lines shown, until they intersect the miter lines, from which intersections draw the middle line, which will be tangent to the quadrant at F. C D and B E show the diameters of the pipe, which are drawn parallel to the lines of the elbow shown.

Fig. 43 shows a four-pieced elbow, to which the same rule is applied. Thus the top and bottom piece equals 2 and the two middle pieces equal 4; total 6. Now divide the quadrant of 90° by 6. $\frac{90}{6} = 15$. Then the first miter line A *a* will equal 15°, the second A *b* 45°, the third A *c* 75°, and the vertical line A C 90°.

The last example is shown in Fig. 44, which shows a five-pieced elbow, in which the top and bottom pieces equal 2, the 3 middle pieces 6; total 8. Divide 90 by 8. $\frac{90}{8} = 11\frac{1}{4}$. Then the first miter line will equal $11\frac{1}{4}°$, the second $33\frac{3}{4}°$, the third $56\frac{1}{4}°$, and the fourth $78\frac{3}{4}°$. By using this method an elbow having any number of pieces may be laid out. When drawing these miter lines it is well to use the protractor shown in Fig. 45, which illustrates how to lay out a three-pieced elbow. From the center point A of the protractor draw lines through $22\frac{1}{2}°$, and $67\frac{1}{2}°$. Now set off A *a*, and the diameter of the pipe *a b*. Draw vertical lines from *a* and *b* to the miter line at *c* and *d*. Lay off similar distances from A to *a'* to *b'* and draw horizontal lines intersecting the $67\frac{1}{2}°$ miter line at *c'* and *d'*. Then draw the lines *d d'* and *c c'* to complete the elbow. In practice, however, it is not necessary to draw out the entire view of the elbow; all that is required is the first miter line, as will be explained in the following problems.

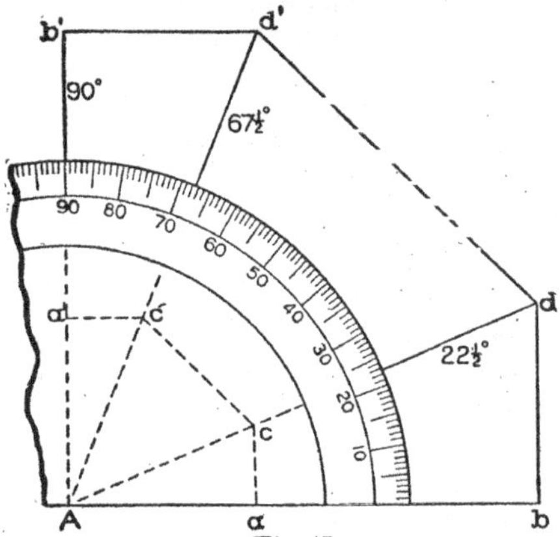

Fig. 45.

EXERCISES FOR PRACTICE.

1. Make the diameter of the pipe $1\frac{3}{4}$ inches and the distances from A to E $1\frac{1}{2}$ inches in Figs. 41 to 44 inclusive.

To obtain the pattern for any elbow, using but the first miter

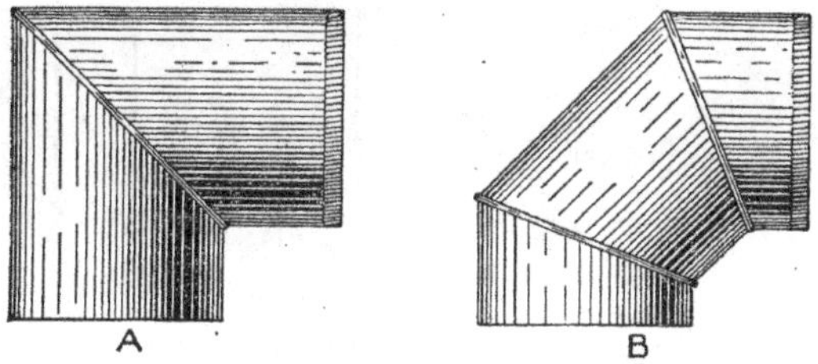

Fig. 46.

line, proceed as follows: In Fig. 46 let A and B represent respectively a two- and three-pieced elbow for which patterns are desired. First draw a section of the elbow as shown at A in Fig. 47 which

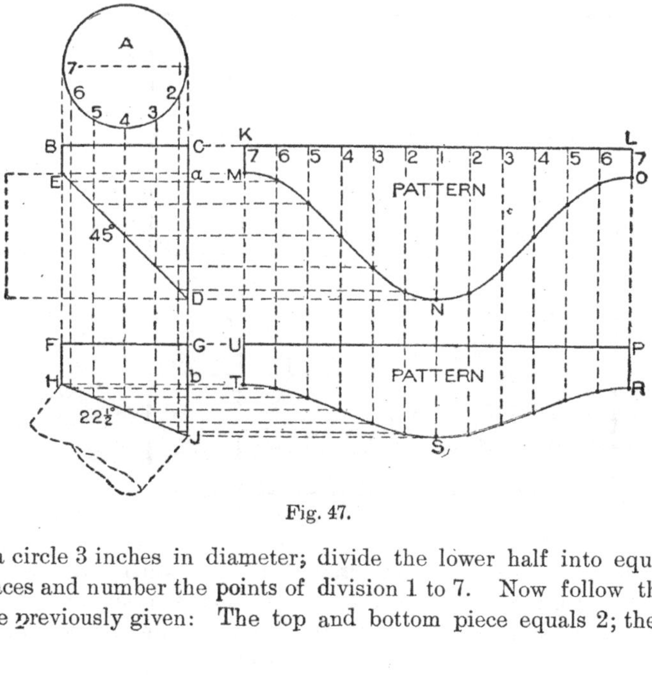

Fig. 47.

is a circle 3 inches in diameter; divide the lower half into equal spaces and number the points of division 1 to 7. Now follow the rule previously given: The top and bottom piece equals 2; then for a two-pieced elbow divide 90 by 2. In its proper position below the section A draw B C D E making E D 45°. From the various points of intersection in A drop vertical lines intersecting E D as

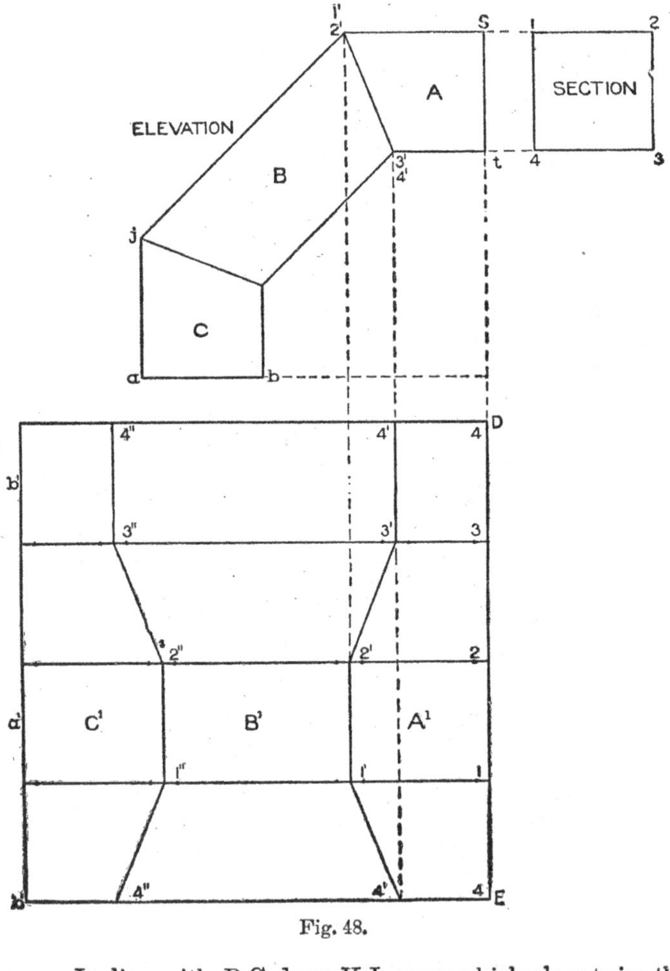

Fig. 48.

shown. In line with B C draw K L upon which place twice the number of spaces contained in the section A as shown by similar figures on K L; from these points drop perpendiculars to intersect with lines drawn from similar intersections on E D, parallel to K L. Trace a line through points shown; then K L O N M will be the pattern. To this laps must be allowed for seaming.

Now to obtain the pattern for a three-pieced elbow, follow the rule. Top and bottom pieces equal 2, one middle piece equals 2; total 4. $\frac{90}{4} = 22\frac{1}{2}$. Therefore in line with the section A below the two-pieced elbow draw F G J H, making H J at an angle of $22\frac{1}{2}°$ to the line H *b*. Proceed as above using the same stretchout lines; then U P R S T will be the desired pattern. It should be understood that when the protractor is used for obtaining the angle as shown in Fig. 45, the heights *a c* and *b* d measured from the horizontal line form the basis for obtaining the heights of the middle pieces, inasmuch as they represent one-half the distance; for that reason the middle pieces count 2 when using the rule. Therefore, the distances F H and G J (Fig. 47), represent one-half of the center piece and U T S R P one-half the pattern for the center piece of a three-pieced elbow.

Fig. 48 shows how the patterns are laid into one another, to prevent waste of metal when cutting. In this example we have a three-pieced elbow whose section is 2 × 2 inches. It is to be laid out in a quadrant whose radius is 5 inches. Use the same principles for square section as for round; number the corners of the section 1 to 4. In line with S *t* draw D E upon which place the stretchout of the square section as shown by similar numbers on D E; from which draw horizontal lines which intersect lines drawn parallel to D E from the intersections 1′ 2′ and 3′ 4′ in A in elevation, thus obtaining similar points in the pattern. Then A^1 will be the pattern for A in elevation. For the pattern for B simply take the distance from 2′ to *j* and place it on the line 4 4′ extended in the pattern on either side as shown by 4′ 4″ on both sides. Now reverse the cut 4′ 2′ 4′ and obtain 4″ 2″ 4″. By measurement it will be found that 4′ 4″ is twice the length of 2′ 2 as explained in connection with Figs. 45 and 47. Make the distance from 1″ to *a′* the same as *j* to *a* in C and draw the vertical line *b′ b′* intersecting the lines 4 4″ extended on both sides. Then A^1, B^1, and C^1 will be the patterns in one piece minus the edges for seaming which must be allowed between these cuts; this would of course make the lengths b′ 4″, 4″ 4′ and 4′ 4 as much longer as the laps would necessitate.

This method of cutting elbows in one piece, from one square is applicable to either round, oval or square sections.

In Figs. 49 and 50 are shown three-pieced elbows such as are

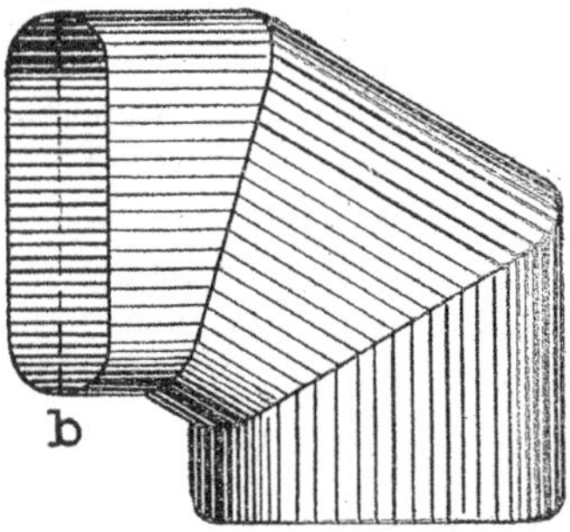

Fig. 49.

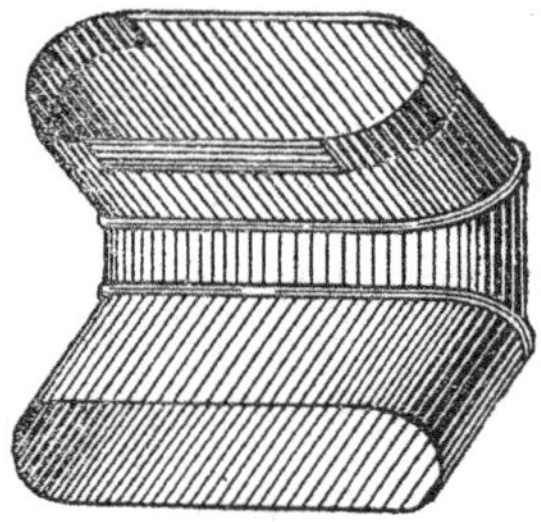

Fig. 50.

used in furnace-pipe work and are usually made from bright tin. Note the difference in the position of the sections of the two elbows. In Fig. 49 *a b* is in a vertical position, while in Fig. 50 it is in a horizontal position. In obtaining the patterns the same rule is employed as in previons problems, care being taken when developing the patterns for Fig. 49 that the section be placed as in Fig. 51 at A; and when developing the patterns for Fig. 50, that the section be placed as shown at A in Fig. 52.

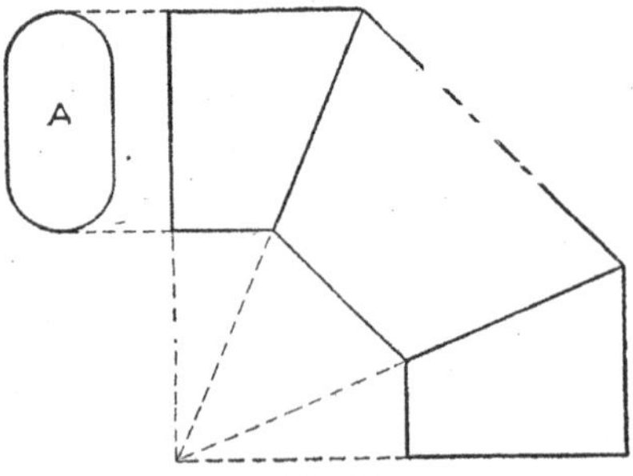

Fig. 51.

Fig. 53 shows a tapering two-pieced elbow, round in section. The method here shown is short and while not strictly accurate, gives good results. It has been shown in previous problems on Intersections and Developments that an oblique section through the opposite sides of a cone is a true ellipse. Bearing this in mind it is evident that if the frustum of the cone H I O N, Fig. 54, were a solid and cut obliquely by the plane J K and the several parts placed side by side, both would present true ellipses of exactly the same size, and if the two parts were placed together again turning the upper piece half-way around as shown by J W M K, the edges

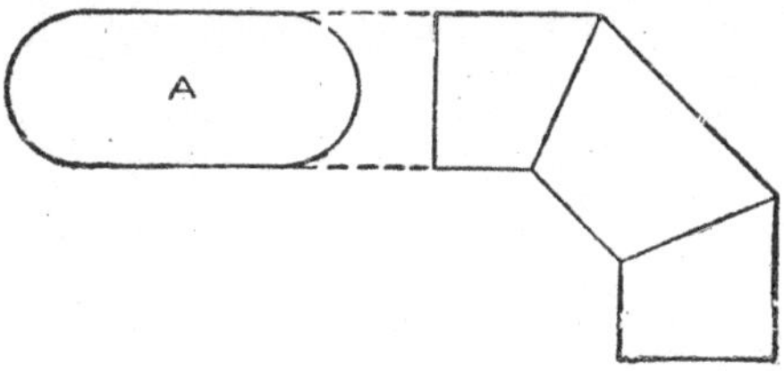

Fig. 52.

of the two pieces from J to K would exactly coincide. Taking advantage of this fact, it is necessary only to ascertain the angle of the line J K, to produce the required angle, between the two pieces of the elbow, both of which have an equal flare. The angle of the miter line, or the line which cuts the cone in two parts, must be found accurately so that when joined together an elbow will be formed having the desired angle on the line of its axis.

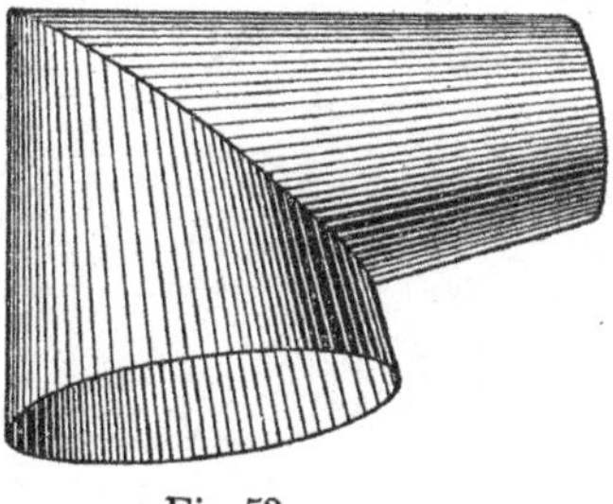

Fig. 53.

Therefore draw any vertical line as A B. With C as a center describe the plan of the desired diameter as shown by E D F B. At right angles to A B draw the bottom line of the elbow H I equal to E F, or in this case, 3 inches. Measuring from the line H I on the line A B the height of the frustum is 5 inches. Through X′ draw the upper diameter O N, $1\frac{1}{2}$ inches. Extend the contour lines of the frustum until they intersect the center line at L. Divide the half plan E D F into a number of equal parts as shown; from these points erect lines intersecting the base line H I from which draw lines to the apex L. As the elbow is to be in two pieces, and the axis at right angles, draw the angle T R S, bisect it at U and draw the line R V. No matter what the angle of the elbow, use this method. Now establish the point J at some convenient point on the cone, and from J, parallel to R V, draw the miter line J K intersecting the radial lines drawn through the cone; from these points and at right angles to the center line A B draw lines intersecting the side of the cone J H from 1 to 7. If it is

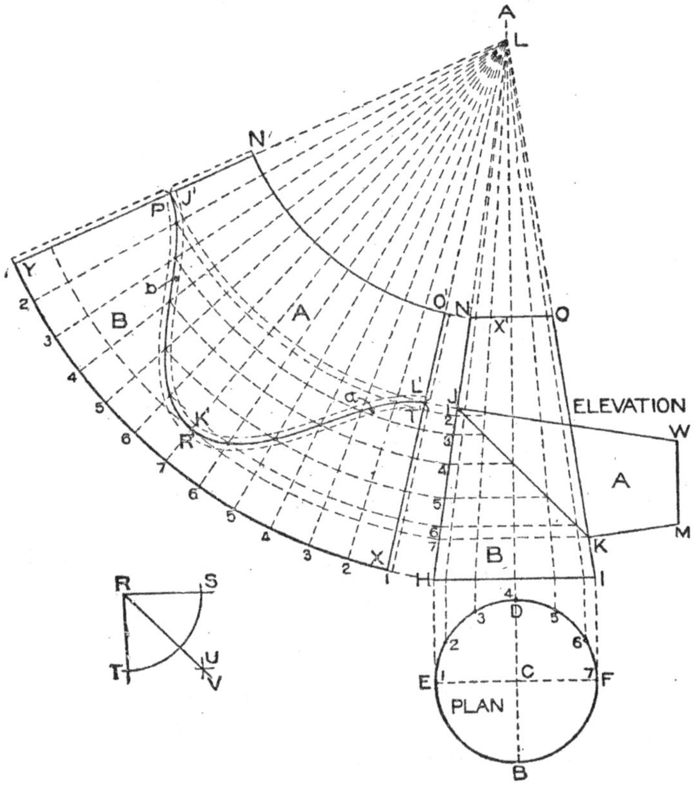

Fig. 54.

desired to know how the side of the tapering elbow would look, take a tracing of N O K J, reverse it and place it as shown by J W M K.

For the pattern proceed as follows: With L as a center and L H as a radius describe the arc 1 1. Starting from 1 set off on this arc twice the stretchout of 1 4 7 in plan, as shown by similar figures on 1 1, from which draw radial lines to the apex L. Again using L as center with radii equal to L N, L 1, L 2 to L 7, draw arcs as shown intersecting radial lines having similar numbers. Through these intersections draw the line J′ L′. Then O′ N′ J′ K′ L′ or A will be the pattern for the upper arm (A) in elevation, and P′ R′ T′ X Y or B the pattern for the lower arm (B) in elevation.

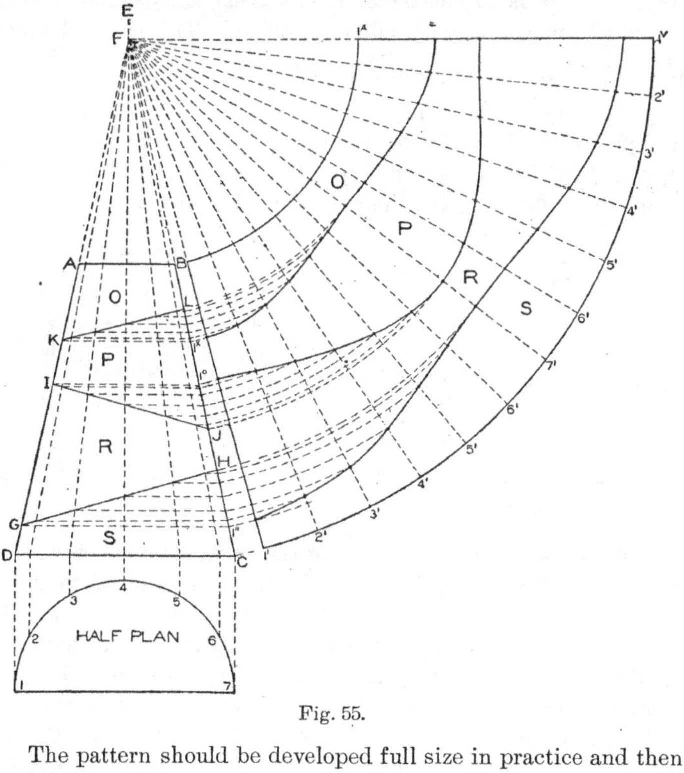

Fig. 55.

The pattern should be developed full size in practice and then pricked from the paper on to the sheet metal, drawing the two patterns as far apart as to admit allowing an edge to A at *a*; also an edge at *b* to B for seaming.

When a pattern is to contain more than two pieces the method of constructing the miter lines in the elevation of the cone is slightly different as shown in Fig. 55. Assume the bottom to be 3 inches in diameter and the top $1\frac{1}{4}$ inches. Let the vertical height be 4 inches. In this problem, as in the preceding, the various pieces necessary to form the elbow are cut from one cone whose dimensions must be determined from the dimensions of the required elbow. The first step is to determine the miter lines, which can be done the same as if regular pieced elbows were being developed. As the elbow is to consist of four pieces in 90°, follow the rule given in connection with elbow drafting. The top and bottom piece equal 2; the two middle pieces equal 4; total 6. $\frac{90}{6} = 15$. Lay off A B C D according to the dimensions given, and draw the half plan below D C; divide it into equal parts as shown. From the points of division erect perpendiculars intersecting D C, from which draw lines meeting the center line E 4 at F.

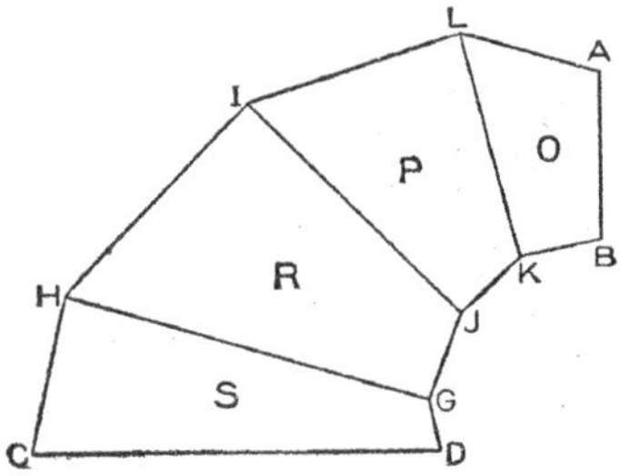

Fig. 56.

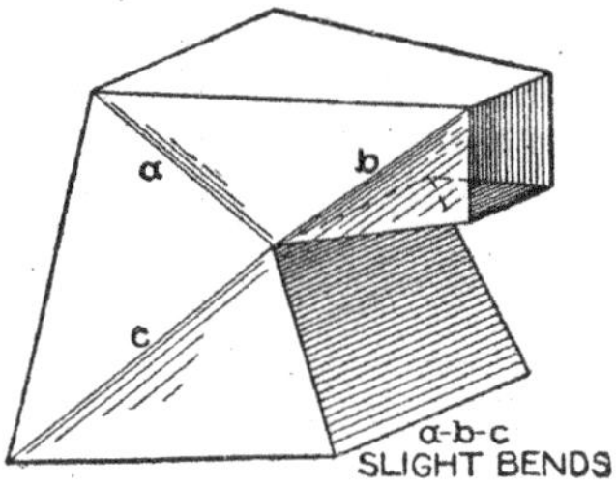

Fig. 57.

We assume that the amount of rise and projection of the elbow are not specified, excepting that the lines of axis will be at right angles. Knowing the angle of the miter line, it becomes a matter of judgment upon the part of the pattern draftsman, what length shall be given to each of the pieces composing the elbow. Therefor establish the points G, I and K, making D G, G I, I K and K A $\frac{1}{2}$, $1\frac{7}{8}$, $\frac{3}{4}$ and 1 inch respectively. From G, I and K draw the horizontal lines G 1″, I 1° and K 1ˣ. To each of these lines draw the lines G H, I J and K L respectively at an angle of 15° intersecting the radial lines in the cone as shown. From these intersections draw horizontal lines cutting the side of the cone. Then using F as a center, obtain the various patterns O, P, R and S in the manner already explained.

In Fig. 56 is shown a side view of the elbow, resulting from preceding operations; while it can be drawn from dimensions obtained in Fig. 55, it would be impossible to draw it without first having these dimensions.

In Fig. 57 is shown a perspective view of a tapering square elbow of square section in two pieces. This elbow may have any given taper. This problem will be developed by triangulation and parallel lines; it is an interesting study in projections as well as in developments. First draw the elevation of the elbow in Fig. 58 making 1–6 equal to $3\frac{1}{2}$ inches, the vertical height 1–2, $4\frac{1}{2}$ inches, and 6–5, $2\frac{1}{2}$ inches; the projection between 1 and 2 should be $\frac{5}{8}$ inch and between 5 and 6, $\frac{3}{8}$ inch. Make the horizontal distance

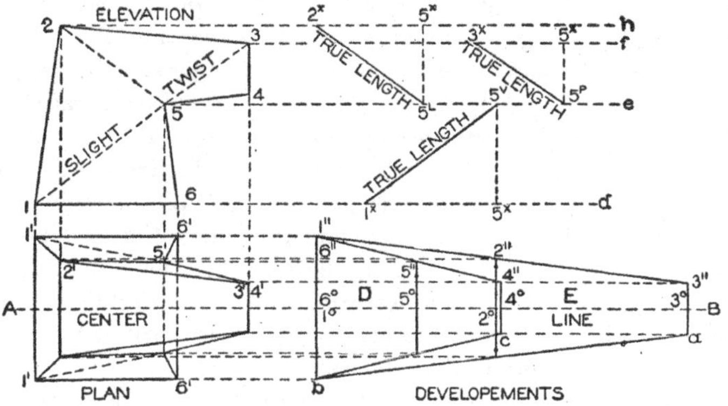

Fig. 58.

from 5 to 4, 2 inches, and the rise at 4 from the horizontal line $\frac{1}{4}$ inch, and the vertical distance from 4 to 3, $1\frac{1}{4}$ inches. Then draw a line from 3 to 2 to complete the elevation.

In its proper position below the line 1–6, draw the plan on that line, as shown by 1′ 1′ 6′ 6′. Through this line draw the center line A B. As the elbow should have a true taper from 1 to 3 and from 4 to 6, we may develop the patterns for the top and bottom pieces first and then from these construct the plan. Therefore, take the distances from 1 to 2 to 3 and from 4 to 5 to 6 in elevation and place them on the line A B in plan as shown respectively from 1° to 2° to 3° and from 4° to 5° to 6°; through these points draw vertical lines as shown. While the full developments

E and D are shown we shall deal with but one-half in the explanation which follows. As the elbow is to have the same taper on either side, take the half distance of the bottom of the elbow 1–6 and place it as shown from 1°–6° to 1″–6″, and the half width of the top of the elbow 3–4 and place it as shown from 3° to 3″ and 4° to 4″. Then draw lines from 3″ to 1″ intersecting the bend 2° at 2″, and a line from 4″ to 6″ intersecting the bend 5° at 5″. Trace these points on the opposite side of the line A B. Then 1″ 3″ *a b* will be the pattern for the top of the elbow and 6″ 4″ *c b* the pattern for the bottom. From these various points of intersection draw horizontal lines to the plan, and intersect them by lines drawn from similarly numbered points in the elevation at right angles to A B in plan. Draw lines through the points thus obtained in plan as shown by 1′, 2′, 3′, 4′, 5′ and 6′ which will represent the half plan view. For the completed plan, trace these lines opposite the line A B as shown. It will be noticed that the line 3–4 in elevation is perpendicular as shown by 3′ 4′ in plan while the points 2′ and 5′ project from it, showing that the piece 2–3–4–5 in elevation must be slightly twisted along the line 5–3 when forming the elbow. Similarly slight bends will be required along the lines 1–5 and 5–2.

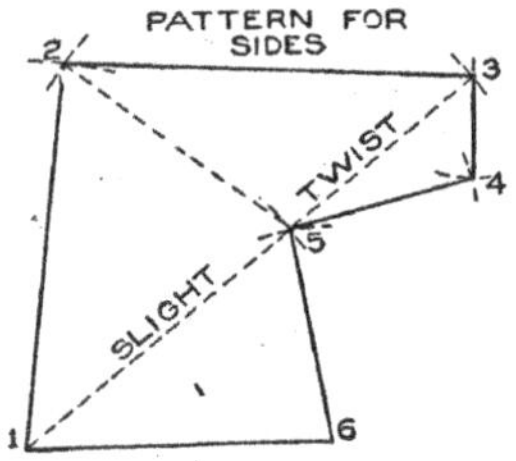

Fig. 59.

It will now be necessary to obtain the true lengths or a diagram of triangles on the lines 1–5, 5–2 and 5–3. Connect similar numbers in plan as shown from 1′ to 5′, 5′ to 2′ and 5′ to 3′, the last two lines being already shown. From similar points in elevation draw horizontal lines as shown by 2–*h*, 3–*f*, 5–*e* and 6–*d*. Take the distances from 1′ to 5′, 5′ to 2′ and 5′ to 3′ in plan and place them on one of the lines having a similar number in elevation, as shown respectively by 1^x 5^x, 5^x 2^x and 5^x 3^x. From the points marked 5^x draw vertical lines intersecting the horizontal line drawn from 5 at 5^v, 5^L and 5^p respectively. Now draw the true lengths 1^x 5^v, 2^x 5^L, and 3^x 5^p. For the pattern draw any line as 1–6 in Fig. 59 equal to 1–6 in Fig. 58. Now with 6″ 5″ in D as a radius and 6 in Fig. 59 as a center, describe the arc 5 which is intersected by an arc struck from 1 as a center and the true length

$1^x 5^v$ in Fig. 58 as radius. Then using the true length $5^L 2^x$ as radius and 5 in Fig. 59 as center, describe the arc 2, which is intersected by an arc struck from 1 as center and 1″ 2″ in E in Fig. 58 as radius. Using the true length $5^p 3^x$ as radius and 5 in Fig. 59 as center, describe the arc 3, and intersect it by an arc struck from 2 as center and 2″ 3″ in E in Fig. 58 as a radius. Now with 5″ 4″ in D as a radius and 5 in Fig. 59 as a center, describe tht arc 4, and intersect it by an arc struck from 3 as center and 3–4 in the elevation in Fig. 58 as a radius. Draw lines from point to point in Fig. 59 to complete the pattern. Laps should be allowed on all patterns, for seaming. Slight bends will take place as shown on the pattern, also as is shown by *a b* and *c* in Fig. 57. If the joint is to be on the line 2–5 in elevation in Fig. 58, the necessary pieces can be joined together.

In Fig. 60 is shown a perspective view of a five-piece tapering elbow, having a round base and an elliptical top. This form is generally known as a *ship ventilator*. The principles shown in this problem are applicable to any form or shape no matter what the respective profiles may be at the base or top. The first step is to draw a correct side view of the elbow as shown in Fig. 61. The outline A B O D E F can be drawn at pleasure, but for practice, dimensions are given. First draw the vertical line A F equal to $4\frac{1}{2}$ inches. On the same line extend measure down $1\frac{1}{4}$ inches to *f* and draw the horizontal line H B. From *f* set off a distance of $1\frac{1}{4}$ inches at G, and using G as a center and G F as a radius describe the arc F E intersecting H B at E, from which draw the vertical line E D equal to 1 inch. Draw D O equal to $1\frac{3}{4}$ inches, then draw C B. From B lay off $5\frac{3}{4}$ inches, and using this point (H) as a center and H B as a radius describe the arc B A. The portion shown B E D C is a straight piece of pipe whose section is shown by I J K L. Now divide the two arcs B A and E F into the same number of parts that the elbow is to have pieces (in this case four) and draw the lines of joint or miter lines as shown by U V, etc.

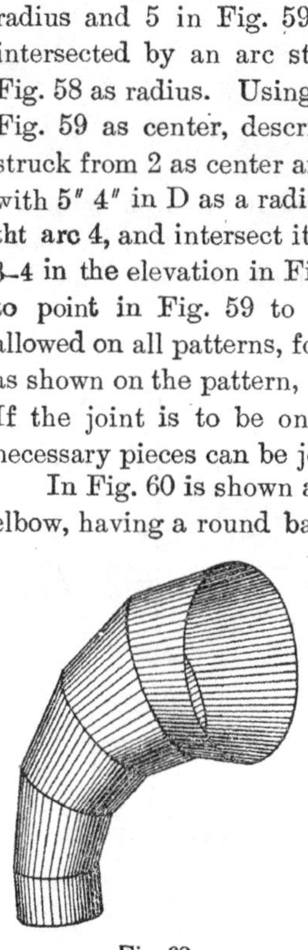

Fig. 60.

Bisect each one of the joint lines and obtain the points *a b c d* and *e*. Then A B C D E F will be the side view.

The patterns will be developed by triangulation, but before this can be done, true sections must be obtained on all of the lines in side elevation. The true sections on the lines B E and C D are shown by I J K L. The length of the sections are shown by the joint lines, but the width must be obtained from a front outline of the elbow, which is constructed as follows: In its proper relation to the side elevation, draw the center line M R upon which draw

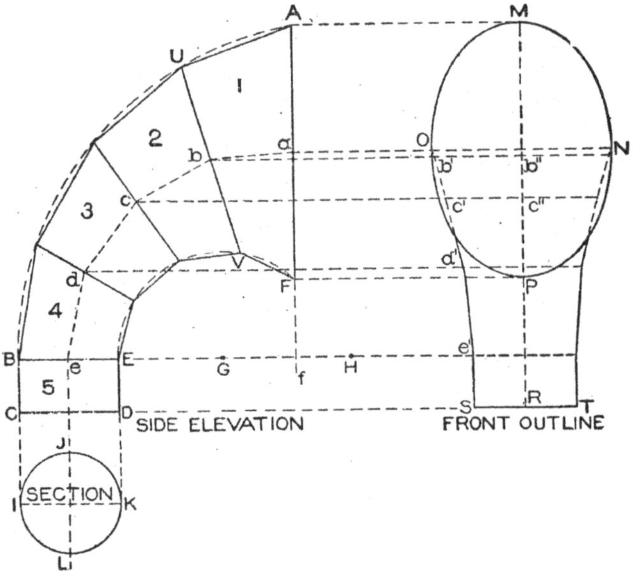

Fig. 61.

the ellipse M N O P (by methods already given in Mechanical Drawing) which represents the section on A F in side. Take half the diameter I K in section and place it on either side of the center line M R as R T or R S. Then draw the outline O S and T N in a convenient location. While this line is drawn at will, it should be understood that when once drawn, it becomes a fixed line. Now from the various intersections *a b c d* and *e* in the side elevation, draw lines through and intersecting the front outline as shown on one side by O, b', c', d' and e'. Then these distances will represent the widths of the sections shown by similar letters in side. For example, the method will be shown for obtaining the true section on U V, and the pattern for piece 1 in side elevation. To avoid a confusion of lines take a tracing of A F V U and place it as shown by 1, 13, 12, O in Fig. 62. On 1–13 place the half profile M N P of Fig. 61. Bisect O–12 in Fig. 62 and obtain the point 6; at a right angle to O–12 from 6 draw the line 6 6′ equal to b' b'' in front outline in Fig. 61. Then through the three points O, 6′ and 12 in Fig. 62, draw the semi-ellipse, which will represent the half section on U V. The other

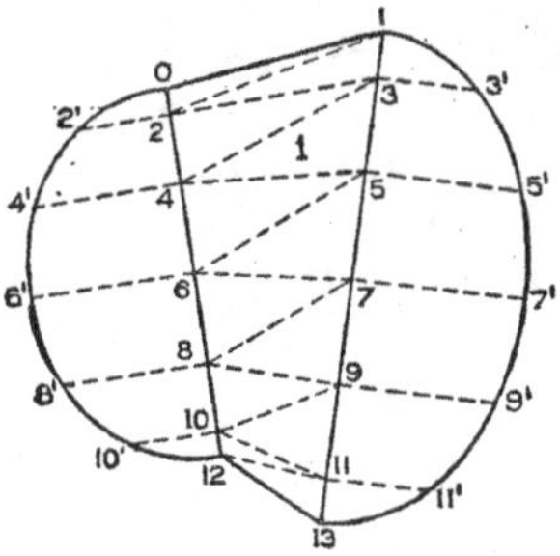

Fig. 62.

sections on the joint lines in side elevation are obtained in the same manner.

If the sections were required for piece 2 in side it would be necessary to use only O 6′ 12 in Fig. 62 and place it on U V in Fig. 61, and on a perpendicular line erected from c, place the width c' c'' shown in front and through the three points obtained again draw the semi-elliptical profile or section. Now divide the two half sections (Fig. 62) into equal parts as shown by the small figures, from which at right angles to 1–13 and O–12 draw lines intersecting these base lines from 1–13. Connect opposite points as 1 to 2 to 3 to 4 to 5, etc., to 12. Then these lines will represent

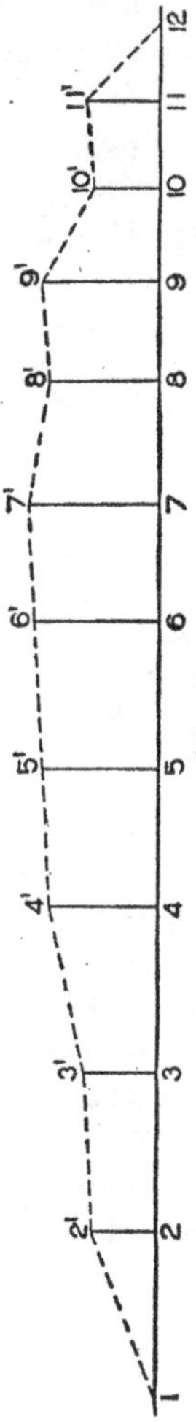

Fig. 63.

the bases of sections whose altitudes are equal to the heights in the half section. For these heights proceed as follows:

Take the various lengths from 1 to 2, 2 to 3, 3 to 4, 4 to 5, etc., to 11 to 12 and place them on the horizontal line in Fig. 63 as shown by similar figures; from these points erect vertical lines equal in height to similar figures, in the half section in Fig. 62 as shown by similar figures in Fig. 63. For example: Take the distance from 7 to 8 in Fig. 62 and place it as shown from 7 to 8 in Fig. 63 and erect vertical lines 7–7′, and 8–8′ equal to 7–7′ and 8–8′ in Fig. 62. Draw a line from 7′ to 8′ in Fig. 63 which is the true length on 7–8 in Fig. 62. For the pattern take the distance of 1–O and place it as shown by 1–O in Fig. 64. Now using O as a center and O 2′ in Fig. 62 as a radius, describe the arc 2 in Fig. 64

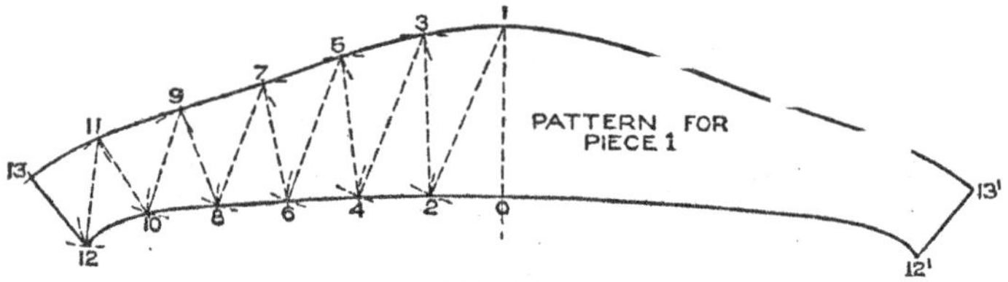

Fig. 64.

and intersect it by an arc struck from 1 as a center with 1–2′ in Fig. 63 as a radius. Now with 1–3′ in Fig. 62 as a radius and 1 in Fig. 64 as a center, describe the arc 3, and intersect it by an arc struck from 2 as center and 2′–3′ in Fig. 63 as a radius. Proceed thus, using alternately as radii, first the divisions in O–6′–12 in Fig. 62, then the proper line in Fig. 63, the divisions in 1–7′–13 in Fig. 62 and again the proper line in Fig. 63, until the line 12–13 in Fig. 64 is obtained, which equals 12–13 in Fig. 62. In this manner all of the sections are obtained, to which laps must be allowed for wiring and seaming.

TABLES.

The following tables will be found convenient for the Sheet-Metal Worker:

WEIGHT OF A SQUARE FOOT OF CAST AND WROUGHT IRON, COPPER, LEAD, BRASS AND ZINC.

FROM $\frac{1}{16}$ INCH TO ONE INCH IN THICKNESS.

THICKNESS.	CAST IRON.	WROUGHT IRON.	COPPER.	LEAD.	BRASS.	ZINC.
Inch.	Lbs.	Lbs.	Lbs.	Lbs.	Lbs.	Lbs.
1-16	2.346	2.517	2.89	3.691	2.675	2.34
1-8	4.693	5.035	5.781	7.382	5.35	4.68
3-16	7.039	7.552	8.672	11.074	8.025	7.02
1-4	9.386	10.07	11.562	14.765	10.7	9.36
5-16	11.733	12.588	14.453	18.456	13.375	11.7
3-8	14.079	15.106	17.344	22.148	16.05	14.04
7-16	16.426	17.623	20.234	25.839	18.725	16.34
1-2	18.773	20.141	23.125	29.53	21.4	18.72
9-16	21.119	22.659	26.016	33.222	24.075	
5-8	23.466	25.176	28.906	36.923	26.75	
11-16	25.812	27.694	31.797	40.604	29.425	
3-4	28.159	30.211	34.688	44.296	32.1	
13-16	30.505	32.729	37.578	47.987		
7-8	32.852	35.247	40.469	51.678		
15-16	35.199	37.764	43.359	55.37		
1	37.545	40.282	46.25	59.061		

NOTE.—The wrought iron and the copper weights are those of hard-rolled plates.

SHEET COPPER.

Official table adopted by the Association of Copper Manufacturers of the United States. Rolled copper has specific gravity of 8.93. One cubic foot weighs 558.125 pounds. One square foot, one inch thick, weighs 46.51 pounds.

Stubs' Gauge, (nearest number).	Thickness in decimal parts of an inch.	Ounces per sq. ft.	Sheets 14" x 48" Weight in pounds.	Sheets 24" x 48" Weight in pounds.	Sheets 30" x 60" Weight in pounds.	Sheets 36" x 72" Weight in pounds.	Sheets 48" x 72" Weight in pounds.
35.......	.00537	4	1.16	2	3.12	4.50	6
33.......	.00806	6	1.75	3	4.68	6.75	9
31.......	.0107	8	2.33	4	6.25	9	12
29.......	.0134	10	2.91	5	7.81	11.25	15
27.......	.0161	12	3.50	6	9.37	13.50	18
26.......	.0188	14	4.08	7	10.93	15.75	21
24.......	.0215	16	4.66	8	12.50	18	24
23.......	.0242	18	5.25	9	14.06	20.25	27
22.......	.0269	20	5.83	10	15.62	22.50	30
21.......	.0322	24	7	12	18.75	27	36
19.......	.0430	32	9.33	16	25	36	48
18.......	.0538	40	11.66	20	31.25	45	60
16.......	.0645	48	14	24	37.50	54	72
15.......	.0754	56	16.33	28	43.75	63	84
14.......	.0860	64	18.66	32	50	72	96
13.......	.095	70	....	35	55	79	105
12.......	.109	81	....	40½	63	91	122
11.......	.120	89	....	44½	70	100	134
10.......	.134	100	....	50	78	112	150
9.......	.148	110	....	55	86	124	165
8.......	.165	123	....	61	96	138	184
7.......	.180	134	....	67	105	151	201
6.......	.203	151	....	75½	118	170	227
5.......	.220	164	....	82	128	184	246
4.......	.238	177	....	88½	138	199	266
3.......	.259	193	....	96	151	217	289
2.......	.284	211	....	105½	165	238	317
1.......	.300	223	....	111½	174	251	335
0.......	.340	253	....	126½	198	285	380

SHEET ZINC.

Numbers		4	5	6	7	8	9	10	11	12	13	14	15	16	17	18	19	20	21	22
Weight per sq. foot		.30	.37	.45	.52	.60	.67	.75	.90	1.05	1.20	1.35	1.50	1.68	1.87	2.06	2.25	2.62	3.00	3.37
Approximate thickness in inches		.008	.010	.012	.014	.016	.018	.020	.024	.028	.032	.036	.040	.045	.050	.055	.060	.070	.080	.090
Size of sheet.	Sq. ft. per sheet.	APPROXIMATE WEIGHT PER SHEET.																		
24 x 84 in.	14.	4.2	5.2	6.3	7.3	8.4	9.4	10.5	12.6	14.7	16.8	18.9	21.	23.5	26.2	28.9	31.5	36.7	42.	47.2
26 x 84	15.2	4.6	5.6	6.9	7.9	9.1	10.2	11.4	13.7	16.	18.3	20.5	22.8	25.6	28.4	31.3	34.2	39.9	45.6	51.2
28 x 84	16.3	4.9	6.	7.4	8.5	9.8	10.9	12.2	14.7	17.1	19.6	22.	24.5	27.4	30.5	33.6	36.7	42.7	48.9	54.9
30 x 84	17.5	5.3	6.5	7.9	9.1	10.5	11.8	13.2	15.8	18.4	21.	23.6	26.2	29.4	32.8	36.1	39.4	45.8	52.5	59.
32 x 84	18.7	5.6	6.9	8.4	9.7	11.2	12.6	14.1	16.9	19.7	22.5	25.3	32.8	31.4	35.	38.5	42.	49.	56.1	63.
34 x 84	19.9	6.0	7.4	9.	10.4	12.	13.4	15.	18.	20.9	23.9	26.9	29.9	33.4	37.2	41.	44.8	52.2	59.7	67.
36 x 84	21.	6.3	7.8	9.5	10.9	12.6	14.1	15.8	18.9	22.	25.2	28.4	31.5	35.3	39.3	43.3	47.2	55.	63.	70.8
36 x 96	24.	7.2	8.9	10.8	12.5	14.4	16.1	18.	21.6	25.2	28.8	32.4	36.	40.3	44.9	49.5	54.	62.8	72.	80.9
36 x 108	27.	8.1	10.	12.2	14.1	16.2	18.1	20.3	24.3	28.4	32.4	36.5	40.5	45.4	50.5	55.6	60.7	70.7	81.	91.
40 x 84	23.4	7.	8.7	10.6	12.2	14.1	15.7	17.6	21.	24.6	28.1	31.6	35.1	39.3	43.8	48.2	52.6	61.3	70.2	78.8
40 x 96	26.8	8.	9.9	12.1	14.	16.1	18.	20.1	24.1	28.1	32.2	36.2	40.2	45.	50.1	55.2	60.3	70.2	80.4	90.3
44 x 84	25.7	7.7	9.5	11.6	13.4	15.4	17.2	19.3	23.1	27.	30.8	34.7	38.6	43.2	48.1	53.	57.8	67.4	77.1	86.6
46 x 90	28.7	8.6	10.6	12.9	14.9	17.2	19.2	21.5	25.8	30.1	34.4	38.7	43.	48.2	53.7	59.1	64.6	75.2	86.1	96.7
48 x 84	28.	8.4	10.4	12.6	14.6	16.8	18.8	21.	25.2	29.4	33.6	37.8	42.	47.	52.4	57.7	63.	73.4	84.	94.4
48 x 96	32.	9.6	11.9	14.4	16.7	19.2	21.5	24.	28.8	33.6	38.4	43.2	48.	53.8	59.9	65.9	72.	83.9	96.	107.8
50 x 108	37.5	11.3	13.9	16.9	19.5	22.5	25.1	28.2	33.8	39.3	45.	50.7	56.3	63.	70.1	77.3	84.4	98.3	112.5	126.4
52 x 84	30.4	9.1	11.3	13.7	15.8	18.3	20.4	22.8	27.4	31.9	36.5	41.	45.6	51.	56.9	62.6	68.4	79.6	91.2	102.5

Casks average about 600 pounds each. No. 4 to No. 17. Boxes average about 500 pounds. No. 18 and heavier.

UNITED STATES STANDARD GAUGE FOR SHEET AND PLATE IRON AND STEEL

COPY [Public—No. 137]

An act establishing a standard gauge for sheet and plate iron and steel.

Be it enacted by the Senate and House of Representatives of the United States of America in Congress assembled, That for the purpose of securing uniformity the following is established as the only standard gauge for sheet and plate iron and steel in the United States of America, namely:

Number of Gauge	THICKNESS		WEIGHT		Number of Gauge
	Approximate thickness in fractions of an inch	Approximate thickness in decimal parts of an inch	Weight per square foot in OUNCES avoirdupois	Weight per square foot in POUNDS avoirdupois	
0000000	1-2	.5	320	20.	0000000
000000	15-32	.46875	300	18.75	000000
00000	7-16	.4375	280	17.5	00000
0000	13-32	.40625	260	16.25	0000
000	3-8	.375	240	15.	000
00	11-32	.34375	220	13.75	00
0	5-16	.3125	200	12.5	0
1	9-32	.28125	180	11.25	1
2	17-64	.265625	170	10.625	2
3	1-4	.25	160	10.	3
4	15-64	.234375	150	9.375	4
5	7-32	.21875	140	8.75	5
6	13-64	.203125	130	8.125	6
7	3-16	.1875	120	7.5	7
8	11-64	.171875	110	6.875	8
9	5-32	.15625	100	6.25	9
10	9-64	.140625	90	5.625	10
11	1-8	.125	80	5.	11
12	7-64	.109375	70	4.375	12
13	3-32	.09375	60	3.75	13
14	5-64	.078125	50	3.125	14
15	9-128	.0703125	45	2.8125	15
16	1-16	.0625	40	2.5	16
17	9-160	.05625	36	2.25	17
18	1-20	.05	32	2.	18
19	7-160	.04375	28	1.75	19
20	3-80	.0375	24	1.5	20
21	11-320	.034375	22	1.375	21
22	1-32	.03125	20	1.25	22
23	9-320	.028125	18	1.125	23
24	1-40	.025	16	1.	24
25	7-320	.021875	14	.875	25
26	3-160	.01875	12	.75	26
27	11-640	.0171875	11	.6875	27
28	1-64	.015625	10	.625	28
29	9-640	.0140625	9	.5625	29
30	1-80	.0125	8	.5	30
31	7-640	.0109375	7	.4375	31
32	13-1280	.01015625	6½	.40625	32
33	3-320	.009375	6	.375	33
34	11-1280	.00859375	5½	.34375	34
35	5-640	.0078125	5	.3125	35
36	9-1280	.00703125	4½	.28125	36
37	17-2560	.0066406	4¼	.265625	37
38	1-160	.00625	4	.25	38

And on and after July first, eighteen hundred and ninety-three, the same and no other shall be used in determining duties and taxes levied by the United States of America on sheet and plate iron and steel. But this act shall not be construed to increase duties upon any articles which may be imported.

SEC. 2. That the Secretary of the Treasury is authorized and required to prepare suitable standards in accordance herewith.

SEC. 3. That in the practical use and application of the standard gauge hereby established a variation of two and one-half per cent either way may be allowed.

Approved, March 3, 1893.

WEIGHTS OF FLAT ROLLED IRON PER LINEAR FOOT.

Iron weighing 480 pounds per cubic foot.

Thickness in Inches.	1″	1¼″	1½″	1¾″	2″	2¼″	2½″	2¾″	12″
1/16	.208	260	.313	.365	.417	.469	.521	.573	2.50
1/8	.417	.521	.625	.729	.833	.938	1.04	1.15	5.00
3/16	.625	.781	.938	1.09	1.25	1.41	1.56	1.72	7.50
1/4	.833	1.04	1.25	1.46	1.67	1.88	2.08	2.29	10.00
5/16	1.04	1.30	1.56	1.82	2.08	2.34	2.60	2.86	12.50
3/8	1.25	1.56	1.88	2.19	2.50	2.81	3.13	3.44	15.00
7/16	1.46	1.82	2.19	2.55	2.92	3.28	3.65	4.01	17.50
1/2	1.67	2.08	2.50	2.92	3.33	3.75	4.17	4.58	20.00
9/16	1.88	2.34	2.81	3.28	3.75	4.22	4.69	5.16	22.50
5/8	2.08	2.60	3.13	3.65	4.17	4.69	5.21	5.73	25.00
11/16	2.29	2.86	3.44	4.01	4.58	5.16	5.73	6.30	27.50
3/4	2.50	3.13	3.75	4.38	5.00	5.63	6.25	6.88	30.00
13/16	2.71	3.39	4.06	4.74	5.42	6.09	6.77	7.45	32.50
7/8	2.92	3.65	4.38	5.10	5.83	6.56	7.29	8.02	35.00
15/16	3.13	3.91	4.69	5.47	6.25	7.03	7.81	8.59	37.50
1	3.33	4.17	5.00	5.83	6.67	7.50	8.33	9.17	40.00
1 1/16	3.54	4.43	5.31	6.20	7.08	7.97	8.85	9.74	42.50
1 1/8	3.75	4.69	5.63	6.56	7.50	8.44	9.38	10.31	45.00
1 3/16	3.96	4.95	5.94	6.93	7.92	8.91	9.90	10.89	47.50
1 1/4	4.17	5.21	6.25	7.29	8.33	9.38	10.42	11.46	50.00
1 5/16	4.37	5.47	6.56	7.66	8.75	9.84	10.94	12.03	52.50
1 3/8	4.58	5.73	6.88	8.02	9.17	10.31	11.46	12.60	55.00
1 7/16	4.79	5.99	7.19	8.39	9.58	10.78	11.98	13.18	57.50
1 1/2	5.00	6.25	7.50	8.75	10.00	11.25	12.50	13.75	60.00
1 9/16	5.21	6.51	7.81	9.11	10.42	11.72	13.02	14.32	62.50
1 5/8	5.42	6.77	8.13	9.48	10.83	12.19	13.54	14.90	65.00
1 11/16	5.63	7.03	8.44	9.84	11.25	12.66	14.06	15.47	67.50
1 3/4	5.83	7.29	8.75	10.21	11.67	13.13	14.58	16.04	70.00
1 13/16	6.04	7.55	9.06	10.57	12.08	13.59	15.10	16.61	72.50
1 7/8	6.25	7.81	9.38	10.94	12.50	14.06	15.63	17.19	75.00
1 15/16	6.46	8.07	9.69	11.30	12.92	14.53	16.15	17.76	77.50
2	6.67	8.33	10.00	11.67	13.33	15.00	16.67	18.33	80.00

WEIGHTS OF FLAT ROLLED IRON PER LINEAR FOOT.

(Continued)

Thickness in Inches.	3″	3¼″	3½″	3¾″	4″	4¼″	4½″	4¾″	12″
1/16	.625	.677	.729	.781	.833	.885	.938	.990	2.50
1/8	1.25	1.35	1.46	1.56	1.67	1.77	1.88	1.98	5.00
3/16	1.88	2.03	2.19	2.34	2.50	2.66	2.81	2.97	7.50
1/4	2.50	2.71	2.92	3.13	3.33	3.54	3.75	3.96	10.00
5/16	3.13	3.39	3.65	3.91	4.17	4.43	4.69	4.95	12.50
3/8	3.75	4.06	4.38	4.69	5.00	5.31	5.63	5.94	15.00
7/16	4.38	4.74	5.10	5.47	5.83	6.20	6.56	6.93	17.50
1/2	5.00	5.42	5.83	6.25	6.67	7.08	7.50	7.92	20.00
9/16	5.63	6.09	6.56	7.03	7.50	7.97	8.44	8.91	22.50
5/8	6.25	6.77	7.29	7.81	8.33	8.85	9.38	9.90	25.00
11/16	6.88	7.45	8.02	8.59	9.17	9.74	10.31	10.89	27.50
3/4	7.50	8.13	8.75	9.38	10.00	10.63	11.25	11.88	30.00
13/16	8.13	8.80	9.48	10.16	10.83	11.51	12.19	12.86	32.50
7/8	8.75	9.48	10.21	10.94	11.67	12.40	13.13	13.85	35.00
15/16	9.38	10.16	10.94	11.72	12.50	13.28	14.06	14.84	37.50
1	10.00	10.83	11.67	12.50	13.33	14.17	15.00	15.83	40.00
1 1/16	10.63	11.51	12.40	13.28	14.17	15.05	15.94	16.82	42.50
1 1/8	11.25	12.19	13.13	14.06	15.00	15.94	16.88	17.81	45.00
1 3/16	11.88	12.86	13.85	14.84	15.83	16.82	17.81	18.80	47.50
1 1/4	12.50	13.54	14.58	15.63	16.67	17.71	18.75	19.79	50.00
1 5/16	13.13	14.22	15.31	16.41	17.50	18.59	19.69	20.78	52.50
1 3/8	13.75	14.90	16.04	17.19	18.33	19.48	20.63	21.77	55.00
1 7/16	14.38	15.57	16.77	17.97	19.17	20.36	21.56	22.76	57.50
1 1/2	15.00	16.25	17.50	18.75	20.00	21.25	22.50	23.75	60.00
1 9/16	15.63	16.93	18.23	19.53	20.83	22.14	23.44	24.74	62.50
1 5/8	16.25	17.60	18.96	20.31	21.67	23.02	24.38	25.73	65.00
1 11/16	16.88	18.28	19.69	21.09	22.50	23.91	25.31	26.72	67.50
1 3/4	17.50	18.96	20.42	21.88	23.33	24.79	26.25	27.71	70.00
1 13/16	18.13	19.64	21.15	22.66	24.17	25.68	27.19	28.70	72.50
1 7/8	18.75	20.31	21.88	23.44	25.00	26.56	28.13	29.69	75.00
1 15/16	19.38	20.99	22.60	24.22	25.83	27.45	29.06	30.68	77.50
2	20.00	21.67	23.33	25.00	26.67	28.33	30.00	31.67	80.00

WEIGHTS OF FLAT ROLLED IRON PER LINEAR FOOT.

(Continued)

Thickness in Inches.	5″	5¼″	5½″	5¾″	6″	6¼″	6½″	6¾″	12″
$\frac{1}{16}$	1.04	1.09	1.15	1.20	1.25	1.30	1.35	1.41	2.50
$\frac{1}{8}$	2.08	2.19	2.29	2.40	2.50	2.60	2.71	2.81	5.00
$\frac{3}{16}$	3.13	3.28	3.44	3.59	3.75	3.91	4.06	4.22	7.50
$\frac{1}{4}$	4.17	4.38	4.58	4.79	5.00	5.21	5.42	5.63	10.00
$\frac{5}{16}$	5.21	5.47	5.73	5.99	6.25	6.51	6.77	7.03	12.50
$\frac{3}{8}$	6.25	6.56	6.88	7.19	7.50	7.81	8.13	8.44	15.00
$\frac{7}{16}$	7.29	7.66	8.02	8.39	8.75	9.11	9.48	9.84	17.50
$\frac{1}{2}$	8.33	8.75	9.17	9.58	10.00	10.42	10.83	11.25	20.00
$\frac{9}{16}$	9.38	9.84	10.31	10.78	11.25	11.72	12.19	12.66	22.50
$\frac{5}{8}$	10.42	10.94	11.46	11.98	12.50	13.02	13.54	14.06	25.00
$\frac{11}{16}$	11.46	12.03	12.60	13.18	13.75	14.32	14.90	15.47	27.50
$\frac{3}{4}$	12.50	13.13	13.75	14.38	15.00	15.63	16.25	16.88	30.00
$\frac{13}{16}$	13.54	14.22	14.90	15.57	16.25	16.93	17.60	18.28	32.50
$\frac{7}{8}$	14.58	15.31	16.04	16.77	17.50	18.23	18.96	19.69	35.00
$\frac{15}{16}$	15.63	16.41	17.19	17.97	18.75	19.53	20.31	21.09	37.50
1	16.67	17.50	18.33	19.17	20.00	20.83	21.67	22.50	40.00
$1\frac{1}{16}$	17.71	18.59	19.48	20.36	21.25	22.14	23.02	23.91	42.50
$1\frac{1}{8}$	18.75	19.69	20.63	21.56	22.50	23.44	24.38	25.31	45.00
$1\frac{3}{16}$	19.79	20.78	21.77	22.76	23.75	24.74	25.73	26.72	47.50
$1\frac{1}{4}$	20.83	21.88	22.92	23.96	25.00	26.04	27.08	28.13	50.00
$1\frac{5}{16}$	21.88	22.97	24.06	25.16	26.25	27.34	28.44	29.53	52.50
$1\frac{3}{8}$	22.92	24.06	25.21	26.35	27.50	28.65	29.79	30.94	55.00
$1\frac{7}{16}$	23.96	25.16	26.35	27.55	28.75	29.95	31.15	32.34	57.50
$1\frac{1}{2}$	25.00	26.25	27.50	28.75	30.00	31.25	32.50	33.75	60.00
$1\frac{9}{16}$	26.04	27.34	28.65	29.95	31.25	32.55	33.85	35.16	62.50
$1\frac{5}{8}$	27.08	28.44	29.79	31.15	32.50	33.85	35.21	36.56	65.00
$1\frac{11}{16}$	28.13	29.53	30.94	32.34	33.75	35.16	36.56	37.97	67.50
$1\frac{3}{4}$	29.17	30.63	32.08	33.54	35.00	36.46	37.92	39.38	70.00
$1\frac{13}{16}$	30.21	31.72	33.23	34.74	36.25	37.76	39.27	40.78	72.50
$1\frac{7}{8}$	31.25	32.81	34.38	35.94	37.50	39.06	40.63	42.19	75.00
$1\frac{15}{16}$	32.29	33.91	35.52	37.14	38.75	40.36	41.98	43.59	77.50
2	33.33	35.00	36.67	38.33	40.00	41.67	43.33	45.00	80.00

WEIGHTS OF FLAT ROLLED IRON PER LINEAR FOOT.

(Continued)

Thickness in Inches.	7″	7¼″	7½″	7¾″	8″	8¼″	8½″	8¾″	12″
1/16	1.46	1.51	1.56	1.61	1.67	1.72	1.77	1.82	2.50
1/8	2.92	3.02	3.13	3.23	3.33	3.44	3.54	3.65	5.00
3/16	4.38	4.53	4.69	4.84	5.00	5.16	5.31	5.47	7.50
1/4	5.83	6.04	6.25	6.46	6.67	6.88	7.08	7.29	10.00
5/16	7.29	7.55	7.81	8.07	8.33	8.59	8.85	9.11	12.50
3/8	8.75	9.06	9.38	9.69	10.00	10.31	10.63	10.94	15.00
7/16	10.21	10.57	10.94	11.30	11.67	12.03	12.40	12.76	17.50
1/2	11.67	12.08	12.50	12.92	13.33	13.75	14.17	14.58	20.00
9/16	13.13	13.59	14.06	14.53	15.00	15.47	15.94	16.41	22.50
5/8	14.58	15.10	15.63	16.15	16.67	17.19	17.71	18.23	25.00
11/16	16.04	16.61	17.19	17.76	18.33	18.91	19.48	20.05	27.50
3/4	17.50	18.13	18.75	19.38	20.00	20.63	21.25	21.88	30.00
13/16	18.96	19.64	20.31	20.99	21.67	22.34	23.02	23.70	32.50
7/8	20.42	21.15	21.88	22.60	23.33	24.06	24.79	25.52	35.00
15/16	21.88	22.66	23.44	24.22	25.00	25.78	26.56	27.34	37.50
1	23.33	24.17	25.00	25.83	26.67	27.50	28.33	29.17	40.00
1 1/16	24.79	25.68	26.56	27.45	28.33	29.22	30.10	30.99	42.50
1 1/8	26.25	27.19	28.13	29.06	30.00	30.94	31.88	32.81	45.00
1 3/16	27.71	28.70	29.69	30.68	31.67	32.66	33.65	34.64	47.50
1 1/4	29.17	30.21	31.25	32.29	33.33	34.38	35.42	36.46	50.00
1 5/16	30.62	31.72	32.81	33.91	35.00	36.09	37.19	38.28	52.50
1 3/8	32.08	33.23	34.38	35.52	36.67	37.81	38.96	40.10	55.00
1 7/16	33.54	34.74	35.94	37.14	38.33	39.53	40.73	41.93	57.50
1 1/2	35.00	36.25	37.50	38.75	40.00	41.25	42.50	43.75	60.00
1 9/16	36.46	37.76	39.06	40.36	41.67	42.97	44.27	45.57	62.50
1 5/8	37.92	39.27	40.63	41.98	43.33	44.69	46.04	47.40	65.00
1 11/16	39.38	40.78	42.19	43.59	45.00	46.41	47.81	49.22	67.50
1 3/4	40.83	42.29	43.75	45.21	46.67	48.13	49.58	51.04	70.00
1 13/16	42.29	43.80	45.31	46.82	48.33	49.84	51.35	52.86	72.50
1 7/8	43.75	45.31	46.88	48.44	50.00	51.56	53.13	54.69	75.00
1 15/16	45.21	46.82	48.44	50.05	51.67	53.28	54.90	56.51	77.50
2	46.67	48.33	50.00	51.67	53.33	55.00	56.67	58.33	80.00

WEIGHTS OF FLAT ROLLED IRON PER LINEAR FOOT.

(Continued)

Thickness in Inches.	9″	9¼″	9½″	9¾″	10″	10¼″	10½″	10¾″	12″
1/16	1.88	1.93	1.98	2.03	2.08	2.14	2.19	2.24	2.50
1/8	3.75	3.85	3.96	4.06	4.17	4.27	4.38	4.48	5.00
3/16	5.63	5.78	5.94	6.09	6.25	6.41	6.56	6.72	7.50
1/4	7.50	7.71	7.92	8.13	8.33	8.54	8.75	8.96	10.00
5/16	9.38	9.64	9.90	10.16	10.42	10.68	10.94	11.20	12.50
3/8	11.25	11.56	11.88	12.19	12.50	12.81	13.13	13.44	15.00
7/16	13.13	13.49	13.85	14.22	14.58	14.95	15.31	15.68	17.50
1/2	15.00	15.42	15.83	16.25	16.67	17.08	17.50	17.92	20.00
9/16	16.88	17.34	17.81	18.28	18.75	19.22	19.69	20.16	22.50
5/8	18.75	19.27	19.79	20.31	20.83	21.35	21.88	22.40	25.00
11/16	20.63	21.20	21.77	22.34	22.92	23.49	24.06	24.64	27.50
3/4	22.50	23.13	23.75	24.38	25.00	25.62	26.25	26.88	30.00
13/16	24.38	25.05	25.73	26.41	27.08	27.76	28.44	29.11	32.50
7/8	26.25	26.98	27.71	28.44	29.17	29.90	30.63	31.35	35.00
15/16	28.13	28.91	29.69	30.47	31.25	32.03	32.81	33.59	37.50
1	30.00	30.83	31.67	32.50	33.33	34.17	35.00	35.83	40.00
1 1/16	31.88	32.76	33.65	34.53	35.42	36.30	37.19	38.07	42.50
1 1/8	33.75	34.69	35.63	36.56	37.50	38.44	39.38	40.31	45.00
1 3/16	35.63	36.61	37.60	38.59	39.58	40.57	41.56	42.55	47.50
1 1/4	37.50	38.54	39.58	40.63	41.67	42.71	43.75	44.79	50.00
1 5/16	39.38	40.47	41.56	42.66	43.75	44.84	45.94	47.03	52.50
1 3/8	41.25	42.40	43.54	44.69	45.83	46.98	48.13	49.27	55.00
1 7/16	43.13	44.32	45.52	46.72	47.92	49.11	50.31	51.51	57.50
1 1/2	45.00	46.25	47.50	48.75	50.00	51.25	52.50	53.75	60.00
1 9/16	46.88	48.18	49.48	50.78	52.08	53.39	54.69	55.99	62.50
1 5/8	48.75	50.10	51.46	52.81	54.17	55.52	56.88	58.23	65.00
1 11/16	50.63	52.03	53.44	54.84	56.25	57.66	59.06	60.47	67.50
1 3/4	52.50	53.96	55.42	56.88	58.33	59.79	61.25	62.71	70.00
1 13/16	54.38	55.89	57.40	58.91	60.42	61.93	63.44	64.95	72.50
1 7/8	56.25	57.81	59.38	60.94	62.50	64.06	65.63	67.19	75.00
1 15/16	58.13	59.74	61.35	62.97	64.58	66.20	67.81	69.43	77.50
2	60.00	61.67	63.33	65.00	66.67	68.33	70.00	71.67	80.00

WEIGHTS OF FLAT ROLLED IRON PER LINEAR FOOT.

(Concluded)

Thickness in Inches.	11″	11¼″	11½″	11¾″	12″	12¼″	12½″	12¾″
1/16	2.29	2.34	2.40	2.45	2.50	2.55	2.60	2.66
1/8	4.58	4.69	4.79	4.90	5.00	5.10	5.21	5.31
3/16	6.88	7.03	7.19	7.34	7.50	7.66	7.81	7.97
1/4	9.17	9.38	9.58	9.79	10.00	10.21	10.42	10.63
5/16	11.46	11.72	11.98	12.24	12.50	12.76	13.02	13.28
3/8	13.75	14.06	14.38	14.69	15.00	15.31	15.63	15.94
7/16	16.04	16.41	16.77	17.14	17.50	17.86	18.23	18.59
1/2	18.33	18.75	19.17	19.58	20.00	20.42	20.83	21.25
9/16	20.63	21.09	21.56	22.03	22.50	22.97	23.44	23.91
5/8	22.92	23.44	23.96	24.48	25.00	25.52	26.04	26.56
11/16	25.21	25.78	26.35	26.93	27.50	28.07	28.65	29.22
3/4	27.50	28.13	28.75	29.38	30.00	30.63	31.25	31.88
13/16	29.79	30.47	31.15	31.82	32.50	33.18	33.85	34.53
7/8	32.08	32.81	33.54	34.27	35.00	35.73	36.46	37.19
15/16	34.38	35.16	35.94	36.72	37.50	38.28	39.06	39.84
1	36.67	37.50	38.33	39.17	40.00	40.83	41.67	42.50
1 1/16	38.96	39.84	40.73	41.61	42.50	43.39	44.27	45.16
1 1/8	41.25	42.19	43.13	44.06	45.00	45.94	46.88	47.81
1 3/16	43.54	44.53	45.52	46.51	47.50	48.49	49.48	50.47
1 1/4	45.83	46.88	47.92	48.96	50.00	51.04	52.08	53.13
1 5/16	48.13	49.22	50.31	51.41	52.50	53.59	54.69	55.78
1 3/8	50.42	51.56	52.71	53.85	55.00	56.15	57.29	58.44
1 7/16	52.71	53.91	55.10	56.30	57.50	58.70	59.90	61.09
1 1/2	55.00	56.25	57.50	58.75	60.00	61.25	62.50	63.75
1 9/16	57.29	58.59	59.90	61.20	62.50	63.80	65.10	66.41
1 5/8	59.58	60.94	62.29	63.65	65.00	66.35	67.71	69.06
1 11/16	61.88	63.28	64.69	66.09	67.50	68.91	70.31	71.72
1 3/4	64.17	65.63	67.08	68.54	70.00	71.46	72.92	74.38
1 13/16	66.46	67.97	69.48	70.99	72.50	74.01	75.52	77.03
1 7/8	68.75	70.31	71.88	73.44	75.00	76.56	78.13	79.69
1 15/16	71.04	72.66	74.27	75.89	77.50	79.11	80.73	82.34
2	73.33	75.00	76.67	78.33	80.00	81.67	83.33	85.00

The weights for 12″ width are repeated on each page to facilitate making the additions necessary for plates wider than 12″. Thus, to find the weight of 15¼″ x ⅞″, add the weights to be found in the same line for 3¼″ x ⅞″ and 12″x⅞″ = 9.48 + 35.00 = 44.48 lbs.

SQUARE AND ROUND IRON BARS.

Thickness or Diameter in Inches.	Weight of □ Bar One Foot long.	Weight of ○ Bar One Foot long.	Area of □ Bar in sq. inches.	Area of ○ Bar in sq. inches.	Circumference of ○ Bar in inches.
0					
1/16	.013	.010	.0039	.0031	.1963
1/8	.052	.041	.0156	.0123	.3927
3/16	.117	.092	.0352	.0276	.5890
1/4	.208	.164	.0625	.0491	.7854
5/16	.326	.256	.0977	.0767	.9817
3/8	.469	.368	.1406	.1104	1.1781
7/16	.638	.501	.1914	.1503	1.3744
1/2	.833	.654	.2500	.1963	1.5708
9/16	1.055	.828	.3164	.2485	1.7671
5/8	1.302	1.023	.3906	.3068	1.9635
11/16	1.576	1.237	.4727	.3712	2.1598
3/4	1.875	1.473	.5625	.4418	2.3562
13/16	2.201	1.728	.6602	.5185	2.5525
7/8	2.552	2.004	.7656	.6013	2.7489
15/16	2.930	2.301	.8789	.6903	2.9452
1	3.333	2.618	1.0000	.7854	3.1416
1/16	3.763	2.955	1.1289	.8866	3.3379
1/8	4.219	3.313	1.2656	.9940	3.5343
3/16	4.701	3.692	1.4102	1.1075	3.7306
1/4	5.208	4.091	1.5625	1.2272	3.9270
5/16	5.742	4.510	1.7227	1.3530	4.1233
3/8	6.302	4.950	1.8906	1.4849	4.3197
7/16	6.888	5.410	2.0664	1.6230	4.5160
1/2	7.500	5.890	2.2500	1.7671	4.7124
9/16	8.138	6.392	2.4414	1.9175	4.9087
5/8	8.802	6.913	2.6406	2.0739	5.1051
11/16	9.492	7.455	2.8477	2.2365	5.3014
3/4	10.21	8.018	3.0625	2.4053	5.4978
13/16	10.95	8.601	3.2852	2.5802	5.6941
7/8	11.72	9.204	3.5156	2.7612	5.8905
15/16	12.51	9.828	3.7539	2.9483	6.0868
2	13.33	10.47	4.0000	3.1416	6.2832
1/16	14.18	11.14	4.2539	3.3410	6.4795
1/8	15.05	11.82	4.5156	3.5466	6.6759
3/16	15.95	12.53	4.7852	3.7583	6.8722
1/4	16.88	13.25	5.0625	3.9761	7.0686
5/16	17.83	14.00	5.3477	4.2000	7.2649
3/8	18.80	14.77	5.6406	4.4301	7.4613
7/16	19.80	15.55	5.9414	4.6664	7.6576
1/2	20.83	16.36	6.2500	4.9087	7.8540
9/16	21.89	17.19	6.5664	5.1572	8.0503
5/8	22.97	18.04	6.8906	5.4119	8.2467
11/16	24.08	18.91	7.2227	5.6727	8.4430

SQUARE AND ROUND IRON BARS.

(Concluded)

Thickness or Diameter in Inches.	Weight of □ Bar One Foot long.	Weight of ○ Bar One Foot long.	Area of □ Bar in sq. inches.	Area of ○ Bar in sq. inches.	Circumference of ○ Bar in inches.
3/4	25.21	19.80	7.5625	5.9396	8.6394
13/16	26.37	20.71	7.9102	6.2126	8.8357
7/8	27.55	21.64	8.2656	6.4918	9.0321
15/16	28.76	22.59	8.6289	6.7771	9.2284
3	30.00	23.56	9.0000	7.0686	9.4248
1/16	31.26	24.55	9.3789	7.3662	9.6211
1/8	32.55	25.57	9.7656	7.6699	9.8175
3/16	33.87	26.60	10.160	7.9798	10.014
1/4	35.21	27.65	10.563	8.2958	10.210
5/16	36.58	28.73	10.973	8.6179	10.407
3/8	37.97	29.82	11.391	8.9462	10.603
7/16	39.39	30.94	11.816	9.2806	10.799
1/2	40.83	32.07	12.250	9.6211	10.996
9/16	42.30	33.23	12.691	9.9678	11.192
5/8	43.80	34.40	13.141	10.321	11.388
11/16	45.33	35.60	13.598	10.680	11.585
3/4	46.88	36.82	14.063	11.045	11.781
13/16	48.45	38.05	14.535	11.416	11.977
7/8	50.05	39.31	15.016	11.793	12.174
15/16	51.68	40.59	15.504	12.177	12.370
4	53.33	41.89	16.000	12.566	12.566
1/16	55.01	43.21	16.504	12.962	12.763
1/8	56.72	44.55	17.016	13.364	12.959
3/16	58.45	45.91	17.535	13.772	13.155
1/4	60.21	47.29	18.063	14.186	13.352
5/16	61.99	48.69	18.598	14.607	13.548
3/8	63.80	50.11	19.141	15.033	13.744
7/16	65.64	51.55	19.691	15.466	13.941
1/2	67.50	53.01	20.250	15.904	14.137
9/16	69.39	54.50	20.816	16.349	14.334
5/8	71.30	56.00	21.391	16.800	14.530
11/16	73.24	57.52	21.973	17.257	14.726
3/4	75.21	59.07	22.563	17.721	14.923
13/16	77.20	60.63	23.160	18.190	15.119
7/8	79.22	62.22	23.766	18.665	15.315
15/16	81.26	63.82	24.379	19.147	15.512
5	83.33	65.45	25.000	19.635	15.708

ANGLE IRON.

Weight Per Linear Foot.

6 x 6 x 5/8 .... 24 Lbs.	2 x 2 x 1/4 .... 3½ Lbs.
5 x 5 x 9/16 .... 16½ "	1¾ x 1¾ x 3/16 .... 2¾ "
4 x 4 x ½ .... 12½ "	1½ x 1½ x 3/16 .... 2 "
3½ x 3½ x 7/16 .... 9 "	1¼ x 1¼ x 3/16 .... 1½ "
3 x 3 x 3/8 .... 7 "	1 x 1 x 1/8 .... 1 "
2½ x 2½ x 5/16 .... 5 "	¾ x ¾ x 1/8 .... 5/8 "
2¼ x 2¼ x ¼ .... 4¼ "	

TEE IRON.

Weight Per Linear Foot.

5 x 8 x 5/8 .... 30 Lbs.	2¼ x 2¼ x ¼ .... 4 Lbs
7 x 6 x 5/8 .... 30 "	2 x 2 x ¼ .... 3¼ "
6 x 3 x ½ .... 16½ "	1¾ x 1¾ x ¼ .... 3 "
4 x 4 x ½ .... 14 "	1½ x 1½ x ¼ .... 2¾ "
3½ x 3½ x ½ .... 12½ "	1¼ x 1¼ x ¼ .... 2¼ "
3 x 3 x 3/8 .... 7¾ "	1 x 1 x 1/8 .... 1 "
2½ x 2½ x ½ .... 8 "	¾ x ¾ x 1/8 .... 3/8 "
2½ x 2½ x 5/16 .... 5 "	

VENTILATION WORK *

In the illustration is shown a system of ventilation, in which the various pipes are led from brick or metal flues to the attic as shown, and connected to the sheet metal drum.

This drum is made in size equal to the combined area of all pipes entering same. The drum is set upon a wooden platform as shown and has a clean-out door made large enough to admit a man's body. Steam coils are placed inside to create a suction, when the heated air rises through the ventilator. The drum is connected to the ventilator as shown, the bracing of the ventilator being fastened to the inside of the curb.

The detail at the right shows the connection joint between the pipe and drum, while that at the left shows the construction of the metal door and frame, with method of fastening to the body of the drum.

---

* The illustration referred to will be found on the back of this page.

CONSTRUCTION DRAWING SHOWING SHEET METAL DRUM AND VENTILATOR IN VENTILATION WORK

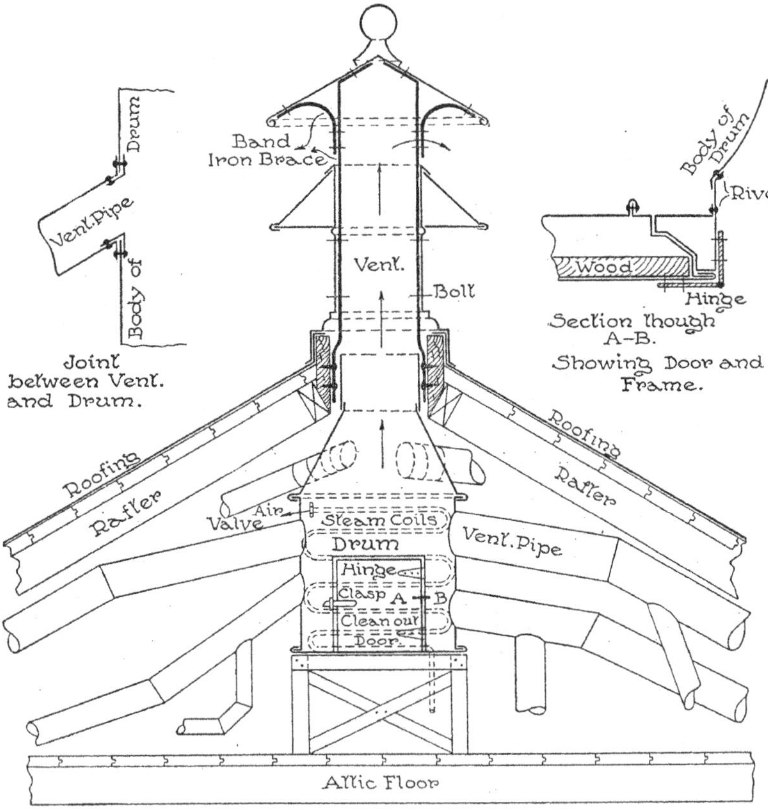

Sectional view showing ventilation pipes connected to drum in attic also steam coils in drum to create suction.

SHEET METAL WORK.

PART II.

PROBLEMS FOR LIGHT GAUGE METAL.

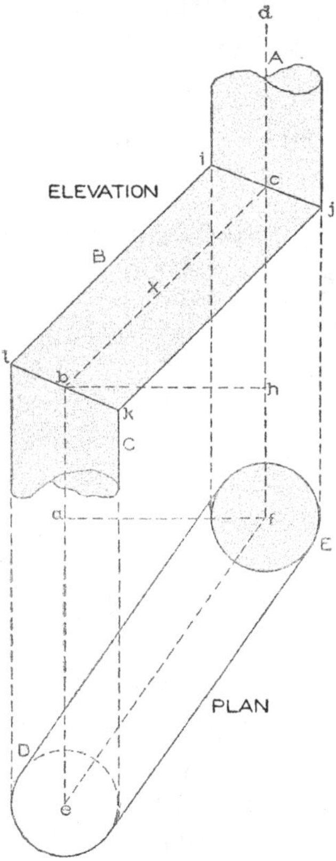

Fig. 65.

It is often the case that the sheet metal worker receives plans for vent, heat, or blower pipes to be constructed, in which the true lengths and angles are not shown but must be obtained from the plans or measurements at the building.

Figs. 65 and 66 show the principles employed for obtaining the true angles and lengths in oblique piping, it being immaterial whether the piping is round, square, or oval in section. The only safe way in obtaining these angles is to use the center line as a basis and after this line has been obtained, build the pipe around it, so to speak. In Fig. 65 let A B C represent the elevation of the elbow shown in plan by D E. Through the center of the pipes draw the center line *a b c d* which intersect the center lines of the pipe in plan at *e* and *f*. In elevation the rise of the middle piece B on the center line is equal to *h c* and projects to the right a distance equal to *b h*, shown in plan by *e f*; this same pipe projects forward in plan a distance equal to *e a*. While the miter lines in elevation *i j* and *k l* have been drawn straight, they would in reality show curved lines; those lines have not been projected as there is no necessity for doing so.

With the various heights and projections in plan and elevation the true length and true angles are obtained as shown in Fig.

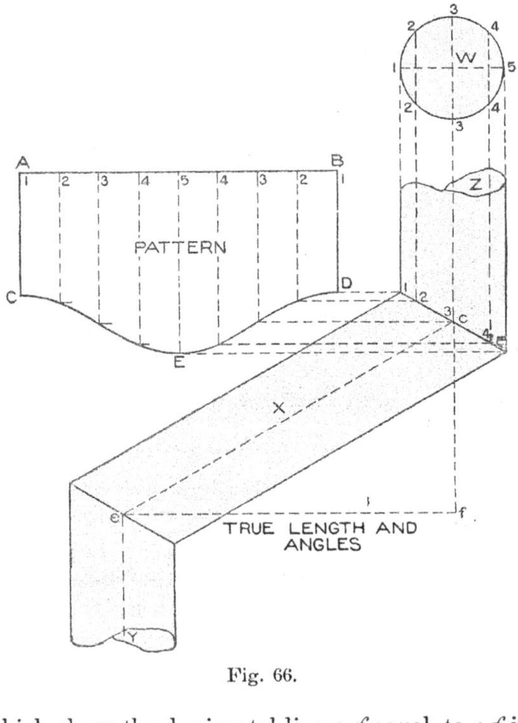

Fig. 66.

66, in which draw the horizontal line *e f* equal to *e f* in plan in Fig. 65. Take the height from *h* to *c* and place it from *f* to *c* in Fig. 66 on a vertical line erected from *f*. Draw a line from *e* to *c* which is the true length on the center line of the pipe shown by B in elevation in Fig. 65. From the points *c* and *e* in Fig. 66 draw perpendicular lines, making Y *e* X and X *c* Z = the true angles shown by *a b* X and X *c d* respectively in Fig. 65. On either side of the center line in Fig. 66 lay off the half diameter of the pipe as shown, and in its proper position draw the profile W.

Divide this into equal spaces and obtain the pattern A B D E C in the usual manner. As both angles are similar the miter cut C E D can be used for all of the patterns. In drawing this problem for practice make the diameter of the pipe 2 inches, the height from *h* to *c* $3\frac{3}{4}$ inches in Fig. 65, the projection *b* to *h* $3\frac{3}{4}$ inches, and the projection in plan *e* to *a* $5\frac{1}{4}$ inches.

Our next problem is that of a rain-water cut-off, a perspective view of which is shown in Fig. 67. While the miter cuts in this problem are similar to elbow work the intersection between the two beveled arms, and the cut-off or slide on the inside require attention. Make the diameter of the three openings each 2 inches; A to B (Fig. 68) $1\frac{1}{2}$ inches. From B at an angle of 45° draw B C $3\frac{1}{4}$ inches and C D 2 inches. From G draw the vertical miter line G *h*. Make the distance from B to T $\frac{1}{2}$ inch. Place the line *d e* of the cut-off $\frac{1}{8}$ inch above the line T U as indicated at *a* and the line *e c* to the right of *h* G, as indicated by *b*, a distance of $\frac{3}{16}$ inch. Parallel to G H draw *c d* giving slight play room between G H, intersecting *e d* and *e c* at *d* and *c* respectively. From *e* at right angles to *d c*, draw a line as shown, intersecting *h* G at *f*, which is the pivot on which the cut-off *c d e* will turn either right or left. The angles of the pipes on opposite sides are constructed in similar manner; A B C D E F G H I J K L M will be the elevation, N, the section on A M and O P R S the section on I J. B T U L shows how far the upper tube projects into the body under which the scoop *e d c* turns right and left to throw the rain water into either elbow as desired. The pattern for the upper piece A T U M is a straight piece of metal whose circumference is equal to N.

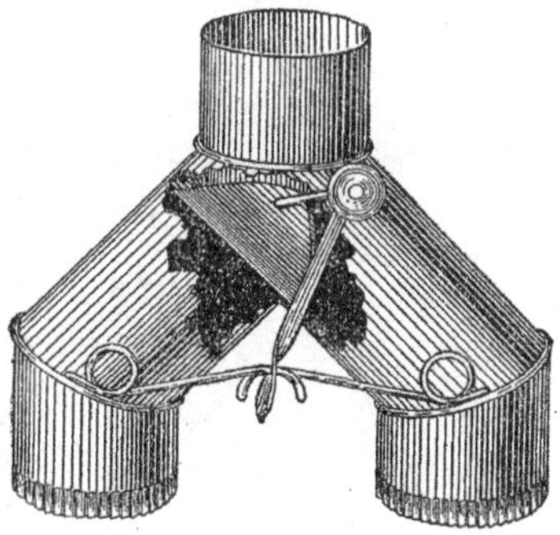

Fig. 67.

For the pattern for (A), divide the half section O P R into equal spaces as shown, from which erect lines intersecting the miter line H K as shown, and from which, parallel to K L and H G, draw lines intersecting the joint lines G *h* L as shown. As none of the lines just drawn intersect the corner *h*, it will be necessary to obtain this point on the half section O P R from which the stretchout of the pattern is taken. Therefore from *h*, parallel to L K draw *h h'* intersecting H K at *h'*, from which, parallel to K J, drop a line intersecting the profile O P R S at *h''*. At right angles to L K draw stretchout of O P R S as shown by similar numbers on T^1 U^1, through which at right angles to T^1 U^1 draw lines which are intersected by lines drawn at right angles to L K from similar in-

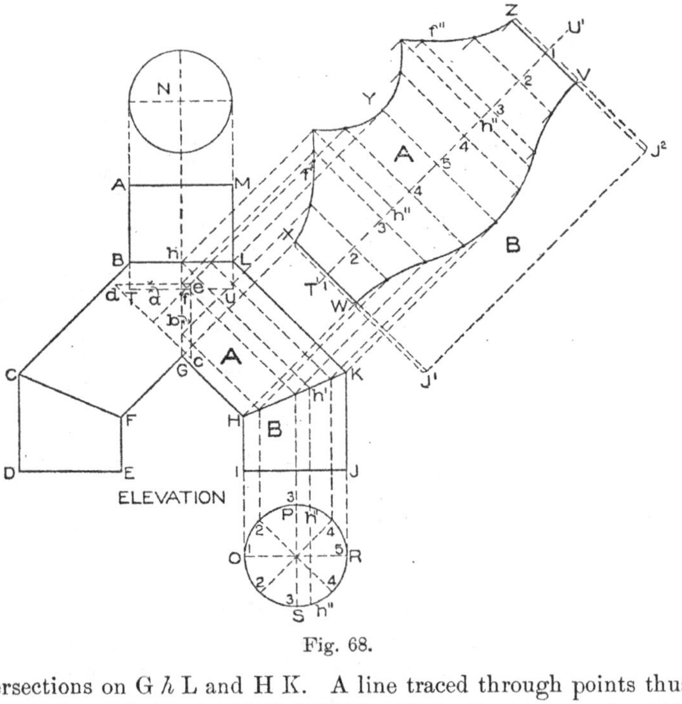

Fig. 68.

tersections on G *h* L and H K. A line traced through points thus obtained as shown by X Y Z V W will be the pattern for (A). From *f* in the elevation at right angles to L K project a line intersecting the miter cut X Y Z at *f'* and *f''*. At *f'* and *f''* holes are to be punched in which the pivot *f* of the scoop *c d e* in elevation will turn.

While the pattern for (B) can be obtained as that for (A) was obtained, a short method is to take the distance K to J and place it as shown from W to J^1 and V to J^2 on the lines of the pattern X W and Z V respectively extended. W V J^2 J^1 will be the pattern for B.

To avoid a confusion of lines in the development of the scoop or cut-off *c d e*, this has been shown in Fig. 69 in which *d e c* is a reproduction of *d e c* in Fig. 68. A true section of the scoop must now be drawn on *x e* in Fig. 69 so that its dimensions will allow it to turn easily inside of the joint line G *h* in elevation in Fig. 68. Therefore draw any horizontal line as 4 5 in Fig. 69, at right angles to which from *f* draw a vertical line intersecting 4 5 at *f*. Now take a distance $\frac{1}{16}$ inch less than one-half the diameter of O R in Fig. 68, and place it in Fig. 69 on either side of the line 4 5 on the vertical line just drawn as shown from *f* to 2 and *f* to 2′. Extend *d c* till it intersects 4 5 at 4. Draw a line from 4 to 2′; by bisecting this line we obtain the line *a b* intersecting 4 5 at *i*. Then with *i* as center and *i* 2′ as radius, describe the arc 2′ 2.

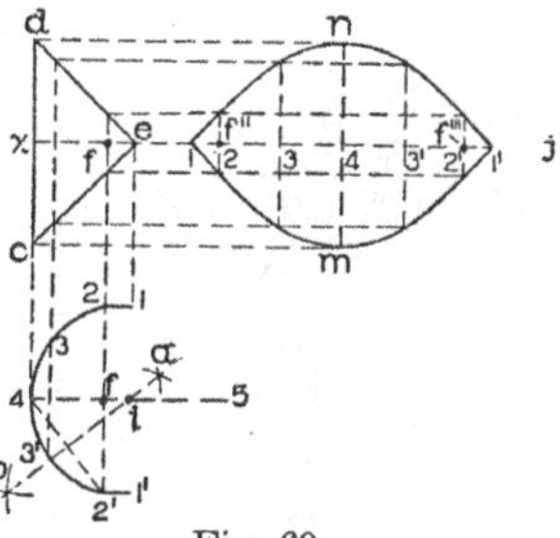

Fig. 69.

From 2 and 2′ draw horizontal lines equal to *f e* as shown by 2 1 and 2′ 1′. Then will 1 4 1′ be the true section on *x e*. Divide the half section into equal spaces as shown from 1 to 4, from which erect lines intersecting *c e* and *e d*. Extend *x e* as *x j*, upon which place the stretchout of 1 4 1′ as shown by similar numbers on *x j*, through which draw vertical lines. These lines intersect with horizontal lines drawn from similar intersections on *d e c*. Through points thus obtained draw the line 1 *n* 1′ *m* which is the desired pattern. As the pivot hole *f* falls directly on line 2, then *f‴ f″* will be the position of the holes in the pattern. Laps must be allowed to all patterns.

In putting up rectangular hot air pipe it is often the case that the pipe will be placed in the partition of one story, then has to fall forward and twist one quarter way around to enter the partition of the upper story which runs at right angles to the lower one. A perspective view showing this condition is shown in Fig. 70, where the upper opening turns one quarter on the lower one and leaning to the right as much as is shown in Fig. 71 in plan. This problem is known as a transition piece in a rectangular pipe.

Full size measurements are given in Fig. 71 which should be drawn one-half size. The height of the transition piece is 1 foot 8 inches, the size of the openings, each 4×10 inches turned as shown, two inches to the left and two inches above the lower section as shown. From the plan construct the front and side elevations as shown by the dotted lines. A B C D and E F G H will then be the front and side elevations of the transition piece respectively

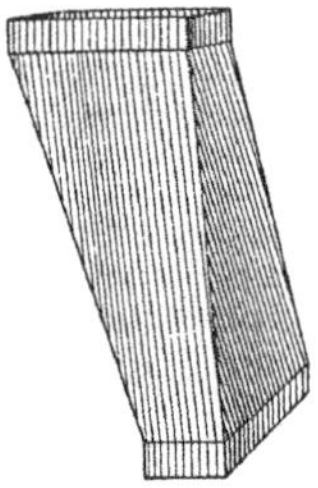

Fig. 70.

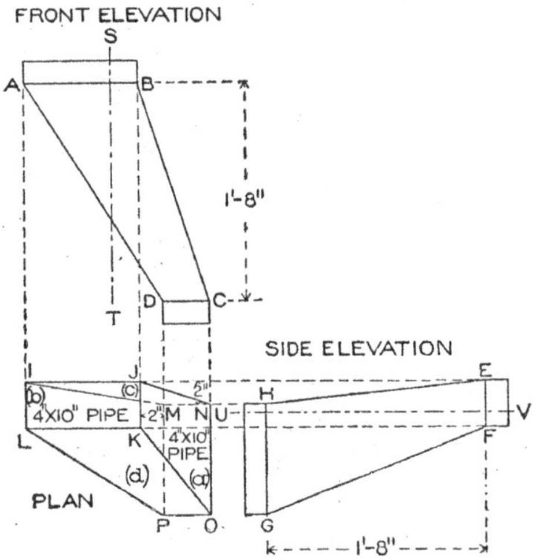

Fig. 71.

equal to 20 inches or 10 inches for practice. Number each side of the plan (*a*), (*b*), (*c*), and (*d*). Through the front and side elevations draw the vertical and horizontal lines S T and U V respectively at pleasure. These lines are only used as bases for measurements in determining the patterns. For the pattern for the side marked (a) in plan take the length of B C and place it on the vertical line B C in Fig. 72. Through the points B and C draw the horizontal lines E F and H G, making B F and B E, and C G and C H equal respectively to the distances measured from the line U V in Fig. 71 to points F, E, G, H. Draw lines from E to H and F to G in Fig. 72, which is the pattern for (a).

For the pattern for (b) in Fig. 71 take the distance of A D, and place it as shown by A D in Fig. 72; through A and D draw E F and H G, making A F and A E, and D G and D H equal

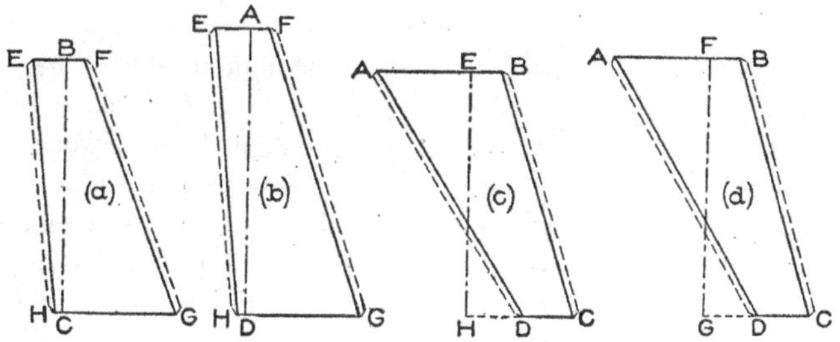

Fig. 72.

respectively to the distances measured from the line U V in side elevation in Fig. 71 to points F, E, G, H. Draw lines from E to H and F to G in Fig. 72, which will be the pattern for (b). In similar manner obtain the patterns for (c) and (d) in plan in Fig. 71. The lengths of E H and F G are placed as shown by similar letters

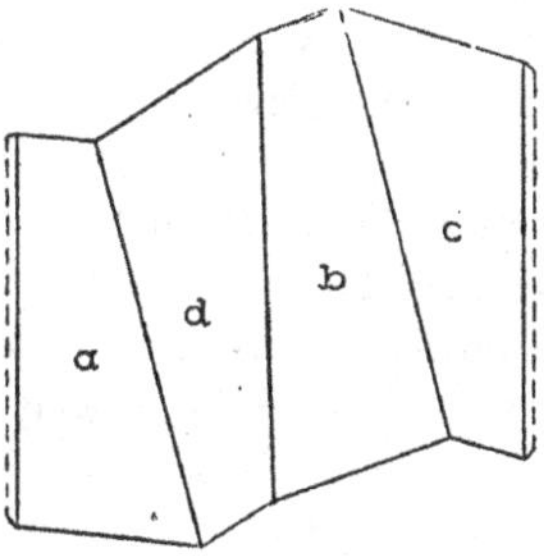

Fig. 73.

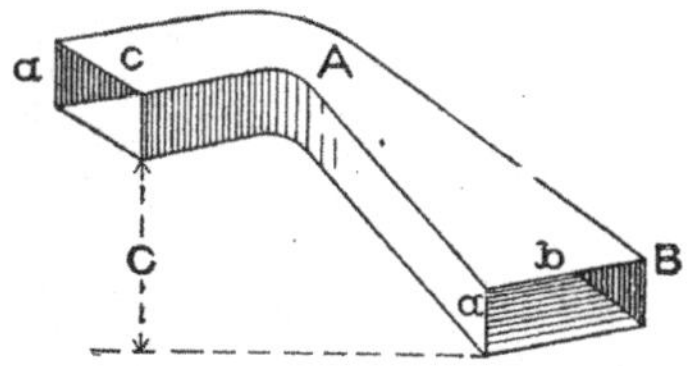

Fig. 74.

in Fig. 72, while the projections to A, B, C, D are obtained from A, B, C, D in front elevation in Fig. 71, measuring in each instance from S T.

If desired the top and lower flange shown in the perspective in Fig. 70 can be added to the patterns in Fig. 72. Laps are allowed to the patterns to allow for double seaming at corners, if, however, the pattern should be required in one piece, it would only be necessary to join the various pieces in their proper positions as shown by *a d b e* in Fig. 73, which would bring the seam on the line J N in plan in Fig. 71.

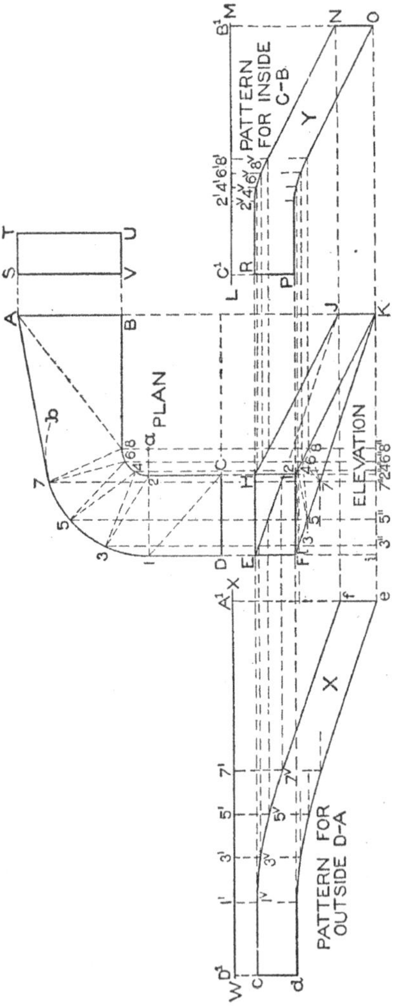

Fig. 75.

In Fig. 74 is shown a perspective view of a curved rectangular chute the construction of which arises in piping and blower work. The problem as here presented shows the sides *a* and *a* in vertical planes having the same height, while the bottom *b* has more width than the top *c*. The top opening is to rise above the bottom opening a given distance equal to C. First draw the plan and elevation as shown in Fig. 75, make A B equal to 2 inches, B 8 $2\frac{1}{2}$ inches; with a radius equal to $\frac{1}{2}$ inch, with *a* as center draw the quarter circle 8 2. From 2 draw the vertical line 2 C equal to $1\frac{3}{8}$ inches and draw C D equal to $1\frac{1}{2}$ inches. Make D 1 equal to C 2 and using *a* as center and *a* 1 as radius draw the arc 1 *b*. From A draw a line tangent to 1 *b* as A 7. A B C D will be the plan of the chute. In line with A B draw the section S T U V. In line with D C draw the section E F I H as shown. Place the desired rise of the chute as shown by F *i* in elevation and from *i* draw a horizontal line as *i* K, which intersect by a line drawn from A B in plan as shown. Make K J equal to F E and draw the lines F K, K I, and E J, J H. F E J K is the elevation of the outside curve, H I K J the inside curve, F I K the bottom, and E H J the top.

Having the plan and elevation in position we will first draw the pattern for the two vertical sides. For the pattern for the side of the chute shown by B C in plan proceed as follows: Divide the inner curve 2 to 8 into equal parts as shown by 2-4-6 and 8, from which points drop lines intersecting the inside of the chute in plan H J K I as shown. At right angles to J K draw L M, upon which place the stretchout of B C in plan as shown by similar letters and numbers on L M, through which draw vertical lines which intersect lines drawn parallel to L M from H J. Through points thus obtained draw the line R 2^V 4^V 6^V 8^VN. The same method can

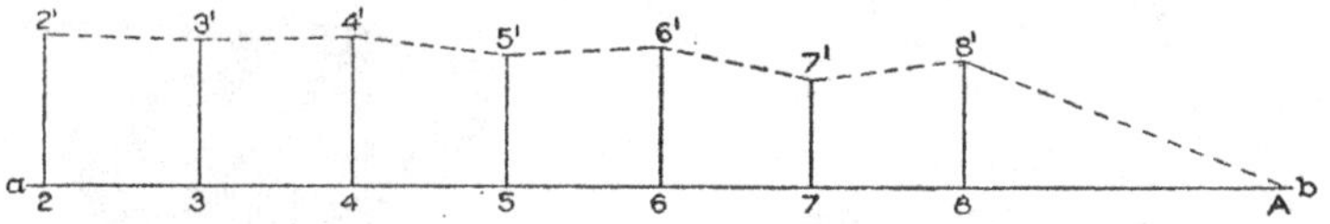

Fig. 76.

be employed for the curve P O, but as the height H I and J K are equal, having a common profile B C, take the height of H I or J K and place it on vertical lines as R P and N O and trace the curve R N as shown by P O. N O P R is the pattern for C B in plan; To obtain the pattern for the outside curve divide the curve 1-7 into equal parts as shown, from which drop vertical lines intersecting similar points in E J K F, in elevation at right angles to E F draw W X, upon which place the stretchout of D A in plan as shown. From the divisions on W X drop vertical lines, which intersect by lines drawn from similar numbered intersections on E J. Trace a line through these points as shown by *c f* and draw *d e* as explained in connection with the inside pattern. *c d e f* is the pattern for the outside of the chute shown in plan by D A.

As both the top and bottom of the chute have the same bevel, the pattern for one will answer for the other. Connect opposite points in plan as shown from C to 1 to 2 to 3 up to 8, then to A. In similar manner connect similar points on the bottom in elevation as shown from 1 to 2 up to K. The lines in plan represent the bases of the sections whose altitudes are equal to the various heights in elevation, measured from *i* K. Take the various lengths from 2 to 3 to 4 to 5 to 6 to 7 to 8 to A in plan and place them as shown by similar numbers on the horizontal line *a b* (Fig. 76); through *a b* draw vertical lines, equal in height to similar numbers in elevation, in Fig. 75, measured from the line *i* K. For example take the distance 4 5 in plan and place it as shown by 4 5 in Fig. 76. Erect perpendiculars 4 4′ and 5 5′ equal to 4″ 4 and 5″ 5 in elevation in Fig. 75. Draw a line from 4′ to 5′ in Fig. 76, which is the true length of 4 5 in plan in Fig. 75. Proceed in similar manner for the balance of the sections. Take a tracing of 1 2 C D in plan and place it as shown by 1, 2, C, D in Fig. 77. Now using 1 as

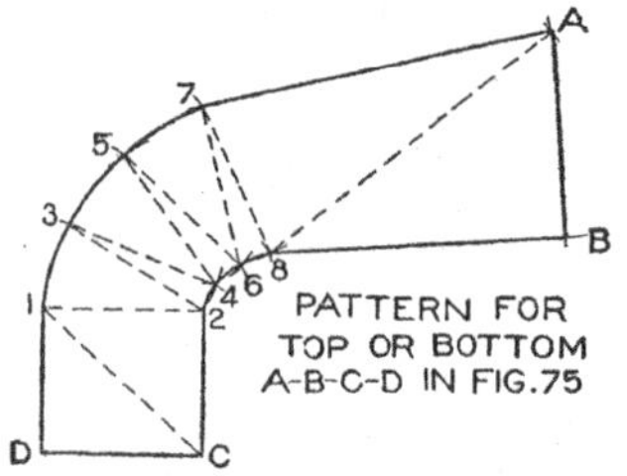

Fig. 77.

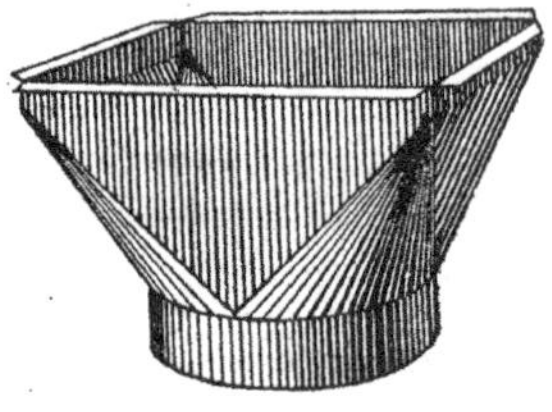
Fig. 78.

center and 1^V 3^V in (*x*), in Fig. 75, as radius, describe the arc at 3, in Fig. 77, which is intersected by an arc, struck from 2 as center, and 2′ 3′, in Fig. 76, as radius. Now with radius equal to 2^V 4^V in (Y) in Fig. 75 and 2 in Fig. 77 as center, describe the arc at 4 which is intersected by an arc, struck from 3 as center and 3′ 4′, Fig. 76, as radius. Proceed in this manner, using alternately as radius, first the divisions in the pattern (X), Fig. 75, then the slant lines in Fig. 76, the divisions in the pattern (Y), Fig. 75, then again the lines in Fig. 76 until the line 7 8, Fig. 77, has been obtained. Then using 7 as center, with a line equal to $7^V f$ in (X), Fig. 75, as radius, describe the arc A, Fig. 77, which is intersected by an arc struck from 8 as center and 8′ A, Fig. 76, as radius. Then with radius, equal to 8^V N in (Y), Fig. 75, and 8, Fig. 77, as center, describe the arc B, which is intersected by an arc, struck from A as center and A B in plan in Fig. 75 as radius. Trace lines through points thus obtained in Fig. 77, and A B C D will be the desired pattern. Laps must be allowed on all patterns for double seaming the corners.

In Fig. 78 is shown a perspective view of a hopper register box usually made from bright tin or galvanized iron in hot air piping. In drawing this problem, the student should first draw the half plan, making the semicircle $3\frac{3}{4}$ inches diameter, and placing it directly in the center of the rectangular top, which is $3\frac{3}{4}$ inches wide and $5\frac{3}{4}$ inches long. Draw the elevation from the plan as shown by A B C D E F G H, making the vertical height V W, $2\frac{1}{4}$ inches, and the flanges at the top and bottom each $\frac{1}{2}$ inch. I K L M in plan is the horizontal section on A B in elevation and O P R the section on E F.

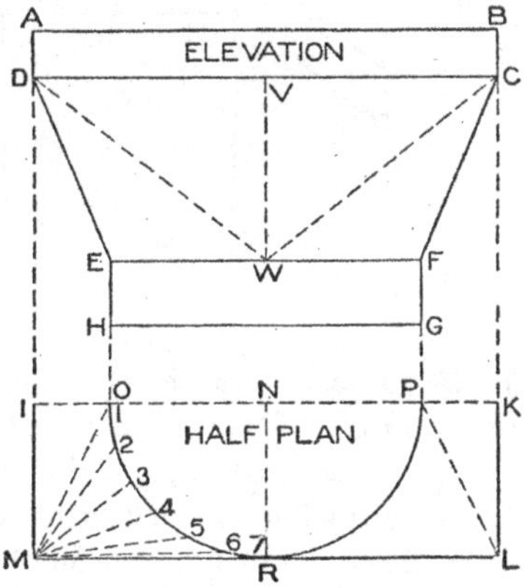

Fig. 79.

The pattern will be developed by triangulation, and the first step is to develop a set of triangles. Divide the quarter circle O R into equal spaces, as shown by the numbers 1 to 7 in plan, from which draw lines to the apex M. These lines represent the bases of triangles whose vertical height is equal to V W in elevation. Therefore, in Fig. 80, draw any horizontal line as T U, upon which place the various lengths M 1, M 2, M 3, etc.) Fig. 79) as shown by similar numbers on T U. From T U erect the line T S equal to the vertical height V W (Fig. 79). Then draw the hypotenuses S 1, S 2, S 3, etc., in Fig. 80, which represent the true lengths of similar numbered lines in plan in Fig. 79.

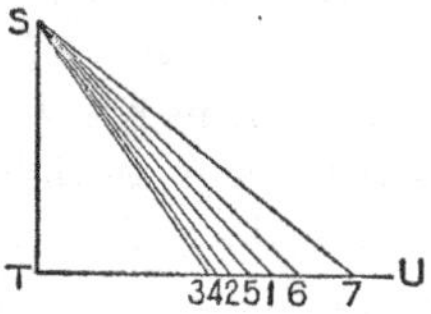

Fig. 80.

For the half pattern with seams on I O and P K in plan, take a tracing of D V W in elevation and place it as shown by D V 7 in Fig. 81. Now using D as center, and with radii equal to the various slant lines in Fig. 80 from S 1 to S 7 strike small arcs as shown from 1 to 7 in Fig. 81. Set the dividers equal to the spaces contained in O R, in Fig. 79, and starting from point 7, in Fig. 81, step from one arc to another until 1 is obtained. Then using 1 as center and E D (Fig. 79) as radius describe the arc D′ in Fig. 81. With D as center and M I in plan in Fig.

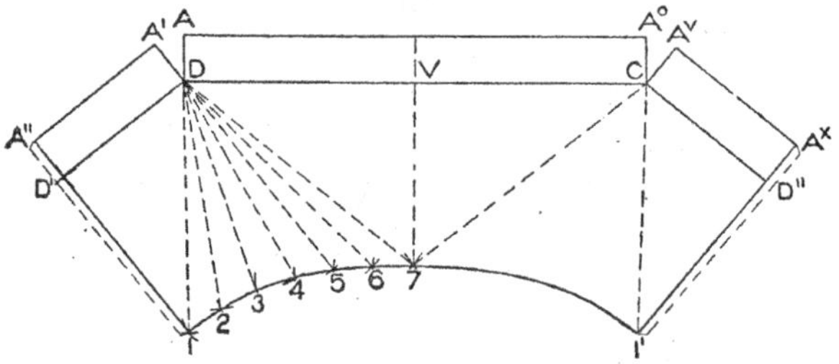

Fig. 81.

79 as radius, draw another arc intersecting the one previously drawn at D′. Draw a line from 1 to D′ to D in Fig. 81, 7 1 D′ D V is the quarter pattern, and the left-hand side of the figure may be made by tracing the quarter pattern reversed as shown by V C D″ 1′ 7. Take the distance of the flange D A in elevation in Fig. 79 and place it at right angles to the line D′ D, D C, C D″ as shown respectively by A″ A′, A A^O and A^V Ax, which completes the half pattern with laps allowed as shown

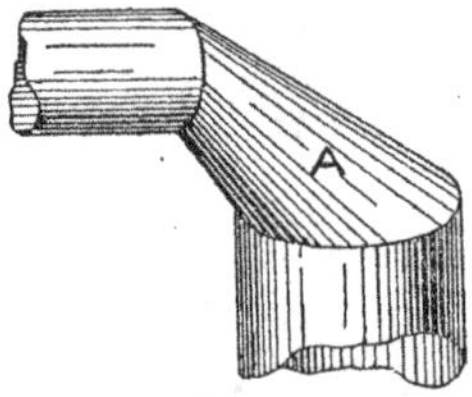

Fig. 82.

The pattern for the collar E F G H in elevation in Fig. 79 is simply a straight strip of metal, equal to the circumference of O P R in plan.

It is often the case that two unequal pipes are to be connected by means of a transition piece as shown by A in Fig. 82, the ends of the pipes being cut at right angles to each other. As the centers of both pipes are in one line when viewed in plan, making both halves of the transition piece equal, the problem then consists of developing a transition piece, from a round base to a round top placed vertically. Therefore in Fig. 83 draw 1 5 equal to $2\frac{1}{4}$ inches, and at an angle of 45° draw 5 6 $1\frac{3}{4}$ inches. At right angles to 1 5 draw 6 10 4 inches long and draw a line from 10 to 1. On 1 5 draw the semicircle 1 3′ 5, and on 6 10 draw the semicircle 6 8′ 10.

Divide both of these into equal spaces as shown, from which draw lines perpendicular to their respective base lines. Connect opposite points as shown by the dotted lines, and construct a diagram of

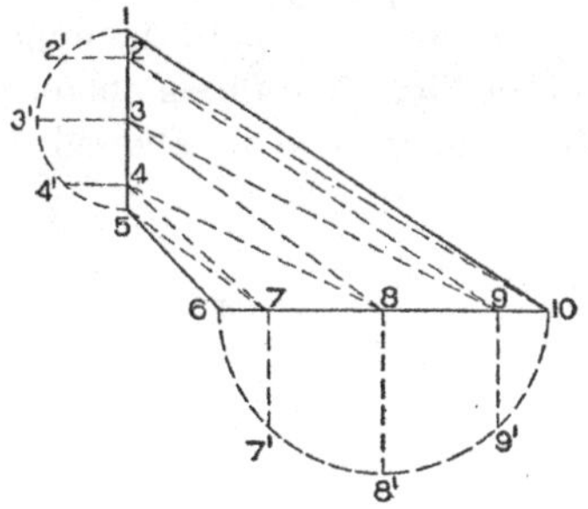

Fig. 83.

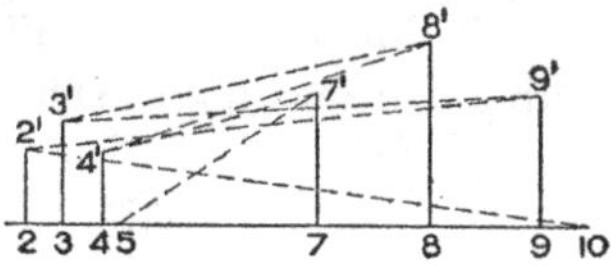

Fig. 84.

sections as shown in Fig. 84 whose bases and heights are equal to similar numbered bases and heights in Fig. 83. For example, take the distance 4 8 and place it as shown by 4 8 in Fig. 84, from which points erect the vertical lines 4 4′ and 8.8′ equal to 4 4′ and 8 8′ in Fig. 83. Draw a line from 4′ to 8′, Fig. 84, which is the true

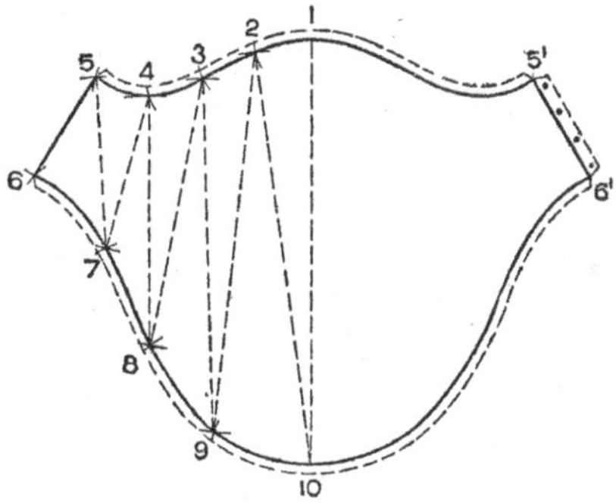

Fig. 85.

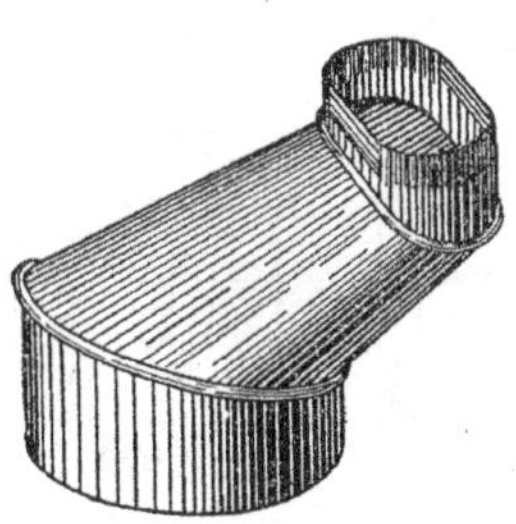
Fig. 86.

length on similar line in Fig. 83. For the pattern take the distance of 1 10 and place it as shown by 1 10 in Fig. 85. Using 1 as center, and 1 2′, Fig. 83, as radius, describe the arc 2 in Fig. 85; intersect it by an arc struck from 10 as center and 10 2′, Fig. 84, as radius. Then using 10 9′ in Fig. 83 as radius, and 10, Fig. 85, as center, describe the arc 9, and intersect it by an arc struck from 2 as center, and 2′ 9′, Fig. 84, as radius. Proceed in this manner using alternately as radii, first the divisions in the half profile 1 3′ 5, Fig. 83, then the length of the proper hypotenuse in Fig. 84, then the divisions in 6 8′ 10 in Fig. 83; then again the hypotenuse in Fig. 84 until the line 5 6 in Fig. 85 has been obtained, which is equal to 5 6 in Fig. 83. Laps should be allowed for riveting and seaming as shown.

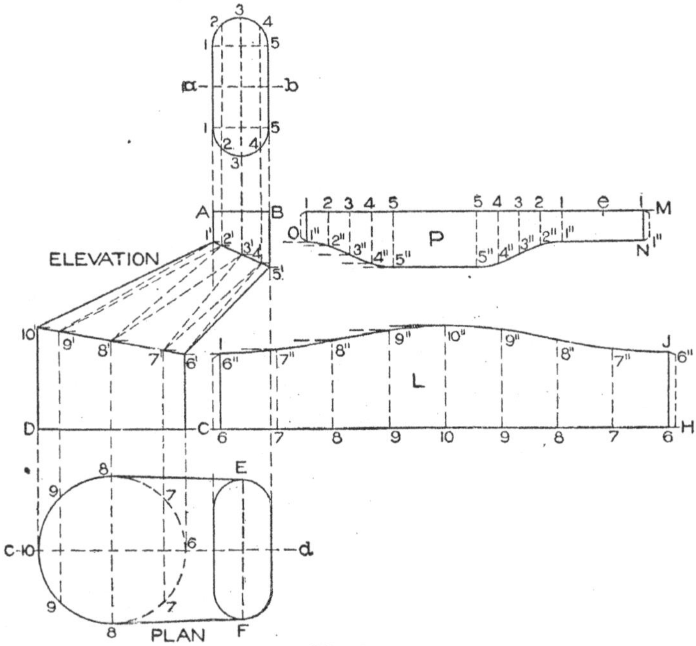

Fig. 87.

In Fig. 86 is shown a perspective of an offset connecting a round pipe with an oblong pipe, having rounded corners.

The first step is to properly draw the elevation and plan as shown in Fig. 87. Draw the horizontal line A B equal to one inch, B 5′ one inch, and from 5′ at an angle of 45° draw 5′ 6′ equal to $2\frac{1}{4}$ inches and 6′ C $1\frac{1}{4}$ inches. Make the diameter C D $2\frac{3}{4}$ inches and D 10′ $1\frac{3}{4}$ inches. Make A 1′ $\frac{1}{2}$ inch and draw a line from 1′ to

10′ which completes the elevation. Directly above the line A B draw the section of the oblong pipe, making the sides 1 1 and 5-5 equal to 1½ inches, to which describe the semicircles on each end as shown. In similar manner draw the section on D C, which is shown by 6 8 10 8. A duplicate of the oblong pipe is also shown in plan by E F, showing that the centers of the pipe come in one line, making both halves symmetrical.

The patterns for the pipes will first be obtained. Divide the semicircular ends of the oblong section into equal parts, in this case four, also each of the semicircles of the round pipe in similar number of parts as shown respectively from 1 to 5 and 6 to 10. Draw vertical lines from these intersections cutting the miter line of the oblong pipe at 1′ 2′ 3′ 4′ 5′ and the miter line of the round pipe at 6′ 7′ 8′ 9′ and 10′. In line with A B draw B M, upon which place the stretchout of the oblong pipe as shown by similar numbers; from B M drop vertical lines intersecting the lines drawn parallel to B M from similarly numbered points on 1′ 5′. Trace a line through points thus obtained, and P N O will be the pattern for the oblong pipe. Now take the stretchout of the round pipe, and place it on C H; erect vertical lines as shown intersecting the lines drawn parallel to C H from similar intersections on 6′ 10′. I J H C is the pattern for the round pipe.

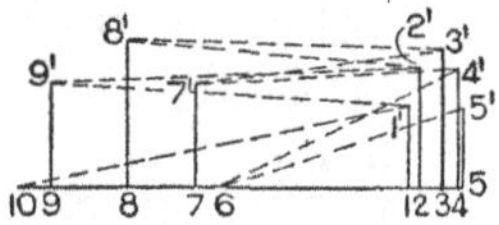

Fig. 88.

The transition piece 1′ 5′ 6′ 10′ will be developed by triangulation, and it is usual to obtain true sections on the lines 1′ 5′ and 6′ 10′; however, in this case it can be omitted because we have the true lengths of the various divisions on the lines 1′ 5′ and 6′ 10′ in the miter cuts in P and L respectively.

The next step is to obtain a diagram of sections giving the true lengths, for which proceed as follows: Connect opposite points in elevation as shown from 1′ to 9′ to 2′ to 8′ to 3′ etc., as shown. For example draw center lines through the oblong and round sections as shown by *a b* and *c d* respectively, and take the length of 1′ 10′ in elevation and place it as shown from 1 to 10 in Fig. 88. From 1 draw the vertical line 1 1′ equal to the height of 1 in the oblong section in Fig. 87 above the center line *a b*. As point 10 in plan has no height, it falls on the center line *c d* in plan, then draw a line from 1′ to 10 in Fig. 88. Now take the distance from 1′ to 9′ in elevation, Fig. 87, and place it as shown from 1 to 9 in Fig. 88. Erect the lines 1 1′ and 9 9′ equal to points 1 and 9 in the oblong and round sections in Fig. 87, measured respectively from the lines *a b* and *c d*. Draw a line from 1′ to 9′ in Fig. 87. Proceed in this manner until all of the sections are obtained. For the pattern proceed as shown in Fig. 89, in which draw any vertical line as *e* 10 equal to 1′ 10′ in elevation in Fig. 87. Now, with one-half of 1 1 in pattern P as *e* 1 as radius, and *e* in Fig. 89 as center, describe the arc 1 which is intersected by an arc struck from 10 as center and 10 1′, in Fig. 88 as radius. With radius equal to 10″ 9″ in pattern L in Fig. 87, and 10 in Fig. 89 as center describe the arc 9, which is intersected by an arc struck from 1 as center and 1′ 9′, in Fig. 88 as radius. Now, using as radius 1″ 2″ in pattern P in Fig. 87 and 1 in Fig. 89 as center, describe the arc 2 which is intersected by an arc struck from 9 as center and 9′ 2′ in Fig. 88 as radius.

Proceed in this manner, using alternately as radii, first the divisions in the pattern cut I J, Fig. 87, then the length of the slant lines in Fig. 88, the divisions in the cut O N in Fig. 87, then again the slant lines in Fig. 88 until the line 5 6 in pattern, Fig. 89, has been obtained. Then using 5 as center and 1 *e* in P, Fig. 87, as radius, describe the arc *e′* in Fig. 89, and intersect it by an arc struck from 6 as center and 6′ 5′ in elevation in Fig. 87 as radius. Draw lines through the various intersections in Fig. 89; 10 *e e′* 6 is the half pattern. By tracing it opposite the line *e* 10, as shown by *e* 1′ 5′ *e″* 6′ 10, the whole pattern, *e′ e e″* 6′ 10 6, is found. Laps should be allowed on all patterns for seaming or riveting both in Figs. 87 and 89.

In Fig. 90 is shown a perspective view of a three-way branch round to round, the inlet A being a true circle, and the outlets B, C, and D also being true circles, the centers of which are in the same vertical plane, thus making both sides of the branch symmetrical.

First draw the elevation and the various sections as shown in Fig. 91. Draw the center line *a b*. From *b* draw the center line of the branch C at an angle of 58° as shown by *b d*. Make the center lines *a b* and *b d* each $3\frac{1}{2}$ inches long. Make the half diameter of the branch B at the outlet $\frac{3}{4}$ inch, and the full diameter of the branch C at the outlet $1\frac{1}{2}$ inches placed on either side of and at right angles to the center lines. Draw a line from *e* to *f*, and with *i* and *h* as centers and radii equal to $\frac{3}{4}$ inch draw arcs intersecting each other at *c*. Draw lines from *i* to *c* to *h*. In similar manner obtain A and the opposite half of B. A B C is the elevation of the three branches whose sections on outlet lines are shown respectively by G F and E and whose section on the inlet line is shown by D.

The next step is to obtain a true section on the miter line or line of joint *b c*. Knowing the height *b c* and the width at the

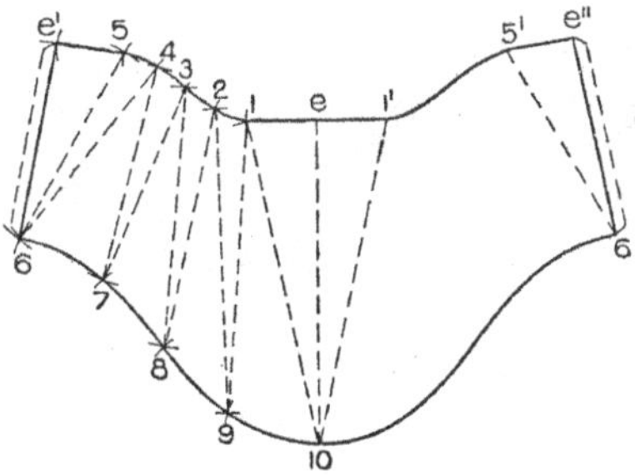

Fig. 89.

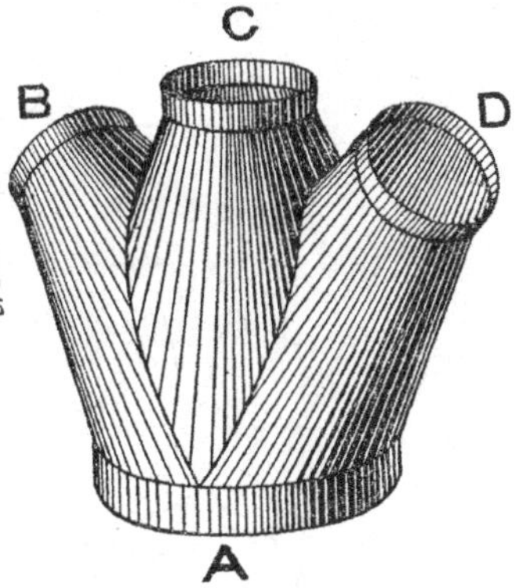

Fig. 90.

bottom, which is equal to the diameter of D, the shape can be drawn at pleasure as shown in Fig. 92, *b c* is drawn equal to *b c*, Fig. 91, while *b d* and *b a* are equal to the half diameter D in Fig. 91. Now through *a c d* in Fig. 92 draw the profile at pleasure as shown, which represents the true section on *c b* in Fig. 91.

As the side branches A and C are alike, only one pattern will be required, also a separate pattern for the center branch both of which will be developed by triangulation. To obtain the measurements for the sections for the center branch B, proceed as shown in Fig. 93 where 1 4 5 8 is a reproduction of one-half the branch B in Fig. 91. As the four quarters of this center branch are alike only one quarter pattern will be developed; then, if desired, the quarter patterns can be joined together, forming one pattern. Now take a tracing of *c b a*, Fig. 92, and place it on the line 5 8 as shown in Fig. 93. Similarly take a tracing of the quarter profile F in Fig. 91 and place it on the line 4 1 in Fig. 93. Divide the two profiles 1' 4 and 5 8' each into the same number of spaces as shown respectively by points 1' 2' 3' 4 and 5 6' 7' 8', from which points at right angles to their respective base lines 1 4 and 5 8 draw lines intersecting the base lines at 1 2 3 4 and 5 6 7 8. Now draw solid lines from 3 to 6 and 2 to 7 and dotted lines from 3 to 5, 2 to 6, and 1 to 7. These solid and dotted lines represent

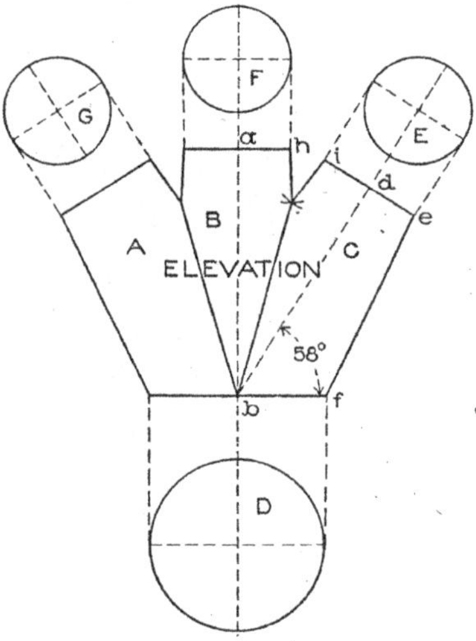

Fig. 91.

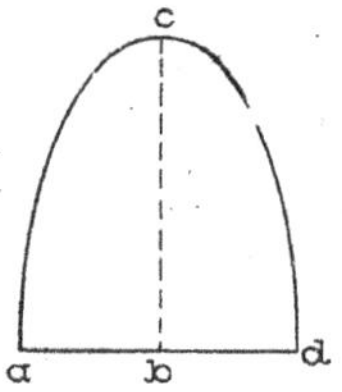

Fig. 92.

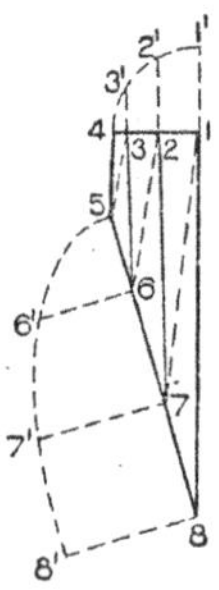

Fig. 93.

the bases of the sections whose altitudes are equal to the various heights of the profiles in Fig. 93. The slant lines in Fig. 94 represent the true distances on similar lines in Fig. 93, as those in Fig. 95 represent the true distances on dotted lines in Fig. 93.

For the pattern take the length of 1' 8', Fig. 94, and place it as shown by 1 8 in Fig. 96, and using 8 as center and 8' 7' in Fig. 93 as radius draw the arc 7, which intersect by an arc struck from 1 as center and 1' 7' in Fig. 95 as radius. Then using 1' 2' in Fig. 93 as radius draw the arc 2, which intersect by an arc struck from 7 as center and 7' 2' in Fig. 94 as radius. Proceed in this manner until the line 4 5 in Fig. 96 has been obtained which equals 4 5 in Fig. 93. Trace a line through points thus obtained in Fig. 96, then will 1 4 5 8 1 give the quarter pattern.

If the pattern is desired in one piece trace as shown by similar figures, to which laps must be allowed for riveting.

As the two branches A and C in Fig. 91 are alike, one pattern will answer for the two. Therefore let 1 7 8 11 14 in Fig. 97 be a reproduction of the branch C in Fig. 91. Now take a tracing of *a b c* in Fig. 92 and place it as shown by 11′ 11 8 in Fig. 97; also take a tracing of the half section E and the quarter section D in Fig. 91 and place them as shown respectively by 1 4′ 7 and

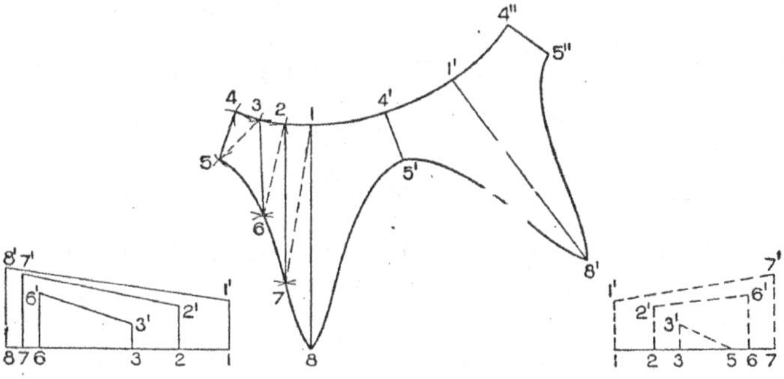

Fig. 94. Fig. 96. Fig. 95.

11 11′ 14 in Fig. 97. Now divide the two lower profiles 8 11 and 11′ 14 each into 3 equal parts, and the upper profile 7 4′ 1 into 6 equal parts as shown by the small figures 8 to 11′, 11′ to 14 and 1 to 7. From these points, at right angles to the various base lines, draw lines, intersecting the base lines as shown by similar numbers. Draw solid and dotted lines as shown, and construct the sections on solid lines as shown in Fig. 98 and the sections on dotted lines as shown in Fig. 99 in precisely the same manner as described in connection with Figs. 94 and 95.

In Fig. 100 is shown the pattern shape (to which laps must be allowed for riveting) obtained as was the development of Fig. 96. First draw the vertical line 1 14, Fig. 100, equal to 1 14 in Fig. 97. Then use alternately as radii, first the divisions in 1 4′ 7 in Fig. 97, the proper slant line in Figs. 98 and 99 and the divisions in 11′ 14 until the line 4 11, Fig. 100, is obtained. Starting from the point 11 use as radii in their regular order the distances marked off between 11′ and 8, Fig. 97, then the proper slant lines in Figs. 98 and 99, the distances shown in the semicircle, 1 4′ 7, Fig. 97, until the line 7 8, Fig. 100, is drawn equal to 7 8, in Fig. 97. Then

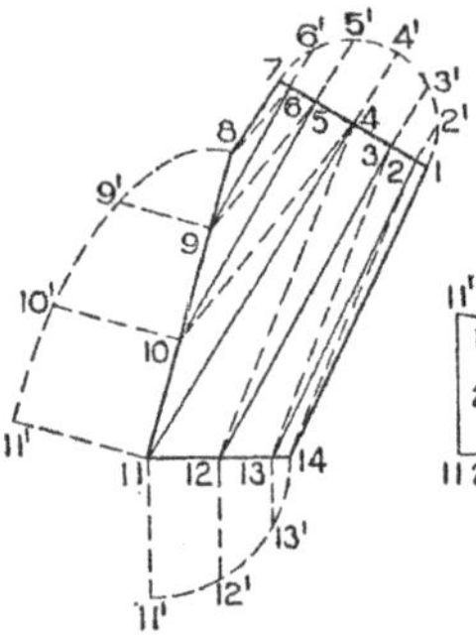

Fig. 97.

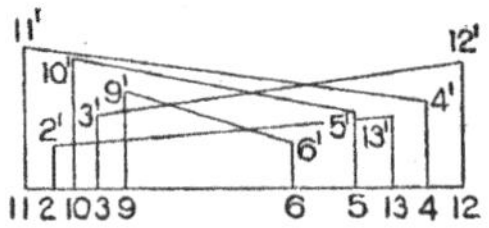

Fig. 98.

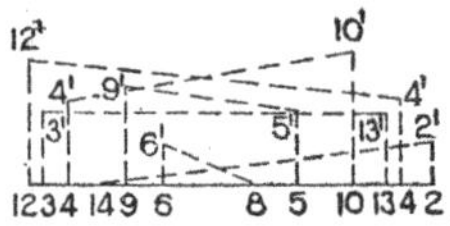

Fig. 99.

1 7 8 11 14, Fig. 100, will be the half pattern. If the pattern is desired in one piece trace 1 7′ 8′ 11′ 14 opposite the line 1 14 as shown.

In Fig. 101 is shown a perspective view of a two-branch fork oval to round, commonly used as breeching for two boilers. As

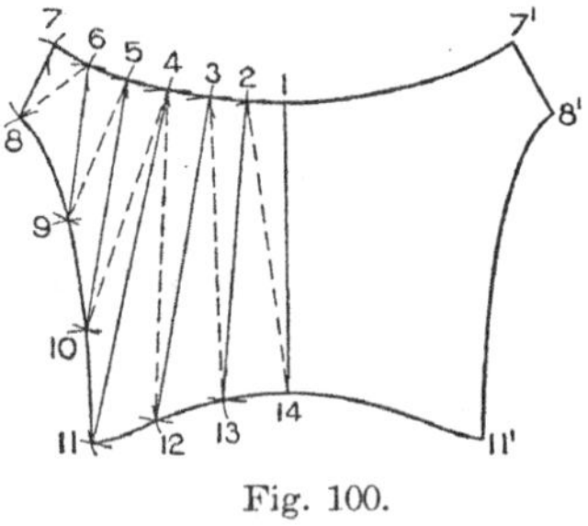

Fig. 100.

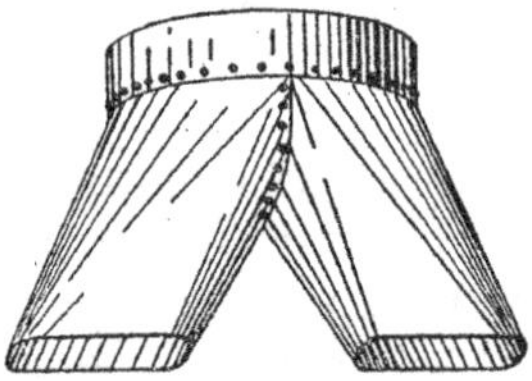

Fig. 101.

both halves of the fork are symmetrical the pattern for one will answer for the other.

While the side elevation shown in Fig. 102 is drawn complete, it is only necessary in practice, to draw one half as follows, and then, if desired, the other half elevation can be traced opposite to the center line E J. First draw J B, $1\frac{1}{2}$ inches, equal to the half diameter of the outlet, and the vertical center height J V, $2\frac{1}{4}$ inches. Establish the height of the joint J E one inch, and the desired projection V D on the base line $1\frac{1}{4}$ inches. Draw the length of the inlet D C $2\frac{3}{4}$ inches, and draw a line from C to B and D to E. Draw a similar figure opposite the line J E, and A B C D E F G shows the side elevation of the fork. In their proper position below A B draw the sections M and N whose semicircular ends are struck from *a b c* and *d* with radii equal to $\frac{1}{2}$ inch. Now draw an end elevation in which the true section on

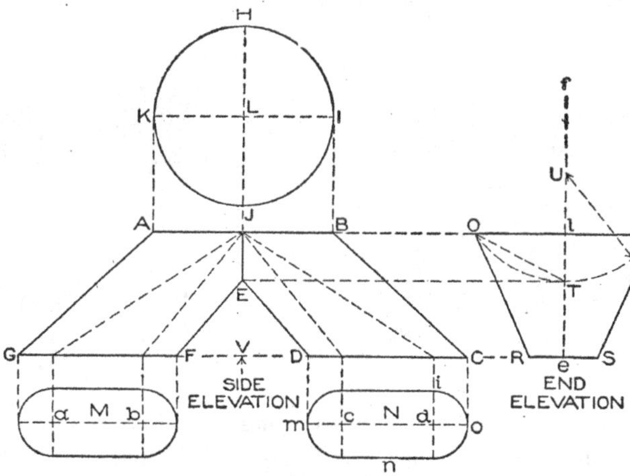

Fig. 102.

J E is obtained. Draw the center line *f e* and extend the lines A B and G C in elevation as A P and G S. Take the half diameter L J and place it on either side of *e f* as shown by O P. In a similar manner take the half diameter of the section N as *d i* and place it on either side of *e f* as shown by R S. Then O P S R shows the end elevation. Draw E T intersecting *e f* at T. Now draw the curve O T P, which in this case is struck from the center U, being obtained by bisecting the line O T. It should be understood that the curve O T P, which represents the true section on J E, can be made any desired shape, but when once drawn, represents a fixed line.

The pattern will be developed by triangulation, for which diagrams of sections must be obtained from which to obtain measurements. These sections are obtained as follows: In Fig. 103 1 4 5 12 13 is a reproduction of J B C D E, Fig. 102. Reproduce the quarter profile H L I, the half profile O T, and the half profile *m n o* as shown by 1′ 1 4, 1″ 13 1 and 12 9′ 8′ 5 in Fig. 103. Divide the round ends in *a* each into 3 parts and the profiles *b* and *c* also each into 3 spaces, as shown by the figures. Drop lines from these figures at right angles to the base lines from 1 to 15 as shown and draw solid and dotted lines in the usual manner. While in some of the previous problems only dotted lines were drawn, we

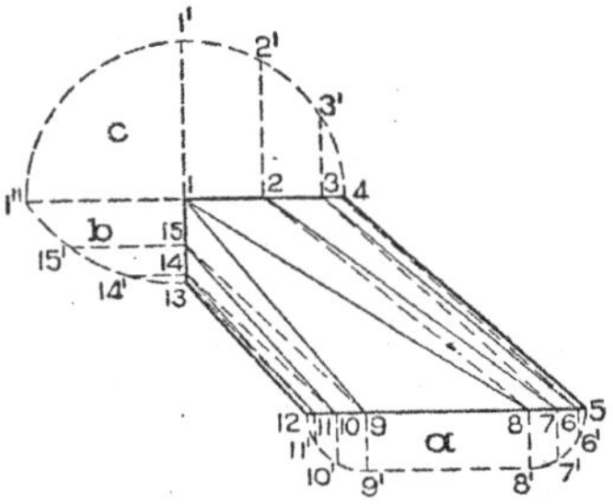

Fig. 103.

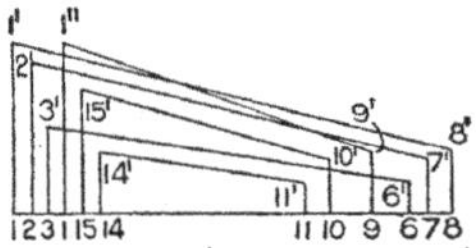

Fig. 104.

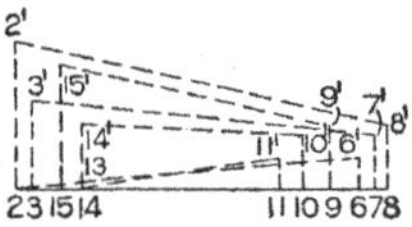

Fig. 105.

have drawn both solid and dotted lines in this case, in order to avoid a confusion of sections. A diagram of sections on solid lines in Fig. 103 is shown in Fig. 104, the figures in both corresponding; while Fig. 105 shows the true sections on dotted lines. The method of obtaining these sections has been described in connection with other problems.

For the pattern draw any vertical line as 4 5, Fig. 106, equal to 4 5 in Fig. 103. Then with 5 6′, Fig. 103, as radius and 5 in Fig. 106 as center draw the arc 6, intersecting it by an arc struck from 4 as center and 4 6′, Fig. 105, as radius. Then using 4 3′, Fig. 103, as radius, and 4 in Fig. 106 as center, describe the arc 3, intersecting it by an arc struck from 6 as center and 6′ 3′ in Fig. 104 as radius. Proceed in this manner, using alternately as radii, first the divisions in *a* in Fig. 103, then the slant lines in Fig. 105; the divisions in *c* in Fig. 103, then the slant lines in Fig.

104, until the line 1 8, Fig. 106, is obtained. Now using 8 as center and 8′ 9′, Fig. 103, as radius draw the arc 9 in Fig. 106, intersecting it by an arc struck from 1 as center and 1″ 9′, Fig. 104, as radius. Then starting at 1 in Fig. 106 use alternately as radii, first the divisions in *b* in Fig. 103, then the slant lines in Fig. 105, the divisions in *a* in Fig. 103, then the length of the slant lines in Fig. 104 until the line 12 13 is obtained in Fig. 106, which equals 12 13 in Fig. 103. Trace a line through points thus obtained in Fig. 106, then will 4 1 13 12 9 8 5 be the half pattern. If the pattern is desired in one piece, trace this half opposite the line 4 5 as shown by 1′ 13′ 12′ 9′ 8′, allowing laps for riveting.

In Fig. 107 is shown a perspective view of a tapering flange around a cylinder passing through an inclined roof, the flange

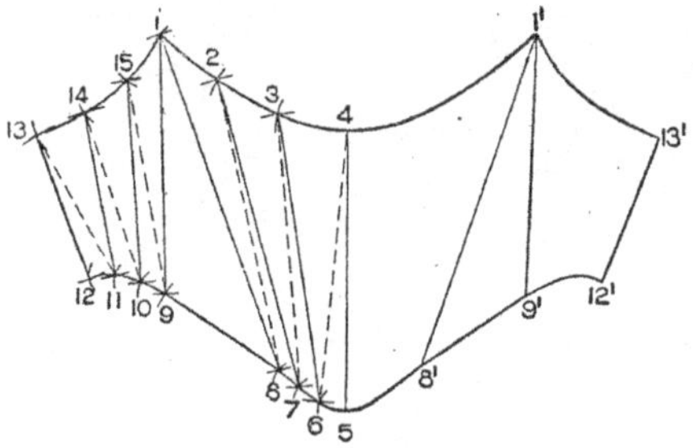

Fig. 106.

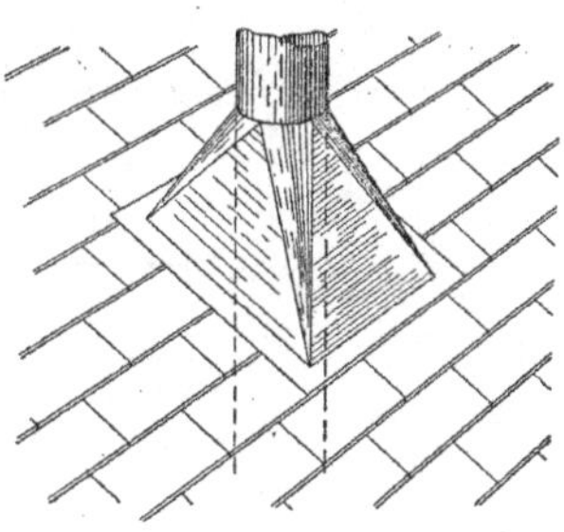

Fig. 107.

being rectangular on the roof line. The problem will be developed by triangulation, a plan and elevation first being required as shown in Fig. 108.

First draw the angle of the roof A B at an angle of 45°, through which draw a center line C D. From the roof line A B on the center line set off *a b* equal to 4 inches and through *b* draw the horizontal line E F, making B F and B E each one inch. Through *d* on the center line draw the horizontal line G H, making *d* H and *d* G each two inches. From H and G erect perpendiculars intersecting the roof line at K and L. Then draw lines from E to K and F to L, completing the elevation. Construct the square in plan making the four sides equal to G H. Bisect H I and draw the center line *c c* intersecting the vertical center at *d′*. Then with radius equal to *b* F or *b* E in elevation and *d′* in plan as center, draw the circle 1 4 7 4′ representing the horizontal section on E F in elevation, while G H I J is the horizontal section on K L in elevation. As the circle in plan is in the center of the square making the two halves symmetrical it is only necessary to divide the semicircle into equal spaces as shown from 1 to 7 and draw lines

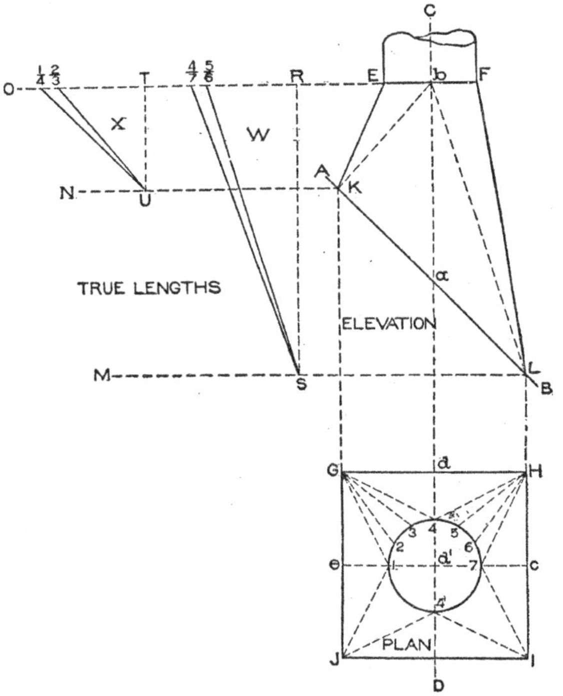

Fig. 108.

from 1, 2, 3 and 4 to G, and 4, 5, 6 and 7 to H. Then will the lines in 1 G 4 and 4 H 7 represent the bases of triangles which will be constructed, whose altitudes are shown respectively by the vertical heights in K E and L F in elevation. Therefore draw horizontal lines through E F, K, and L as shown by F O, K N, and L M. From any point as R and T on F O, draw the perpendiculars R S and T U respectively, meeting the horizontal lines drawn from L and K. Now take the various lengths in plan as G1, G2, G3, and

G4 and place them on the line F O as shown by T1, T2, T3 and T4, from which points draw lines to U which will represent the true lengths on similar lines in plan. In similar manner take the distances in plan from H to 4, to 5, to 6, to 7, and place them on the line F O, from R to 4, to 5, to 6, to 7, from which points draw lines to S which represent the true lengths on similar lines in plan.

For the pattern take the distance F L in elevation and place it on the vertical line 7′ L in Fig. 109. At right angles to 7′ L draw L S equal to *c* H or *c* I in plan, Fig. 108. Draw the dotted

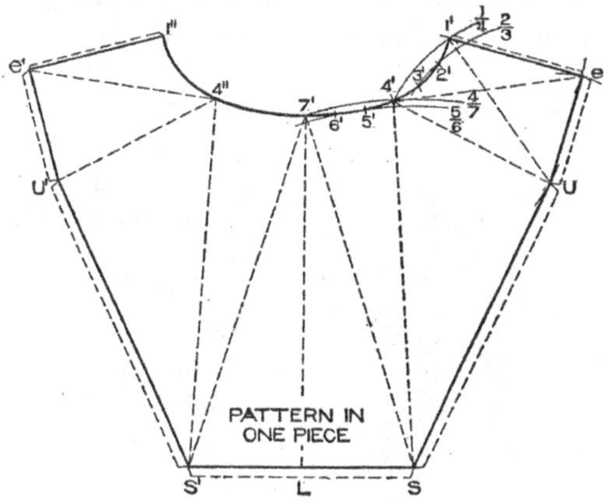

Fig. 109.

line from 7′ to S in Fig. 109, which should be equal to S 7 in W in Fig. 108. Now with radii equal to S $\frac{4}{7}$, and S $\frac{5}{6}$ and S, Fig. 109, as center, draw the arcs indicated by similar numbers. The dividers should equal the spaces in the semicircle in plan in Fig. 108, and starting at 7′ in Fig. 109, step from arc to arc of corresponding numbers as shown by 6′, 5′, 4′. Draw a dotted line from 4′ to S. Then using S as center and L K in elevation, Fig. 108, as radius, describe the arc U in Fig. 109, intersecting it by an arc struck from 4′ as center and U 4, Fig. 108, as radius. Now using U $\frac{1}{4}$, and U $\frac{2}{3}$ in X as radii, and U, Fig. 109, as center, describe arcs having similar numbers. Again set the dividers equal to the spaces in plan in Fig. 108, and starting from 4′ in Fig. 109 step to corresponding numbered arcs as shown by 3′, 2′, 1′.

Draw a dotted line from 4′ to U to 1′. With K E in elevation, Fig. 108, as radius, and 1′ in Fig. 109 as center, describe the arc *e* intersecting it by an arc struck from U as center and G *e* in plan in Fig. 108 as radius. Draw a line connecting S, U, *e*, and 1′. 7′ 4′ 1′ *e* U S L 7′ shows the half pattern, which can be traced opposite the line 7′ L to complete the full pattern as shown by 7′ 4″ 1″ *e*′ U′ S′ L.

One of the difficult problems often encountered by the sheet metal worker is that of a cylinder joining a cone furnace top at any angle. The following problem shows the principle to be applied, no matter what size the furnace top has, or what size pipe is used, or at what angle the pipe is placed in plan or elevation, the principles being applicable under any conditions.

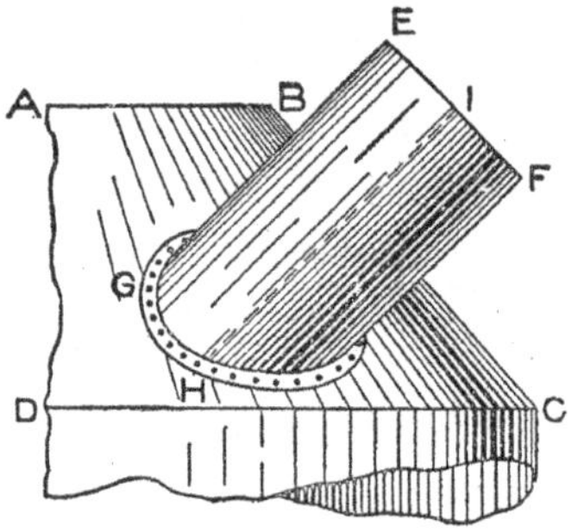

Fig. 110.

Fig. 110 shows a view of a cylinder intersecting a conical furnace top, the top being placed to one side of the center of the top. A B C D represents a portion of the conical top, intersected by the cylinder E F G H, the side of the cylinder H I to intersect at a given point on the conical top as at H. This problem presents an interesting study in projections, intersections, and development, to which close attention should be given.

In Fig. 111 first draw the center line A X. Then draw the half elevation A B C D, making A B $1\frac{5}{8}$ inches, C D $3\frac{1}{2}$ inches and the vertical height A D $2\frac{1}{8}$ inches. Draw the line from B to C. Directly below C D draw the one-quarter plan using Z as center, as shown by Z C^1 D^1 and in line with A B of the elevation draw the quarter plan of the top as Z B^1 A^1. Let *a* in the elevation represent the desired distance that the side of the cylinder is to meet the cone above the base line as H in Fig. 110. From *a*, parallel to C D in Fig. 111, draw *a b*. Then from *a* drop a vertical line intersecting the line Z C^1 in plan at *a*′. Then using Z as center and Z *a*′ as radius, describe the quarter circle *a*′ *b*′. Z *a*′ *b*′

in plan represents the true section on the horizontal plane *a b* in elevation. Now locate the point where the side of the cylinder as H in Fig. 110 shall meet the arc *a′ b′* in plan, Fig. 111, as shown

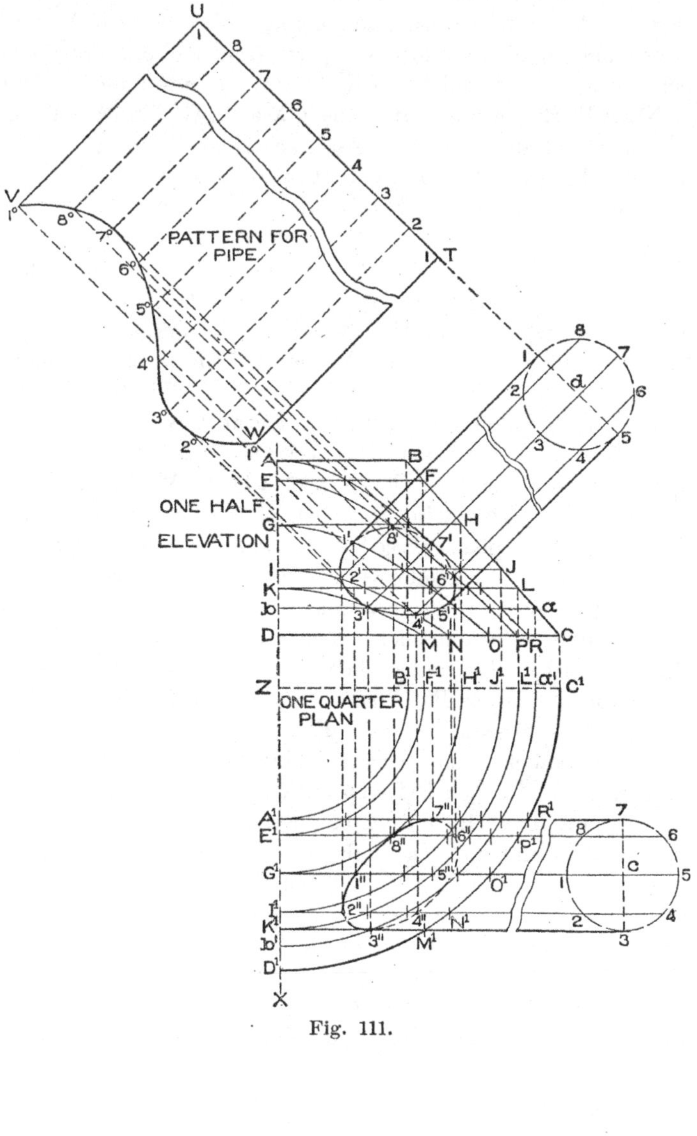

Fig. 111.

at 3″. Through 3″ draw the horizontal line intersecting the center line at K^1, the outer arc at M^1 and extend it indefinitely to 3. From 3 erect the perpendicular equal to the diameter of the cylinder, or $1\frac{5}{8}$ inches, bisect it and obtain the center *c*. Using *c* as center with *c* 7 as radius, describe the profile of the cylinder as shown, and divide it into equal parts from 1 to 8. From these points draw lines parallel to 3 K^1, intersecting the outer arc D^1 C^1 at N^1 O^1 P^1 R^1 and the center line Z X at I^1, G^1, E^1, A^1. With Z as center and the various intersections from K^1 to A^1 as radii, describe the arcs K^1 L^1, I^1 J^1, G^1 H^1, E^1 F^1, and A^1 B^1. From the intersection B^1, F^1, H^1, J^1, L^1 erect vertical lines into the elevation intersecting the side of the cone B C as shown by similar letters B F H J L. From these points draw horizontal lines through the elevation as shown respectively by A B, E F, G H, I J, and K L. These lines represent a series of horizontal planes, shown in plan by similar letters. For example, the arc E^1 F^1 in plan represents the true section on the line E F in elevation, while the arc G^1 H^1 is the true section on the line G H in elevation, etc.

The next step is to construct sections of the cone as it would appear, if cut by the lines shown in plan by K^1 M^1, I^1 N^1, G^1 O^1, E^1 P^1, and A^1 R^1. To obtain the section of the cone in elevation on the line A^1 R^1 in plan, proceed as follows: At right angles to the line A^1 R^1 and from the intersections on the various arcs, draw lines upward (not shown) intersecting similar planes in elevation corresponding to the arcs in plan. A line traced through intersections thus obtained in elevation as shown from A to R, will be the true section on the line A^1 R^1 in plan. For example, the line K^1 M^1 of the cylinder intersects the arcs at K^1 3″ and M^1 respectively. From these intersections, erect vertical lines intersecting K L, *b a*, and D C in elevation at K, 3′, and M respectively. Trace a curve through these points, then will K 3′ M be the section of the cone if cut on the line K^1 M^1 in plan. In similar manner obtain the other sections. Thus the section line E P, G O, and I N in elevation, represent respectively the sections if cut on the lines E^1 P^1, G^1 O^1, and I^1 N^1 in plan. Now from the given point 3″ in plan erect a line which must meet the intersection of the plane *b a* and section K M in elevation at 3′. From 3′ at its desired angle, in this case 45°, draw the line 3′ 7. At any point as *d* at right angles to 3′ 7 draw the line 1 5 through *d*, making *d* 5 and *d* 1 each equal to half the diameter of the cylinder shown in plan. With *d* 5 as radius and *d* as center draw the profile of the cylinder in elevation, and divide it into the same number of parts as shown in C in plan, being careful to allow the circle *d* in elevation to make a quarter turn, bringing the number 1 to the top as shown.

The next operation is to obtain the miter line or line of joint between the cylinder and cone in elevation. By referring to the plan it will be seen that the point 7 in the profile *c* lies in the plane of the section A^1 R^1. Then a line from the point 7 in the profile *d* in elevation, drawn parallel to the lines of the cylinder, must cut the section A R which corresponds to the plane A^1 R^1 in plan as shown by 7′ in elevation. The points 6 and 8 in the profile *c* in plan, are in the plane at the section E^1 P^1, then must the corresponding points 6 and 8 in the profile *d* in elevation, intersect the section E P as shown by 6′ and 8′. As the points 1 5, 2 4, and 3 in the profile *c* in plan, are in the planes of the sections G^1 O^1, I^1 N^1, and K^1 M^1 respectively, the corresponding points 1 5, 2 4, and 3 in the profile *d* in elevation must intersect the sections G O, I N, and K M respectively at points 1′ 5′, 2′ 4′, and 3′ as shown. Trace a line through these points, which will show the line of intersection between the cone and cylinder.

For the pattern for the cylinder, proceed as follows: At right angles to the line of the cylinder in elevation, draw the line T U upon which place the stretchout of the profile *d* as shown by similar figures on T U. In this case the seam of the pipe has been placed at 1 in *d*. Should the seam be desired at 3, 5 or 7, lay off the stretchout on T U starting with any of the given numbers. At right angles to U T from the small figures 1 to 1 draw lines which intersect with lines drawn from similar numbered intersections in the miter line in elevation at right angles to 1′ 1, resulting in the intersections 1 to 5° to 1° in the pattern. Trace a line through points thus obtained, then will U V W T be the development for the cylinder to which laps must be allowed for riveting to the cone as shown in Fig. 110 and seaming the joint T W in pattern in Fig. 111.

While the pattern for the cone is obtained the same as in ordinary flaring ware, the method will be described for obtaining the pattern for the opening to be cut into the cone. Before this can be done a plan view of the intersection between the pipe and cone must first be obtained as follows: From the various intersections 1′ to 8′ in elevation drop vertical lines intersecting lines drawn from similar numbers in the profile *c* in plan, thus obtaining the intersections 1″ to 8″ through which a line is traced which is the desired plan view.

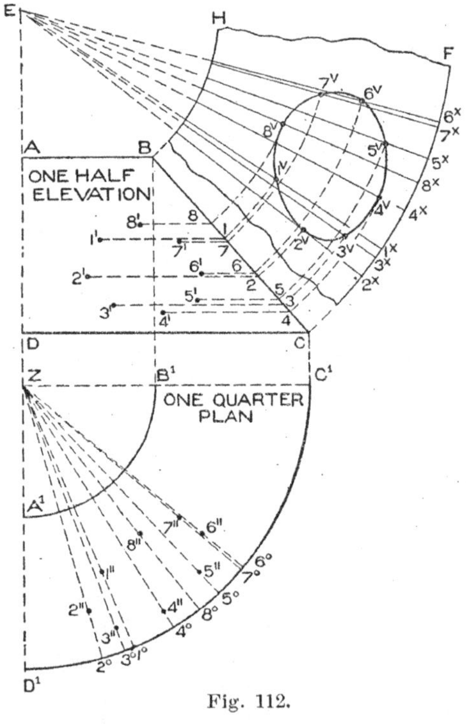

Fig. 112.

For the pattern for the opening in the cone, the outline of the half elevation and one-quarter plan with the various points of intersections both in plan and elevation in Fig. 112 is a reproduction of similar parts in Fig. 111, and has been transferred to avoid a confusion of lines which would otherwise occur in obtaining the pattern. Parallel to D C in Fig. 112 from the various intersections 1′ to 8′ draw lines intersecting the side of the cone B C from 1 to 8. Through the various intersections 1″ to 8″ in plan from the apex Z draw lines intersecting the outer curve from 1° to 8° as shown. Extend the line C B in elevation until it meets the center line D A extended at E. Then using E as center, with E C and E B as radii draw the arcs C F and B H respectively. At any point as 2^X on the arc C F lay off the stretchout of the various points on D^1 C^1 in plan from 2° to 6° as shown by similar figures on C F as shown from 2^x to 6^x. From these points draw radial lines to the apex E, and intersect them by arcs struck from E as center whose radii are equal to the various intersections on B C having similar numbers. Thus arc 4 intersects radial line 4^x at 4^v; arcs 3, 5, and 2 intersect radial lines 3^x, 5^x, and 2^x at 3^v, 5^v, and 2^v, and so on. Trace a line through points thus obtained as shown from 1^v to 8^v which is the desired shape. If a flange is desired to connect with the cylinder, a lap must be allowed along the inside of the pattern.

COPPERSMITH'S PROBLEMS.

In the five problems which will follow, particular attention is given to problems arising in the coppersmith's trade. While all the previous problems given in the course can be used by the coppersmith in the development of the patterns where similar shapes are desired, the copper worker, as a rule, deals mostly with hammered surfaces, for which flaring patterns are required. The principles which will follow, for obtaining the blanks or patterns for the various pieces to be hammered, are applicable to any size or shape of raised work. The copper worker's largest work occurs in the form of brewing kettles, which are made in various shapes, to suit the designs of the different architects who design the work. In hammering large brewing kettles of heavy copper plate, the pieces are developed, hammered, and fitted in the shop, then set together in the building, rope and tackle being used to handle the various sections for hammering, as well as in construction at the building. While much depends upon the skill the workman has with the hammer, still more depends upon the technical knowledge in laying out the patterns.

In all work of this kind the patterns are but approximate, but no matter what size or shape the work has, the principles contained in the following problems are applicable to all conditions.

In Fig. 113 is shown a perspective of a sphere which is to be constructed of horizontal sections as shown in Fig. 114, in which for practice draw the center line A B, on which, using *a* as center, and with radius equal to $2\frac{1}{4}$ inches, describe the elevation of the sphere B C D E. Divide the quarter circle D C into as many spaces as the hemi-sphere is to have sections, as shown by C F G D. From these points draw horizontal lines through the elevation, as shown by C E, F H, and G I. Now through the extreme points as E H, H I, and I D draw lines intersecting the center line B A at J, K, and D respectively. For the pattern for the first section Z, take D I as radius, and using D^1 in Z^1 as center, describe the circle shown. For the pattern for the second section Y, use K I and K H as radii, and with K^1 as center draw the arcs I^1 I^2 and H^2

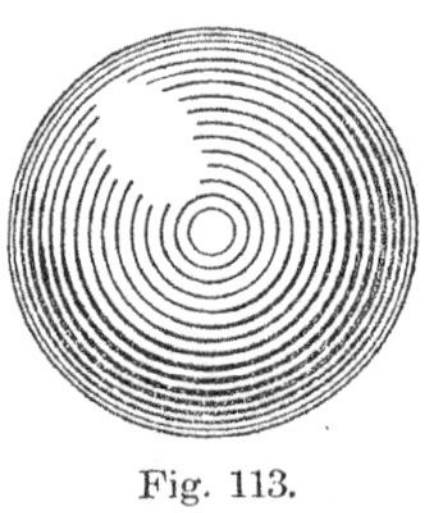

Fig. 113.

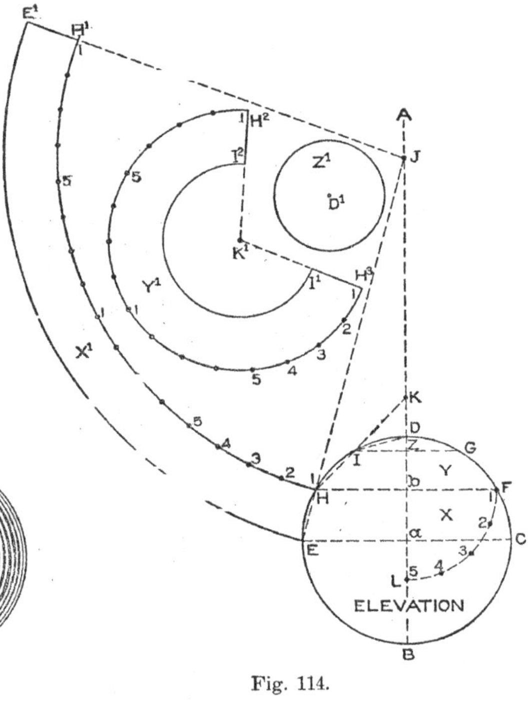

Fig. 114.

H^3. From any point as H^3 draw a line to the center K^1. It now becomes necessary to draw a section, from which the true length of the patterns can be obtained. Therefore with *b* F as radius, describe the quarter circle F L, which divide into equal spaces, as shown by the figures 1 to 5. Let the dividers be equal to one of those spaces and starting at H^3 on the *outer arc* in Y^1 step off four times the amount contained in the quarter section F L, as shown from 1 to 5 to 1 to 5 to 1 in Y^1. From 1 or H^2 draw a line to K^1. Then will H^2 I^2 I^1 H^3 be the pattern for the section Y in elevation.

For the pattern for the third section, use J as center, and with radii equal to J H and J E draw the arcs H H^1 and E E^1. Now set the dividers equal to one of the equal spaces in F L and starting from H set off four times the amount of L F as shown from 1 to 5 to 1 to 5 to 1 on the *inner curve* H H^1. From the apex J through H^1 draw a line intersecting the outer curve at E^1. E E^1 H^1 H shows the pattern for the center section. It will be noticed in the pattern X^1 we space off on the inner curve, while on the pattern Y^1 we space off on the outer curve. These two curves must contain the same amount of material as they join together when the ball is raised. To all of the patterns laps must be allowed for brazing or soldering. The patterns shown are in one piece; in practice where the sphere is large they are made in a number of sections.

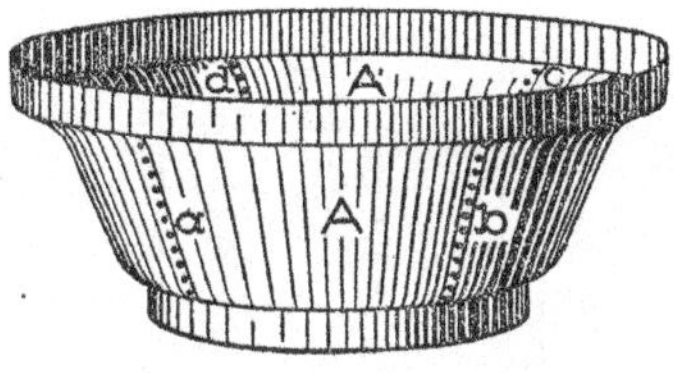

Fig. 115.

In Fig. 115 is shown the perspective view of a circular tank whose outline is in the form of an ogee. The portion for which the patterns will be described is indicated by A A, made in four sections, and riveted as shown by *a b c d*.

Fig. 116 shows how the pattern is developed when the center of the ogee is flaring as shown from 3 to 4 in elevation. First draw the elevation A B C D, making the diameter of A B equal to 7 inches, the diameter of D C 4 inches, and the vertical height of the ogee $1\frac{3}{4}$ inches. Through the center of the elevation draw the center line *f h*, and with any point upon it as *i*, draw the half plan through A B and C D in elevation as shown respectively by E F and H G. Now divide the curved parts of the ogee into equal spaces as shown from 1 to 3 and 4 to 6. Draw a line through the flaring portion until it meets the center line *f h* at *j*. *j* will, therefore, be the center with which to strike the pattern. Take the stretchout of the curve from 3 to 1 and 4 to 6 and place it on the flaring line from 3 to 1′ and 4 to 6′ as shown by the figures. Then will 1′ 6′ be the stretchout for the ogee. It should be understood that no hammering is done to that part shown from 3 to 4. The portion shown from 3 to 1′ is stretched to meet the required profile 3 2 1, while the lower part 4 to 6′ is raised to conform with the lower curve 4 5 6. Therefore, knowing that the points 3 and 4 are fixed points, then from either of these, in this case point 4,

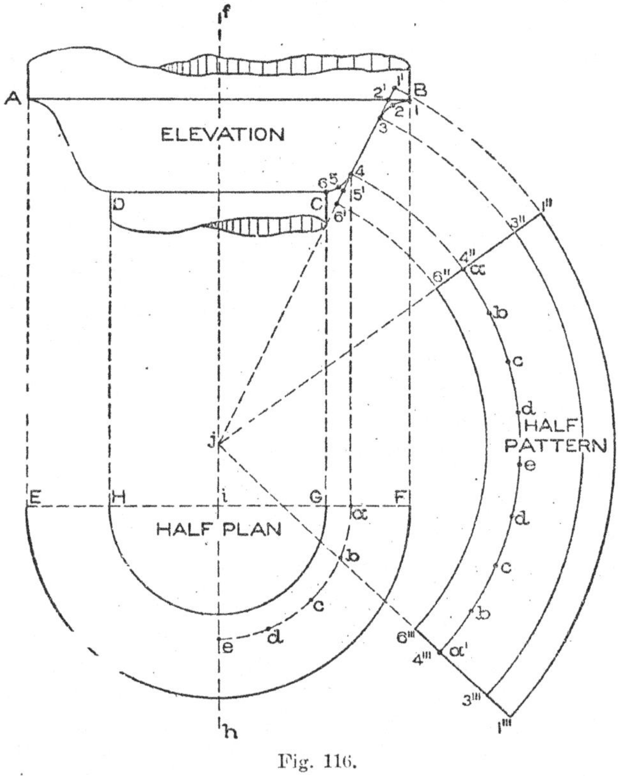

Fig. 116.

drop a vertical line intersecting the center line E F in plan at *a*. Then with *i* as center and *i a* as radius, describe the quarter circle *a e*, and space it into equal parts as shown by *a*, *b*, *c*, *d*, *e*, which represent the measuring line in plan on the point 4 in elevation. Using *j* as center, and *j* 6′, *j* 4, *j* 3 and *j* 1′ as radii, draw the arcs 1″-1‴, 3″-3‴, 4″-4‴ and 6″-6‴ as shown. From 1″ draw a radial line to *j* intersecting all the arcs as shown. Now starting at 4″ step off on the arc 4″-4‴ twice the stretchout of the quarter circle *a e* as shown by similar letters *a* to *e* to *a′* in pattern. From *j* draw a line through *a′* intersecting all of the arcs as shown. 1″-1‴-6‴-6″ shows the half pattern for the ogee.

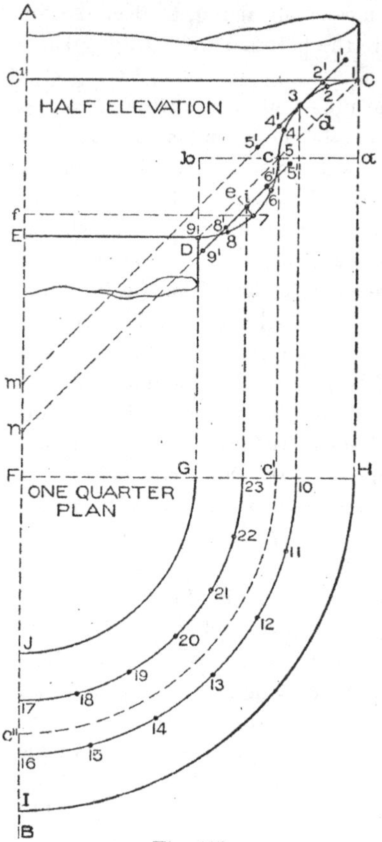

Fig. 117.

While in the previous problem the greater part of the ogee was flared, occasion may arise where the ogee is composed of two quarter circles struck from centers as shown in Fig. 117. First draw the center line A B, then draw the half diameter of the top C¹ C equal to $3\frac{1}{4}$ inches and the half diameter E D $1\frac{3}{4}$ inches. Make the vertical height of the ogee $1\frac{1}{2}$ inches, through the center of which draw the horizontal line *a b*. From C and D draw vertical lines intersecting the horizontal line *a b*, at *a* and *b* respectively. Then using *a* and *b* as centers with radii equal respectively to *a* C and *b* D draw the quarter circles shown completing the ogee. In the quarter plan below which is struck from the center F, G J and H I are sections respectively on D E and C C¹ in elevation. The methods of obtaining the patterns in this case are slightly different than those employed in the previous problems. The upper curve shown from C to *c* will have to be stretched, while the lower curve shown from *c* to D will have to be raised. Therefore in the stretchout of the pattern of the upper part from 1′ to 3 and 3 to 5′ the edges must be stretched so as to obtain more material to allow the metal to increase in diameter and conform to the desired shape shown from 1 to 3 and 3 to 5. In the lower curve the opposite method must be employed. While in the upper curve the edges had to be stretched to increase the diameters, in the lower curve the edges must be drawn in by means of raising, to decrease the diameter, because the diameters to the points 5″ and 9′ are greater than to points *c* and *d*.

To obtain the pattern for the upper curve C *c* which must be stretched, draw a line from C to *c*; bisect it and obtain *d*, from which erect the perpendicular *d* 3 intersecting the curve at 3. Through 3 draw a line parallel to C *c* intersecting the center line A B at *m*. Now divide the curve C *c* into equal spaces as shown from 1 to 5 and starting from the point 3 set off on the line just drawn on either side of 3 the stretchout shown from 3 to 1′ and 3 to 5′. 1′ 5′ shows the amount of material required to form the curve C *c*. In this case 3 represents the stationary point of the blank on which the pattern will be measured. Therefore from 3 drop a vertical line intersecting the line F H at 10. Then using F as center and F 10 as radius, describe the arc 10 16, and divide it into equal spaces as shown from 10 to 16. Now with radii equal to *m* 5′, *m* 3 and *m* 1′, Fig. 117, and with *m* in Fig. 118 as center, describe the arcs 5 5′, 3 3′ and 1 1′. Draw the radial line *m* 1 intersecting the two inner arcs at 3 and 5. As the arc 3 3′ represents the stationary point 3 in elevation in Fig. 117, then set the dividers equal to the spaces 10 16 in plan and step off similar spaces in Fig. 118 on the arc 3 3′, starting at 3 as shown by similar numbers 16 to 10. Through 10 draw a line to the apex *m*, intersecting the inner curve at 5′ and the outer curve at 1′. 1 1′ 5′ 5 is the quarter pattern for the upper curve or half of the ogee, to which laps must be allowed for riveting and brazing.

For the pattern for the lower curve in elevation in Fig. 117 draw a line from *c* to D; bisect it at *e* and from *e* erect a perpendicular intersecting the curve at 7. From 7 draw a horizontal line intersecting the center line at *f*. Now the rule to be followed in "raising" is as follows: Divide the distance from *e* to 7 into as many parts, as the half diameter F 7 is equal to inches. In this case 7 *f* equals 2¼ inches; (any fraction up to the ½ inch is not taken into consideration, but over $\frac{1}{2}$ inch one is added). Therefore for $2\frac{1}{4}$ inches use 2. Then divide the distance from *e* to 7 into two parts as shown at *i* and through *i* parallel to *c* D draw a line as shown intersecting the center line at N. Now divide the curve *c* to D into equal spaces as shown by the figures 5 to 9. Let off on either side of *i* the stretchout from 5 to 9 as shown from 5″ to

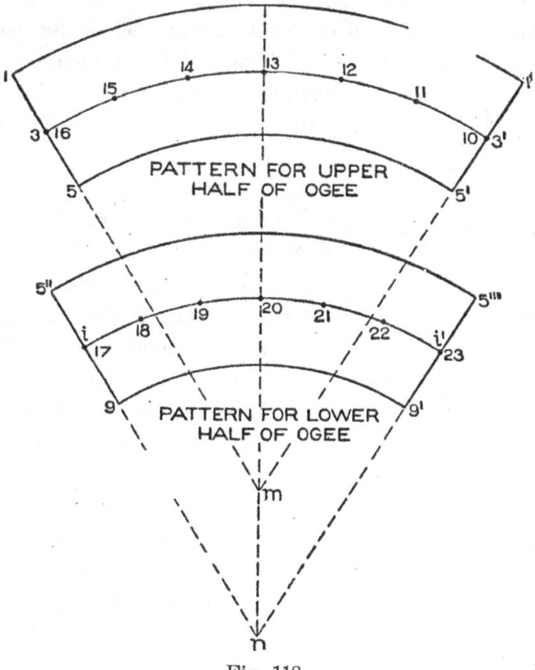

Fig. 118.

9′. From *i* drop a vertical line intersecting F H in plan at 23. Then using F as center draw the arc 23 17 as shown, which represents the measuring line in plan on *i* in the stretchout.

The student may naturally ask, why is *i* taken as the measuring line in plan, when it is not a stationary point, for when "raising" *i* will be bulged outward with the raising hammer until it meets the point 7. In bulging the metal outward, the surface at *i* stretches as much as the difference between the diameter at *i* and

7. In other words, if the measuring point were taken on 7 it would be found that after the mould was "raised" the diameter would be too great. But by using the rule of dividing *e* 7 into as many parts as there are inches in *f* 7 the diameter will be accurate while this rule is but approximate. In this case *e* 7 has only been divided into two equal parts, leaving but one point in which a line would be drawn througn parallel to *c* D. Let us suppose that the semi-diameter 7 *f* is equal to eleven inches. Then the space from *e* to 7 would be divided into just so many parts, *and through the first part nearest the cove* the line would be drawn parallel to *c* D and used as we have used *i*. Now with radii equal to *n* 9′, *n i*, and *n* 5″ and *n* in Fig. 118 as center, describe the arcs 5″ 5‴ *i i′* and 9 9′. From any point as 5″ draw a line to *n* intersecting all the arcs shown. Now take the stretchout from 17 to 23 in plan, Fig. 117, and starting from 17 in Fig. 118 mark off equivalent distances on the arc *i i′* as shown. Draw a line through 23 to the apex *n*, intersecting the inner and outer arcs at 9′ and 5‴. Then will 9 5″ 5‴ 9′ be the greater pattern for the lower part of the ogee.

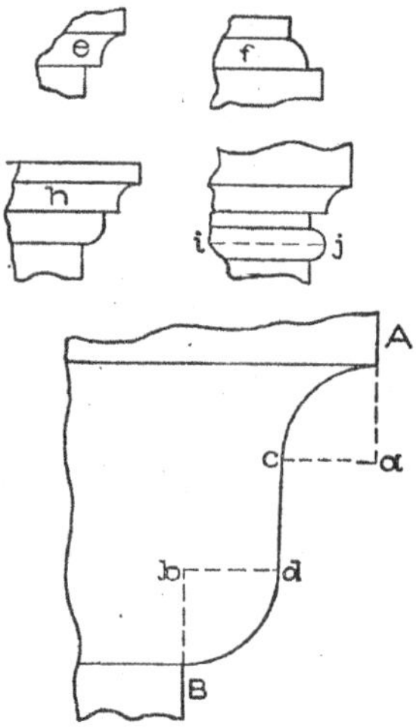

Fig. 119.

Another case may arise where the center of the ogee is vertical as shown from *c* to *d* in Fig. 119 in A B. In this case the same principles are applied as in Fig. 117; the pattern for *c d* in Fig. 119 being a straight strip as high as *c d* and in length equal to the quarter circumference *c′ c″* in plan in Fig. 117 which is the section on *c* in elevation. These rules are applicable to any form of mould as shown in Fig. 119, by *e*, *f*, *h*, and *j*. The bead *i* in *j* would be made in two pieces with a seam at *i* as shown by the dotted line, using the same method as explained in connection with *c* D in elevation in Fig. 117.

The coppersmith has often occasion to lay out the patterns for curved elbows. While the sheet metal worker lays them out in pieces, the coppersmith's work must form a curve as shown in Fig. 120 which represents a curved elbow of 45°.

In Fig. 121 is shown how an elbow is laid out having 90°, similar principles being required for any degree of elbow. First draw the side elevation of the elbow as shown by A B C D, mak-

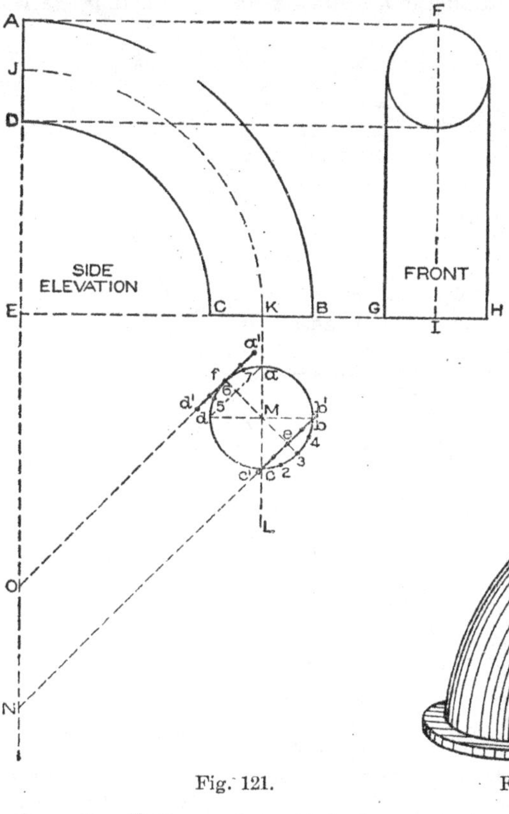

Fig. 121.

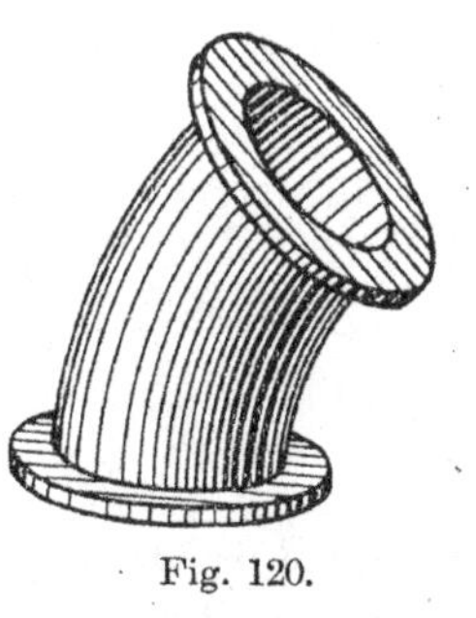

Fig. 120.

ing the radius E B equal to $4\frac{1}{2}$ inches and the diameter B C 2 inches. Bisect C B at K. Then with E as center and E K as radius draw the arc K J representing the seam at the sides. Draw the front view in its proper position as F G H, through which draw the center line F I representing the seam at back and front, thus making the elbow in four pieces. Directly below C B draw the section of the elbow as shown by *a b c d* struck from M as center. Through M draw the diameters *b d* and *a c*. The inner curve of the elbow *a d c* in plan will be stretched, while the outer curve *a b c* in plan will be raised. Through M draw the diagonal 3 6 intersecting the circle at 3 and *f* respectively. Now draw *a d;* through *f* parallel to *a d* draw a line intersecting the center

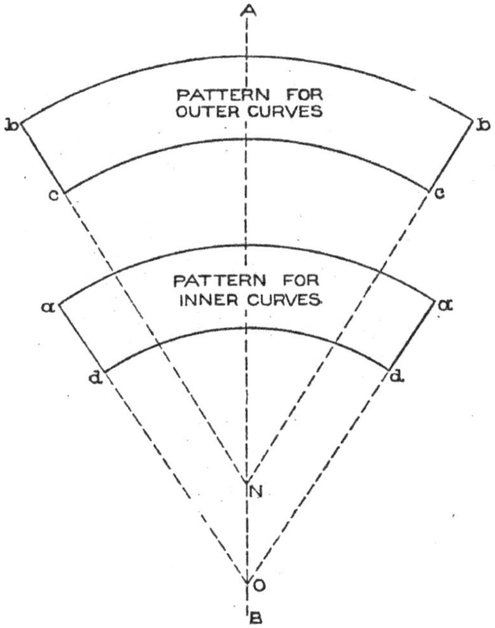

Fig. 122.

line A E extended at O. On either side of *f* place the stretchout of 6 *a* and 6 *d* as shown by *f a′* and *f d′*. Then with radii equal to O *d′* and O *a′* and with O on the line A B, Fig. 122, as center describe the arcs *d d* and *a a*. Make the length of *d d* equal to the inner curve D C in Fig. 121. From *a* and *d* in Fig. 122 draw lines to the apex O extending them to meet the outer curve at *a* and *a*. Then will *a d d a* be the half pattern for the inner portion of the elbow for two sides. The radius for the pattern for the outer curve is shown in Fig. 121 by N *c′*, N *b′*, placing the stretchout of the curve on either side of the point *e*. *b b c c* in Fig. 122 shows the pattern for the outer curve, the length *b b* being obtained from A B in elevation in Fig. 121.

In work of this kind the patterns are made a little longer, to allow for trimming after the elbow is brazed together. Laps must be allowed on all patterns for brazing.

Fig. 123 shows a perspective view of a brewing kettle, made in horizontal sections and riveted. The same principles which were employed for obtaining the patterns for a sphere in Fig. 114 are applicable to this problem. Thus in Fig. 124, let A B C represent a full section of a brewing kettle as required according to architect's design. Through the middle of the section draw the center line D E. Now divide the half section B to C into as many parts as the kettle is to have pieces as shown by *c*, *d*, *e*, *f*. From these small letters draw horizontal lines through the section, as shown by *c* A, *d d'*, *e e'*, and *f f'* and in its proper position below the section, draw the plan views on each of these horizontal lines in elevation, excepting *d' d*, as shown respectively by I F G H, *e'' e'''* and *f'' f'''*, all struck from the center *a*. Now through the points *c d* draw a line which if extended would meet the center line. Then this intersection would be the center with which to draw the arcs *c c'* and *d d''*; the flange *c b* would be added to the pattern as shown by *b'*. The stretchout for this pattern 1^1 would be obtained from the curved line F G H I in plan and stepped off on the outer arc *c c'*. In similar manner through *d e*, *e f*, and *f* C draw the lines intersecting the center line D E at K, L, and C. Then using the points as center, describe the patterns 2^1, and 3^1, and the full circle 4^1.

Fig. 123.

The stretchout for the patterns 2^1 and 3^1 is obtained from the circle *e'' e'''* in plan and placed on the inner curve of the pattern 2^1, and on the outer curve of the pattern 3^1. If desired the stretchout could be taken from *f'' f'''* in plan, and placed on the inner curve of 3^1 which would make the pattern similar as before.

In large kettles of this kind, the length of the pattern is guided by the size of the sheets in stock, and if it was desired that each ring was to be made in 8 parts then the respective circle in plan from which the stretchout is taken would be divided into 8 parts, and one of these parts transferred to the patterns, to which laps must be allowed for seaming and riveting.

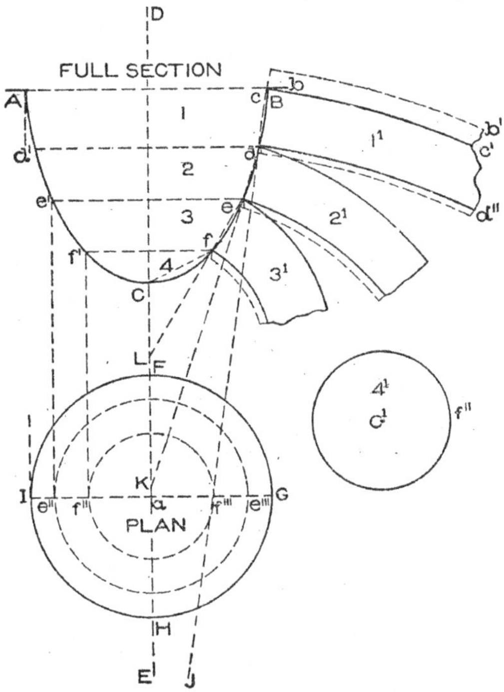

Fig. 124.

PROBLEMS FOR WORKERS IN HEAVY METAL.

While all of the problems given in this course are applicable to developments in heavy metal as well as in that of lighter gauge, the following problems relate to those forms made from boiler plate.

When using metal of heavier gauge than number 20, for pipes, elbows, or any other work, it is necessary to have the exact inside diameter. It is customary in all shops working the heavier metal, to add a certain amount to the stretchout to make up for the loss incurred in bending, in order that the inside diameter of the article (pipe, stack, or boiler shell) may be kept to a uniform and desired size. This amount varies according to different practice of workmen, some of whom allow 7 times the thickness of the metal used, while others add but 3 times the thickness. Theoretically the amount is 3.1416 times the thickness of the metal.

For example, suppose a boiler shell or stack is to be made 48 inches in diameter out of $\frac{1}{2}$-inch thick metal. If this shell is to measure 48 inches on the inside, add the thickness of the metal, which is $\frac{1}{2}$ inch, making $48\frac{1}{2}$ inches. Multiply this by 3.1416 and the result will be the width of the sheet. If, on the other hand, the outside diameter is to measure 48 inches, subtract the thickness of the metal, which would give $47\frac{1}{2}$ inches and multiply that by 3.1416 which would give the proper width of the sheet. It is well to remember that no matter what the thickness of the plate may be, if it is not added, the diameter of the finished article will not be large enough; for where no account is taken of the thickness of the metal, the diameter will measure from the center of the thickness of the sheet. While this rule is theoretically correct there is always a certain amount of material lost during the forming operations. It is, therefore, considered the best practice to use seven times the thickness of the metal in question. The circumference for a stack 48 inches in diameter inside using $\frac{1}{2}$ inch metal would be, on this principle, $3.1416 \times 48 + (7 \times \frac{1}{2})$ to which laps would have to be allowed for riveting. Where the stack has both diameters equal a butt joint is usually employed with a collar as shown at either *a* or *b* in Fig. 125, but where one end of the stack is to fit into the other, a tapering pattern must be obtained which will be described as we proceed.

In putting up large boiler stacks it is usual to finish at the top with a moulded cap, and while the method of obtaining the patterns is similar to parallel line developments, the method of developing such a pattern will be given showing how the holes are punched for a butt joint.

In Fig. 126 a view of the moulded cap on a stack is shown. On a large size stack the cap is often divided into as many as 32 pieces. If the stack is to be made in horizontal sections the rules given in the problems on coppersmithing apply. While in obtaining the patterns for a cap in vertical sections, the plan is usually divided into 16 to 32 sides, according to the size of the stack; we have shown in Fig. 127 a quarter plan so spaced as to give 8 sides to the full circle. This has been done to make each step distinct, the same principles being applied no matter how many sides the plan has.

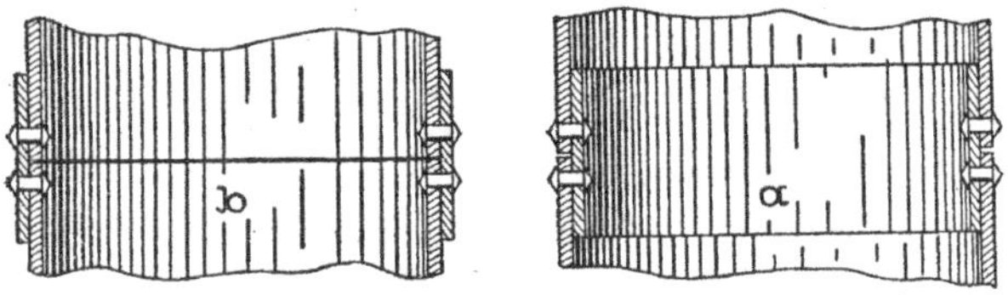

Fig. 125.

First draw the center line A B and with any point as C with radius equal to $4\frac{1}{2}$ inches draw the quadrant D E. Now tangent to D and E, draw the line D F and E G, and at an angle of 45°, tangent to the curve at Y, draw G F intersecting the previous lines drawn at G and F. C D F G E shows the plan view of the extreme outline of the cap. Directly above the plan draw a half section of the cap, the curve 5 8 being struck from *b* as center and with a radius equal to *b* 8 or $1\frac{3}{4}$ inches. Then using the same radius with *a* as center describe the quarter circle 5 2. Make 2 1 equal to $\frac{5}{8}$ inch, and 8 9 one inch. From the corners F and G in plan draw the miter lines F C, C G. Divide the profile of the cap into equal spaces as shown by the figures 1 to 9, from which drop vertical lines, intersecting the miter line F C as shown. On C D extended as C H place the stretchout of the profile of the cap as shown by similar numbers. At right angles to D H draw lines as shown, and intersect them by lines drawn parallel to D H from the intersections on C F. Trace a line through points thus obtained as shown by J I and trace this outline on the opposite side of the

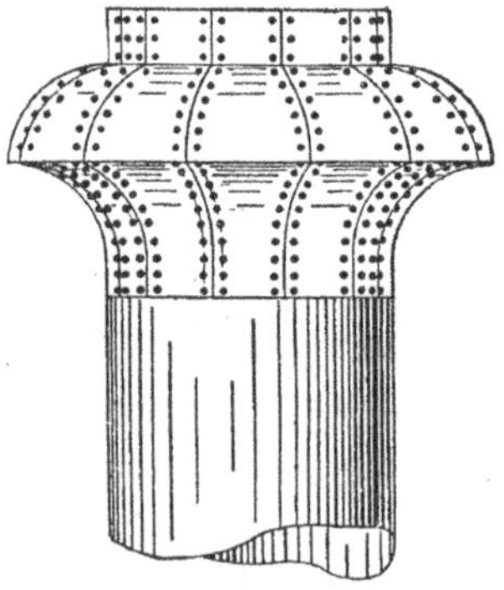

Fig. 126.

line D H as shown by J^1 I^1. Then will J I I^1 J^1 be the complete pattern for one side.

When riveting these pieces together an angle is usually placed on the inside and the miters butt sharp, filing the corners to make a neat fit. This being the case the holes are punched in the pattern before bending as shown by X X X etc. Assuming that the

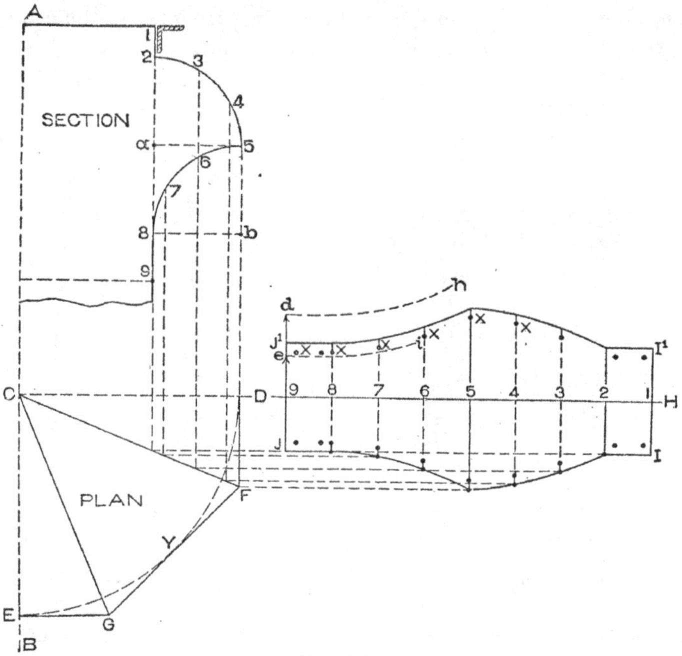

Fig. 127.

stack on which the cap is to fit is 48 inches in diameter, obtain the circumference as previously explained and divide by 8 (because the plan is composed of 8 pieces) placing one-half of the distance on either side of the center line D H in pattern. Assuming that $\frac{1}{16}$ of the circumference is equal to 9 *e*, trace from *e* the entire miter cut, as partly shown by *e i* to the line I^1 I. If the $\frac{1}{16}$ circumference were equal to 9 *d*, the cut would then be traced as shown in part by *d h* until it met the line I I^1. This, of course, would be done on the half pattern 9 J I I before tracing it opposite the center line D H. Should the plan be divided into 32 parts, divide the circumference of the stack by 32 and place $\frac{1}{64}$ of the circumference on 9 J in pattern, measuring from the center line D H, and after obtaining the proper cut, trace opposite the line D H.

In constructing a stack where each joint tapers and fits inside of the other, as shown in Fig. 128, a short rule is employed for obtaining the taper joints without having recourse to the center. In the illustration *a b* represents the first joint, the second C slip-

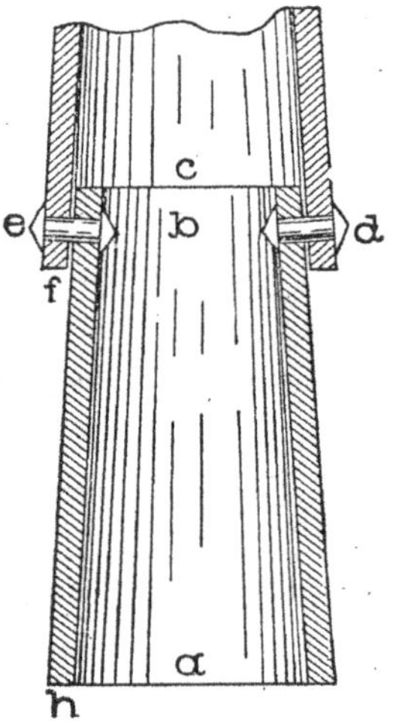

Fig. 128.

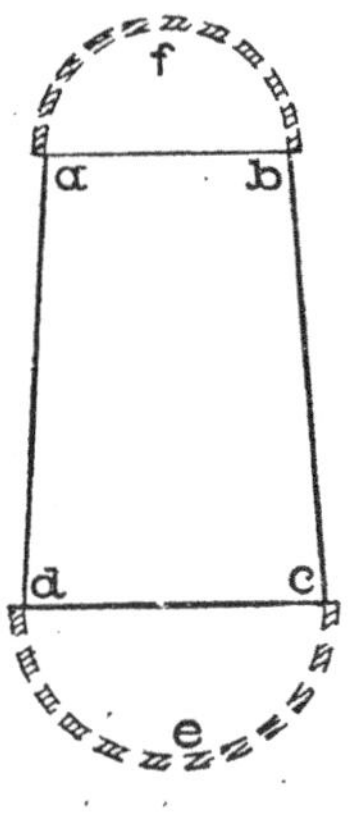

Fig. 129.

ping over it with a lap equal to *f*, the joint being riveted together at *e* and *d*. When drawing the first taper joint *a b*, care must be taken to have the diameter at *f* on the outside, equal to the inside diameter at the bottom at *h*. This allows the second joint to slip over a certain distance so that when the holes are punched in the sheets before rolling, the holes will fit over one another after the pipe is rolled.

In Fig 129 *a b c d* is a taper joint drawn on the line of its inside diameter, as explained in Fig. 128 *f*, and *e* in Fig 129 represents respectively the half sections on *a b* and *d c*. By the short rule the radial lines of the cone are produced without having recourse to the apex, which, if obtained in the full-size drawings, would be so far away as to render its use impracticable. A method similar to the following is used for obtaining the arcs for the pattern in all cases where the taper is so slight as to render the use of a common apex impracticable.

Let *a b c d*, Fig. 130, be a reproduction of *a b c d* in Fig. 129. On either side of *a d* and *b c*, in Fig. 130, place duplicates of *a b c d* as shown by *b' c'* and *a' d'*. This can be done most accurately by using the diagonals *d b* and *c a* as radii, and with *d* and *c* as centers describe the arcs *b b'* and *a a'* respectively, and intersect

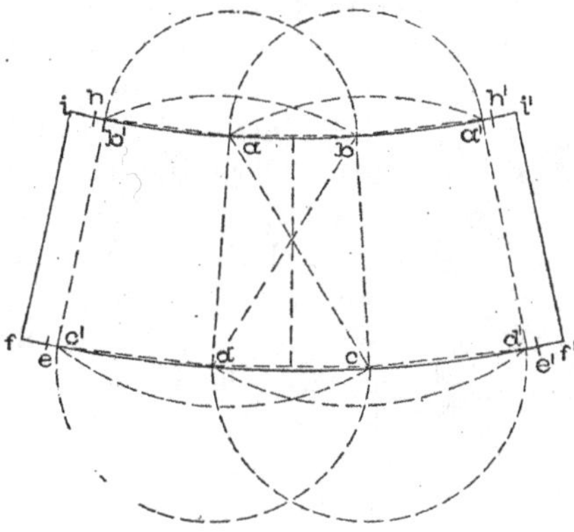

Fig. 130,

them by arcs struck from *a* and *b* as centers, with radii equal respectively to *a b* and *b a* as shown. In precisely the same manner obtain the intersection *c'* and *d'* at the bottom. Now through the intersections *b' a b a'* and *d' c d c'* draw the curve as shown by bending the straight-edge or any straight strip of wood placed on edge and brought against the various intersections, extending the curves at the ends and top and bottom indefinitely. Since the circumference of the circle is more than three times the diameter, and as we only have three times the diameter as shown from *c'* to *d'* and *b'* to *a'*, then multiply .1416 times the bottom and top diameter *d c* and *a b* respectively, and place one-half of the amount on either side of the bottom and top curves as shown by *e*, *e'*, and *h*, *h'*. Now take one-half of seven times the thickness of the metal in use and place it on either side on the bottom and top curves as shown by *f*, *f'* and *i*, *i'*, and draw a line from *i* to *f* and *i'* to *f'*. To this lap must be allowed for riveting. The desired pattern is shown by *i i' f' f*.

Fig. 131 shows a three-pieced elbow made from heavy metal, the two end pieces fitting into the center pieces, to which laps are allowed for riveting. The principles which shall be explained to cut these patterns and make the necessary allowance for any thickness of metal is applicable to any elbow.

In Fig. 132 draw as previously described the elbow A B C, below G H draw the section of the inside diameter as D which is struck from *a*, and divide into equal spaces as shown by the figures 1 to 5 on both sides. Through these figures draw vertical lines intersecting the miter line *b c*, and from these intersections parallel to *c d* draw lines intersecting the line *d e* as shown.

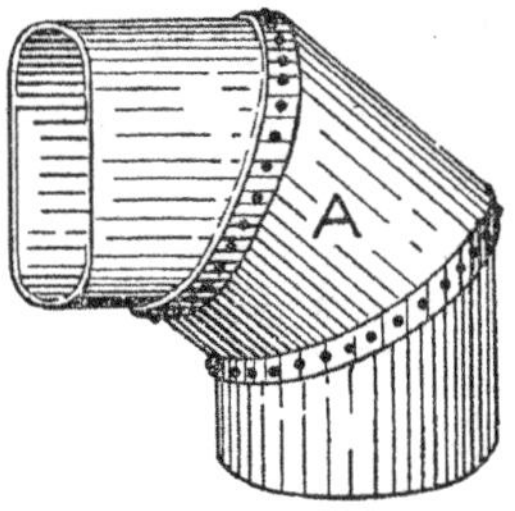

Fig. 131.

Before obtaining the stretchout for these elbows, a preliminary drawing must be constructed, in which an allowance is made for the thickness of the material that is to be used. This drawing makes practical use of a principle well known to draughtsmen from its application to the proportional division of lines and is clearly shown at (R). In allowing for the thickness of the metal in use, it is evident that we cannot allow it at one end, but must distribute it uniformly throughout the pattern. In (R) draw any horizontal line as E F, upon which place the stretchout of the inside diameter of the pipe D, as shown by similar figures on E F. From 1° on E F lay off the distance 1° *m* equal to 7 times the thickness of the metal in use as before explained. Then using E as center and E *m* as radius, draw the arc *m* 1' intersecting the vertical line drawn from 1°, and from the various intersections from 1 to 1° on E F erect perpendiculars intersecting the slant line 1 1' at 2' 3' 4', etc., as shown. The slant line 1 1, with the various intersections is now the correct stretchout for the elbow made of such heavy material called for by the specifications. On G H extended, as H I, place the stretchout of the slant line 1 1' as shown from 1 to 1' on H I. At right angles to H I and from the various intersections, erect lines, which are intersected by lines drawn parallel to H I from similar numbered intersections on the miter line *b c*. Trace the curve L M. L M I H shows the pattern for the two end pieces of the elbow.

As the middle section A in Fig. 131 is to overlap the two end pieces, it is unnecessary to allow for any additional thickness on

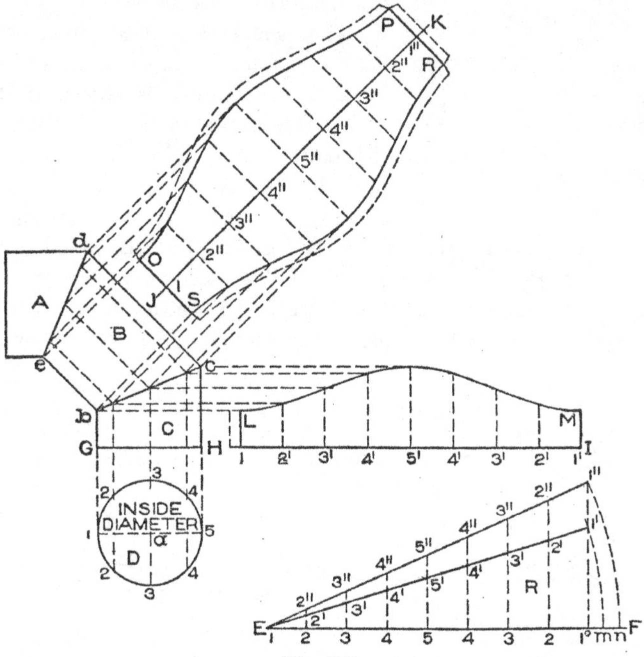

Fig. 132.

account of this lap when suitable flanging machines are available; but since it is desirable, in some instances, to make an allowance in the pattern for riveting, the method of allowing for this lap will be explained.

In (R), Fig. 132, lay off on the line E F the distance *m n* equal to 7 times the thickness of the metal in use, and with radius equal to E *n* draw an arc intersecting the line 1° 1′ extended at 1″. Draw the slant line from 1″ to 1 and extend all the vertical lines to intersect 1 1″ at 2″ 3″ 4″, etc. The slant line 1 1″ is the correct stretchout for the middle section B. At right angles to *d c* draw J K equal to 1 5″ 1″ in (R), as shown by similar figures in J K, through which draw lines at right angles to J K, and intersect them by lines drawn at right angles to *d c* as shown. Trace the curved lines to produce O P R S, which is the pattern for the middle section, to which flanges are allowed as shown by dotted lines.

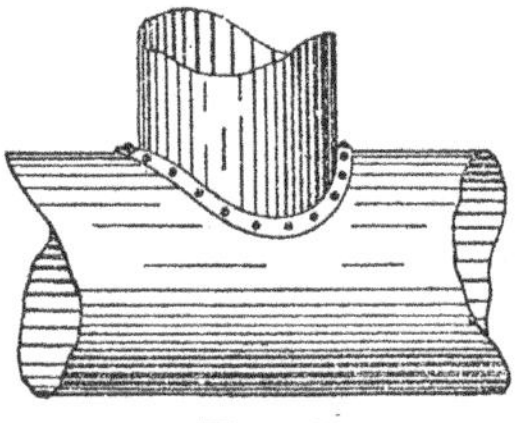

Fig. 133.

The perspective of an intersection between pipes having different diameters in boiler work is shown in Fig. 133. While the method of obtaining the patterns is similar in principle to parallel line developments, a slight change is required in obtaining the allowance in the stretchout for the thickness of the metal in use.

Let A B, Fig. 134, represent the part section of a boiler struck with a radius equal to $3\frac{3}{8}''$ and let 1 7 7° 1° be the elevation of the intersecting pipe, whose inside diameter is $4\frac{7}{8}''$, as shown by 1 7.

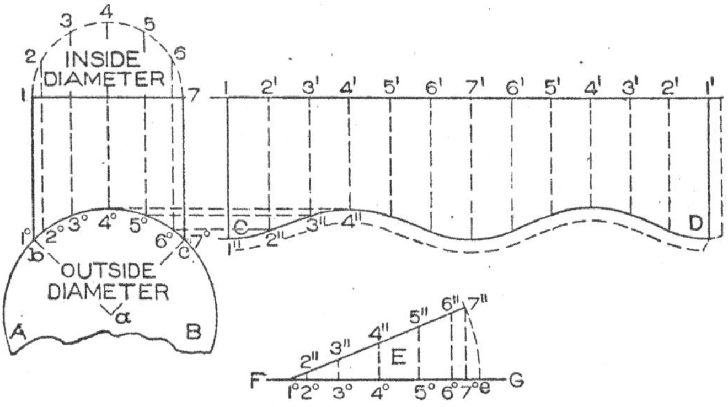

Fig. 134.

Divide the half section 1 4 7 into an equal number of spaces, as numbered, from which drop vertical lines intersecting the outside line of the boiler at 1° to 7° as shown. A true stretchout must now be obtained in which allowance has been made for the thickness of the metal in use. Therefore, in Fig. 135, on the horizontal line A B lay off the stretchout of twice the inside section of the pipe in Fig. 134, as shown by similar figures on A B in Fig. 135, adding 1^x *a*, equal to 7 times the thickness of the metal in use. For example, supposing $\frac{1}{4}$-inch steel was used; the distance 1^x *a* would then be equal to 7 × $\frac{1}{4}$, or $1\frac{3}{4}$ inches. Now draw the arc *a* 1′, using 1 as center, which is intersected by the vertical line drawn from 1^x. From 1′ draw a line to 1, and from the various points on A B erect perpendiculars intersecting 1 1′ at 2′ 3′ 4′, etc. 1 1′ shows the true stretchout to be be laid off on the line 1 7 extended in Fig. 134 as 1 1′, and from the various intersections on 1 1′ drop vertical lines and intersect them by lines drawn parallel to 1 1′ from similar intersections on the curve 1° 7° as shown. Trace a curved line as shown from C to D. 1 C D 1′ shows the pattern for the vertical pipe to which a flange must be allowed for riveting as shown by the dotted line.

It is now necessary to obtain the pattern for the shape to be cut out of the boiler sheet, to admit the mitering of the vertical pipe. In some shops the pattern is not developed, only the vertical pipe is flanged, as shown in Fig. 133, then set in its proper position on the boiler and line marked along the inside diameter of the pipe, the pipe is then removed and the opening cut into the boiler with a chisel. We give, however, the geometrical rule for obtaining the pattern, and either method can be used.

As A B in Fig. 134 represents the outside diameter of the boiler, to which 7 times the thickness of the metal used must be added to the circumference in laying out the sheet, and as the vertical pipe intersects one-quarter of the section as shown by *a b c*, take the stretchout from 1° to 7° and place it from 1° to 7° on F G in (E), to which add 7° *e*, equal to $\frac{1}{4}$ of 7 times the thickness of the plate used. Draw the arc *e* 7″, using 1° as center, intersecting it by the vertical line drawn from 7°. Erect the usual vertical lines and draw 7″ 1°, which is the desired stretchout. Now place this stretchout on the line A B in Fig. 136, erecting vertical lines as shown. Measuring in each and every instance from the line 1 7 in Fig. 134, take the various distances to points 2, 3, 4, 5, and 6 and place them in Fig. 136 on lines having similar numbers, measuring in each instance from A B on either side, thus obtaining the points 2, 3, 4, 5, and 6. Trace the curve 1° 4 7″ 4, which is the desired shape.

Fig. 137 shows a perspective of a gusset sheet A on a locomotive, the method of obtaining this pattern in heavy metal is shown in Fig. 138. First draw the end view A B C, the semicircle 4 1 4 being struck from *a* as center with a radius equal to 2

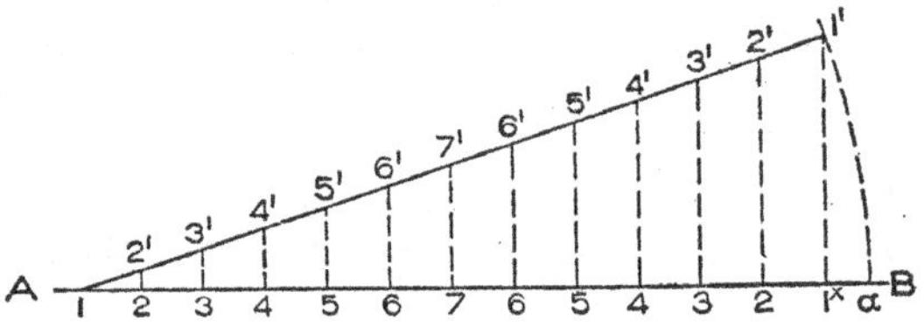

Fig. 135.

inches. Make the distance 4 to C and 4 to B both $3\frac{3}{4}$ inches and draw C B. Draw the center line A F, on which line measure up $2\frac{1}{4}$ inches and obtain *b*, which use as center with radius equal to *a* 4, draw the section of the boiler D E F G. In its proper position draw the side view H I J K L M N. H I L M N H shows the side view of the gusset sheet shown in end view by G A E D G.

Divide the semicircle 4 1 4 in end view into equal spaces as shown, from which draw horizontal lines intersecting H N in side

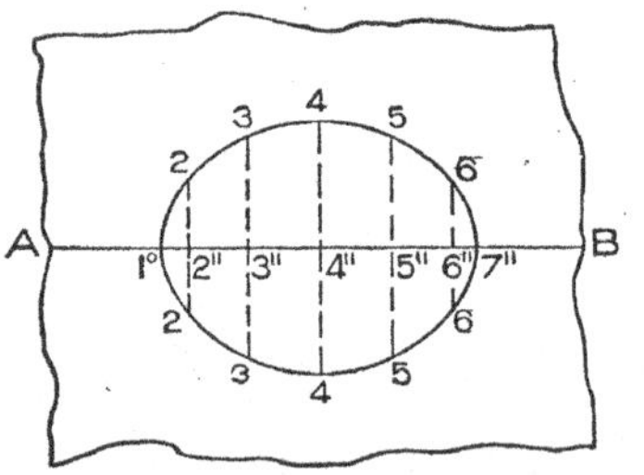

Fig. 136.

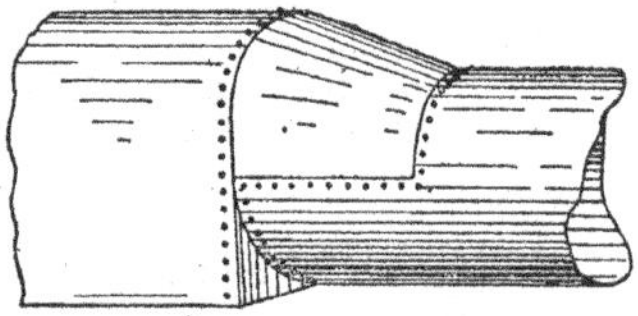

Fig. 137.

view from 1′ to 4′. From these intersections parallel to H I, draw lines indefinitely intersecting I L from 1″ to 4″. At right angles to N L produced draw the line at *c d*, on which a true section must be obtained at right angles to the line of the gusset sheet. Measuring from the line A D in end view, take the various distances to points 2, 3, and 4 and place them on corresponding lines measuring from the line *c d* on either side, thus obtaining the intersections 1° to 4°, a line traced through these points will be the true section. In (Y) on any line as O P lay off the stretch-out of the true section as shown from 4°, 1°, 4°. As the gusset sheet only covers a portion equal to a half circle, add the distance 4° *e* equal to ½ of 7 times the thickness of the metal in use and

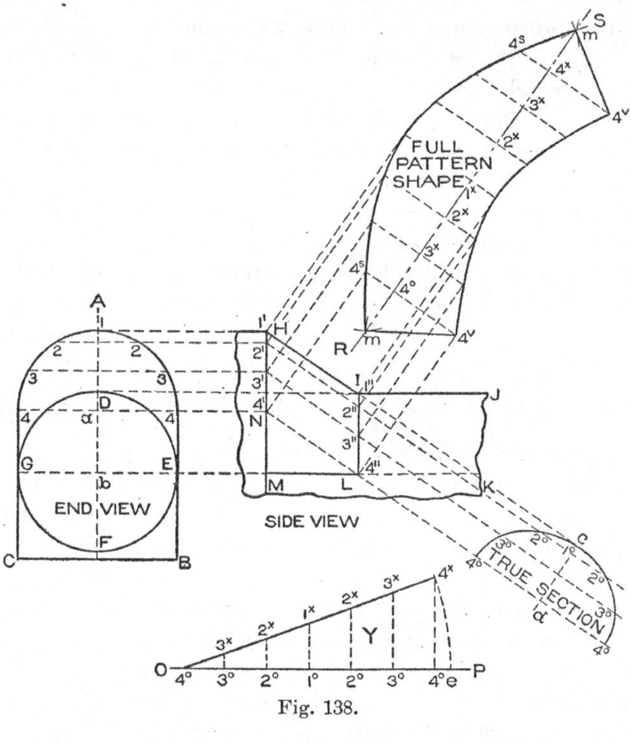

Fig. 138.

using 4° at the left, as center with 4° *e* as radius, describe the arc *e* 4^x, intersecting it at 4^x by the vertical line drawn from 4°. From O P erect vertical lines intersecting the line drawn from 4^x to 4° at 3^x, 2^x, 1^x, etc. 4° 4^x is the true stretchout, and should be placed on the line R S drawn at right angles to H I. Through the numbers on R S and at right angles draw the lines shown and intersect them by lines drawn from similarly numbered intersections on H N and I L at right angles to H I. Through points thus obtained trace a curved line 4^S, 4^S, and 4^V, 4^V. It now becomes necessary to add the triangular piece shown by L M N in side view, to the pattern which can be done as follows: Using L M in side view as radius and 4^V at either end of the pattern as centers, describe the arcs *m* and *n*; intersect them by arcs struck from 4^S and 4^S as centers, and M N in side view as radius. Then draw lines from 4^S to *m* to 4^V in the pattern on either side. The full pattern shape for the gusset sheet will then be shown by *m* 4^S 4^S *m* 4^V 4^V, to which laps must be allowed for riveting.

Fig. 139 shows a conical piece connecting two boilers with the flare of A such that the radial lines can be used in developing the pattern. In all such cases this method should be used in preference to that given in connection with Fig. 130. Thus in Fig. 139 the centers of the two boilers are on one line as shown by *a b*. While the pattern is developed the same as in flaring work, the method of allowing for the metal used is shown in Fig. 140.

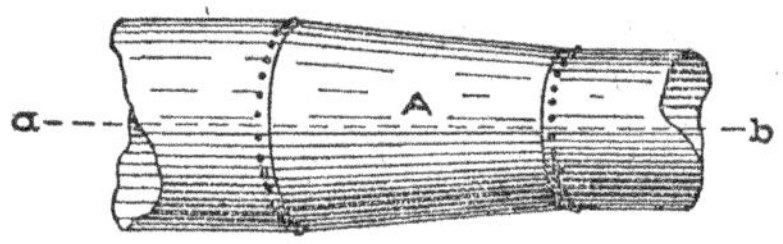

Fig. 139.

A B C D is the elevation of the conical piece, the half inside section being shown by 1 4 7 which is divided into equal spaces. 1 7 1 in (E) is the full stretchout of the inside section A 4 D in elevation, and 1 *e* is equal to 7 times the thickness of the metal used. The line 1 1′ is then obtained in the usual manner as are the various intersections 2′ 3′ 4′, etc. Now extend the lines A B and D C in elevation until they meet the center line *a b* at *a*. Then using *a c* and *a d* draw the arcs 1′ 7′ and 1″ 7″. From 1′ draw a radial line to *a*, intersecting the inner arc at 1″. Now set the dividers equal to the spaces on 1 1′ in (E) and starting from 1′ in the pattern step off 6 spaces and draw a line from 7′ to *a* intersecting the inner arc at 7″. 1′ 7′ 1″ 7″ shows the half pattern to which flanges must be allowed for riveting.

Fig. 141 shows a view of a scroll sign, generally made of heavy steel, heavy copper, or heavy brass. So far as the sign is concerned it is simply a matter of designing, but what shall be given attention here is the manner of obtaining the pattern and elevation of the scroll. As these scrolls are usually rolled up in form of a spiral, the method of drawing the spiral will first be shown.

Establish a center point as *a'* in Fig. 142, and with the desired radius describe the circle shown, which divide into a polygon of

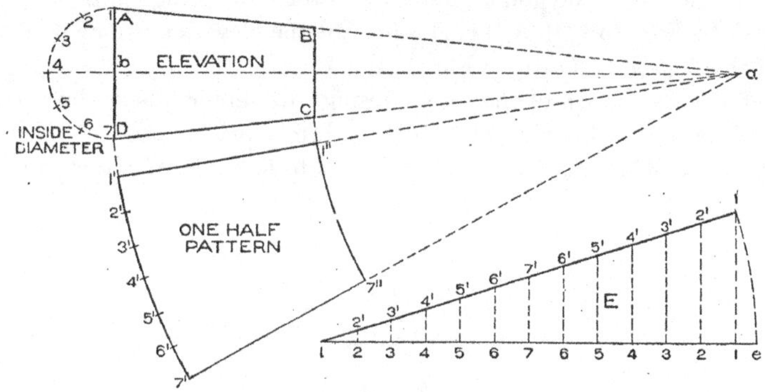

Fig. 140.

any number of sides, in this case being 6 sides or a hexagon. The more sides the polygon has, the nearer to a true spiral will the figure be. Therefore number the corners of the hexagon 1 to

Fig. 141.

5 and draw out each side indefinitely as 1 *a*, 2 *b*, 3 *c*, 4 *d*, 5 *e*, and 6 *f*. Now using 2 as center and 2 1 as radius, describe the arc 1 A; then using 3 as center and 3 A as radius, describe the arc

A B, and proceed in similar manner using as radii 4 B, 5 C, 6 D, and 1 E, until the part of the spiral shown has been drawn. Then using the same centers as before continue until the desired spiral is obtained, the following curves running parallel to those first drawn. The size of the polygon *a'*, determines the size of the spiral.

In Fig. 143 let A B C D represent the elevation of one corner of the flag sign shown in Fig. 141. In its proper position in Fig. 143 draw a section of the scroll through its center line in elevation as shown by *a* 17 to 1, which divide into equal spaces as shown from 1 to 17. Supposing the scroll is to be made of $\frac{1}{8}$ inch thick

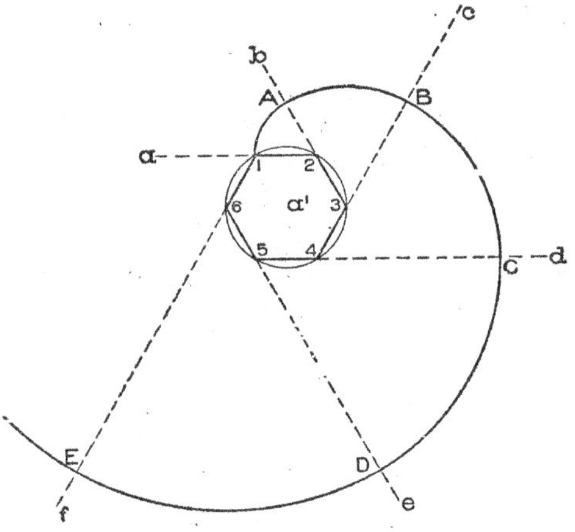

Fig. 142.

metal, and as the spiral makes two revolutions then multiply $\frac{1}{8}$ by 14, which would equal $1\frac{3}{4}$ inches. Then on E F in Fig 144 place the stretchout of the spiral in Fig. 143, as shown by similar numbers, to which add 17 E equal to 14 times the thickness of metal in use, and draw the arc E 17′ in the usual manner and obtain the true stretchout with the various intersections as shown. Through the elevation of the corner scroll in Fig. 143 draw the center line E F, upon which place the stretchout of 17′ E, Fig. 144, as shown by similar numbers on E F in Fig. 143. At right angles to E F, through 1′ and 17′, draw 17° 17° equal to A B and 1° 1° equal to the desired width of the scroll at that point. Then at pleasure draw the curve 1° 17° on either side, using the straight-

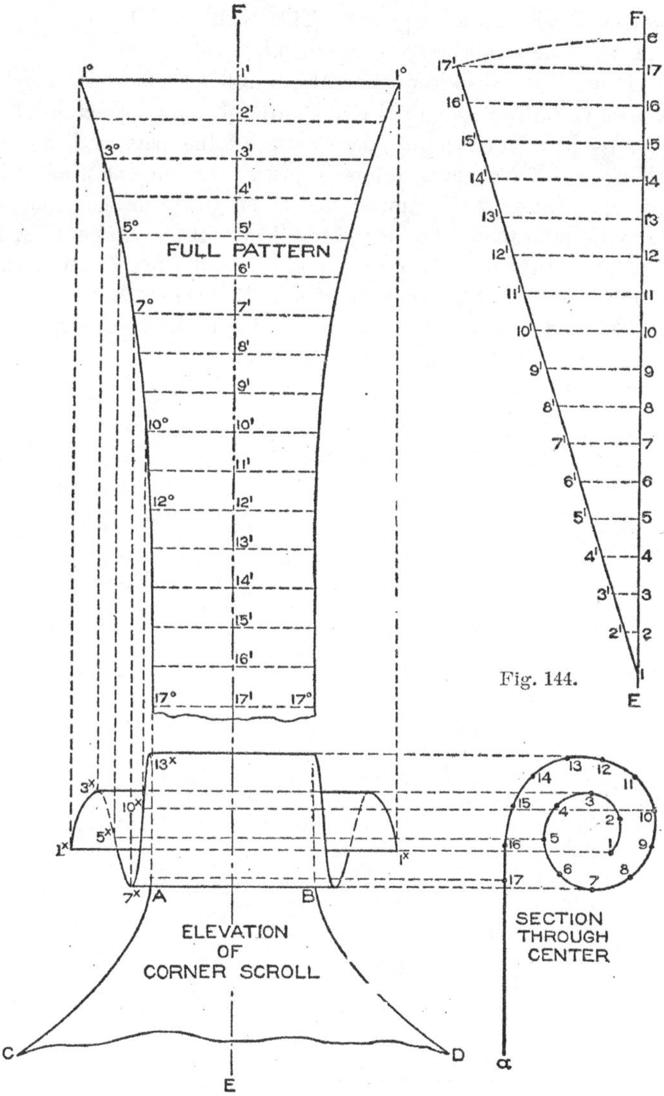

Fig. 144.

Fig. 143.

edge and bending it as required. Then will 1° 1° 17° 17° be the pattern for the scroll using heavy metal.

If it is desired to know how this scroll will look when rolled up, then at right angles to E F and through the intersections $1'$ to $17'$ draw lines intersecting the curves of the pattern 1°-17° on both sides. From these intersections, shown on one side only, drop lines intersecting similar numbered lines, drawn from the intersections in the profile of the scroll in section parallel to A B. To avoid a confusion of lines the points 1^x, 3^x, 5^x, 7^x, 10^x, 12^x, and 17^x have only been intersected. A line traced through points thus obtained as shown from 1^x to 17^x in elevation gives the projections at the ends of the scroll when rolled up.

SKYLIGHT WORK *

The upper illustration shows the layout of a flat pitched skylight whose curb measures 6′ – 0″ × 7′ – 6″, the run of the rafter or length of the glass being 6′ 0″ on a horizontal line. Five bars are required, making the glass 15 inches wide A working section through AB and CD is shown below.

It will be noticed in the section through AB that the flashing is locked to the roofing and flanged around the inside of the angle iron construction; over this the curb of the skylight rests, bolted through the angle iron as shown, the bolt being capped and soldered to avoid leakage.

The same construction is used in the section through CD, with the exception, that when the flashing cannot be made in one piece, a cross lock is placed in the manner indicated, over the fireproof blocks.

* The illustration referred to will be found on the back of this page.

CONSTRUCTION DRAWING SHOWING LAY OUT OF FLAT SKYLIGHT AND METHOD OF FASTENING FLASHING ON ANGLE IRON CONSTRUCTION.

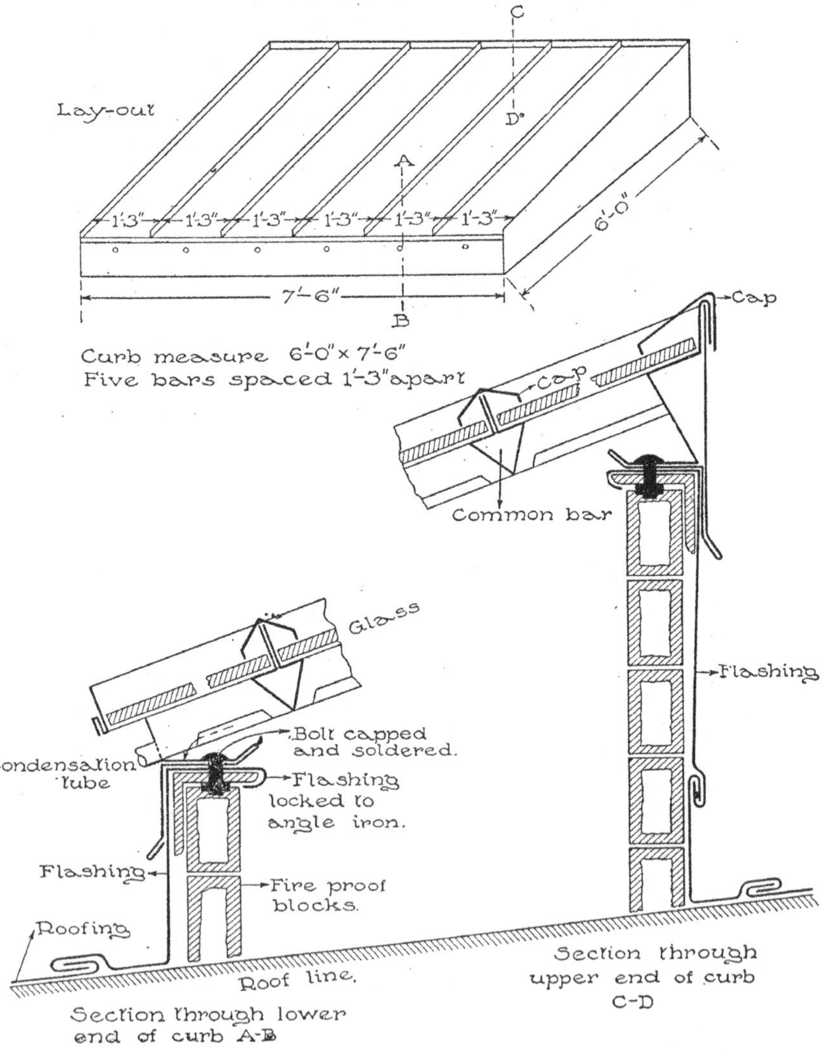

SHEET METAL WORK

PART III

SKYLIGHT WORK

Where formerly skylights were constructed from wrought iron or wood, to-day in all the large cities they are being made of galvanized sheet iron and copper. Sheet metal skylights, having by their peculiar construction lightness and strength, are superior to iron and wooden lights; superior to iron lights, inasmuch as there is hardly any expansion or contraction of the metal to cause leaks or breakage of glass; and superior to wooden lights, because they are fire, water and condensation-proof, and being less clumsy, admit more light.

The small body of metal used in the construction of the bar and curb and the provisions which can be made to carry off the inside condensation, make sheet metal skylights superior to all others constructed from different material.

CONSTRUCTION

The construction of a sheet metal skylight is a very simple matter, if the patterns for the various intersections are properly developed. For example, the bar shown in Fig. 145 consists of a piece of sheet metal having the required stretchout and length, and bent by special machinery, or on the regular cornice brake, into the shape shown, which represents strength and rigidity with the least amount of weight. A A represent the condensation gutters to receive the condensation from the inside when the warm air strikes against the cold surface of the glass, while B B show the rabbets or glass-rest for the glass.

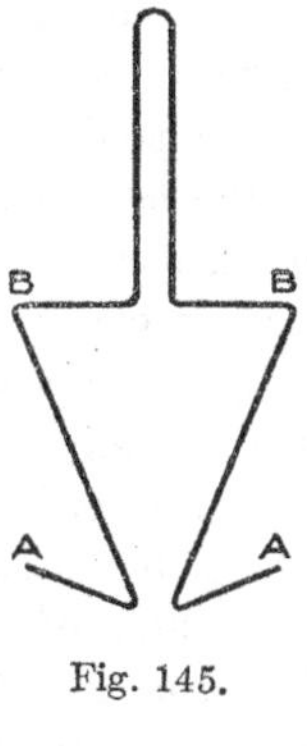

Fig. 145.

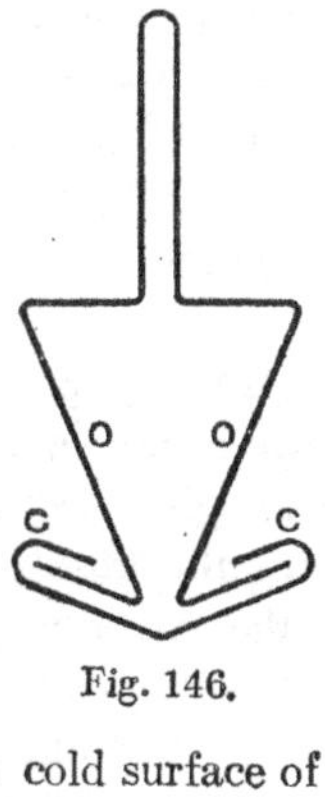

Fig. 146.

In Fig. 146, C C is a re-enforcing strip, which is used to hold the two walls O O together and impart to it great rigidity. When skylight bars are required to bridge long spans, an internal core is made of sheet metal and placed as shown at A in Fig. 147, which adds to its weight-sustaining power. In this figure B B shows the glass laid on a bed of putty with the metal cap C C C, resting snugly against the glass, fastened in position by the rivet or bolt D D. Where a very large span is to be bridged a bar similar to that shown in Fig. 148 is used. A heavy core plate A made of ¼-inch thick metal is used, riveted or bolted to the bar at B and B. In construction, all the various bars terminate at the curb shown at A B C in Fig. 149, which is fastened to the wooden frame D E.

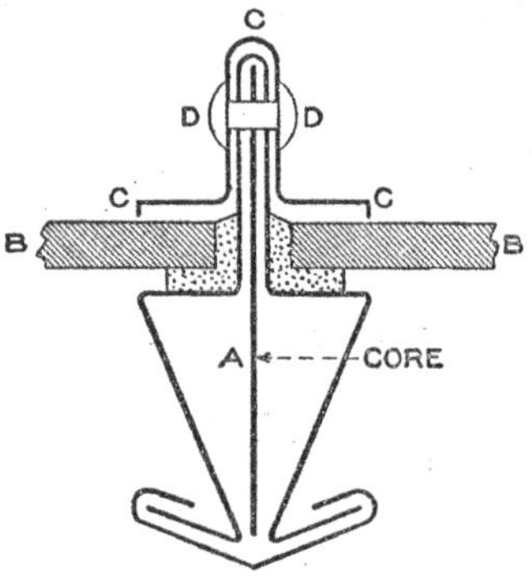

Fig. 147.

The condensation gutters C C in the bar *b*, carry the water into the internal gutter in the curb at *a*, thence to the outside through holes provided for this purpose at F F. In Fig. 150 is shown a sectional view of the construction of a double-pitched skylight. A shows the ridge bar with a core in the center and cap attached over the glass. B shows the cross bar or clip which is used in large skylights where it is impossible to get the glass in one length, and where the glass must be protected and leakage prevented by means of the cross bar, the gutter of which conducts the water into the gutter of the main bar, thence outside the curb as before explained. C is the frame generally made of wood or angle iron and covered by the metal roofer with flashing as shown at F. D shows the skylight bar with core showing the glass and cap in position. E is the metal curb against which the bars terminate, the condensation being let out through the holes shown.

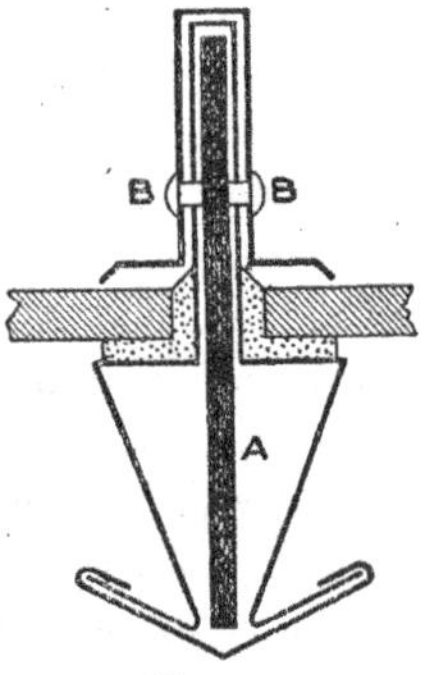

Fig. 148.

In constructing pitched skylights having double pitch, or being hipped, the pitch is usually one-third. In other words it is one-third of the span. If a skylight were 12 feet wide and one-third pitch were required, the rise in the center would be one-third of 12, or 4 feet. When a flat skylight is made the pitch is usually built in the wood or iron frame and a flat skylight laid over it. The glass used in the construction of metallic skylights is usually $\frac{1}{4}$-inch rough or ribbed glass; but in some cases heavier glass is used.

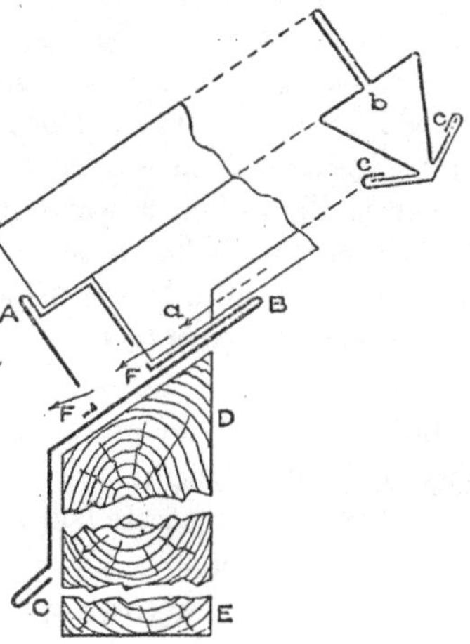

Fig. 149.

If for any reason it is desired to know the weight of the various thickness of glass, the following table will prove valuable.

Weight of Rough Glass Per Square Foot.

Thickness in inches.	$\frac{1}{8}$.	$\frac{3}{16}$.	$\frac{1}{4}$.	$\frac{3}{8}$.	$\frac{1}{2}$.	$\frac{5}{8}$.	$\frac{3}{4}$.	1.
Weight in pounds.	2.	$2\frac{1}{2}$.	$3\frac{1}{2}$.	5.	7.	$8\frac{1}{2}$.	10.	$12\frac{1}{2}$.

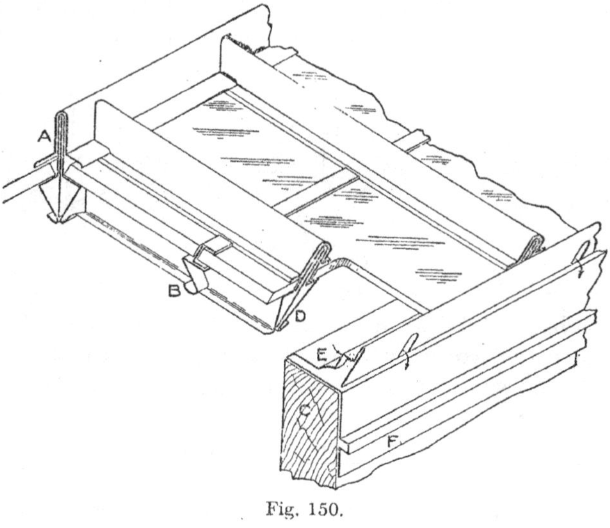

Fig. 150.

SHOP TOOLS

In the smaller shops the bars are cut with the hand shears and formed up on the ordinary cornice brake. In the larger shops, the strips required for the bars or curbs are cut on the large squaring shears, and the miters on the ends of these strips are cut on what is known as a miter cutter. This machine consists of eight foot presses on a single table, each press having a different set of dies for the purpose of cutting the various miters on the various bars. The bars are then formed on what is known as a Drop Press in which the bar can be formed in two operations to the length of 10 feet.

METHOD EMPLOYED IN OBTAINING THE PATTERNS

The method to be employed in developing the patterns for the various skylights is by parallel lines. If, however, a dome, conservatory or circular skylight is required, the blanks for the various curbs, bars, and ventilators, are laid out by the rule given in Sheet Metal Work, Part IV, under "Circular Work".

VARIOUS SHAPES OF BARS

In addition to the shapes of bars shown in Figs. 145 to 148 inclusive, there is shown in Fig. 151 a plain bar without any condensation gutters, the joint being at A. B B represents the glass resting on the rabbets of the bar, while C shows another form of cap which covers

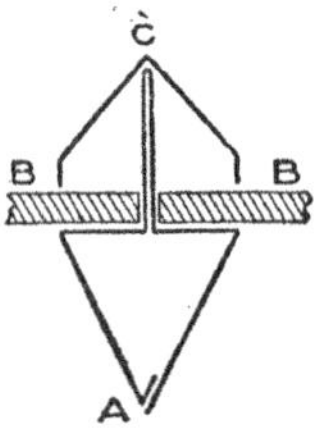

Fig. 151.

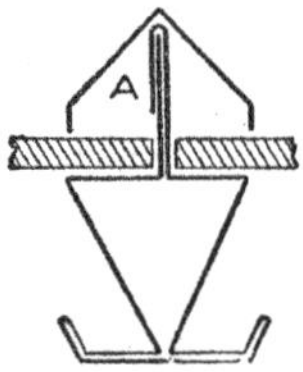

Fig. 152.

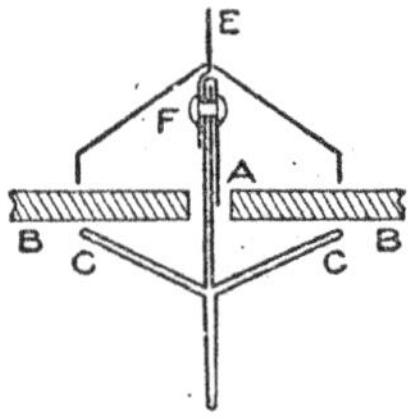

Fig. 153.

the joint between the bar and glass. Fig. 152 gives another form of bar in which the condensation gutters and bar are formed from one piece of metal with a locked hidden seam at A. Fig. 153 shows a bar on which no putty is required when glazing. It will be noticed that it is bent from one piece of metal with the seam at A, the glass B B resting on the combination rabbets and gutters C C. D is the cap which is fastened by means of the cleat E. These cleats are cut about ½-inch wide from soft 14-oz. copper, and riveted to the top of the bar at F; then a slot is cut into the cap D as shown from *a* to *b* in Fig. 154; then the cap is pressed firmly onto the glass and the cleat E turned down which holds the cap in position.

When a skylight is constructed in which raising sashes are required, as shown in Fig. 155, half bars are required at the sides A and B, while the bars on each side of the sash to be raised are so constructed that a water-tight joint is obtained when closed. This is shown in Fig. 156, which is an enlarged section through A B in Fig. 155. Thus in Fig. 156, A A represents the two half bars with condensation gutters as shown, the locked seam taking place at B B. C C represent the two half bars for the raising sash with the caps D D attached to same, as shown, so that when the sash C C is closed, the caps

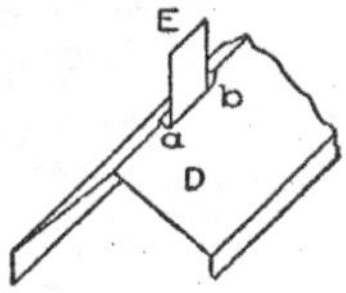

Fig. 154.

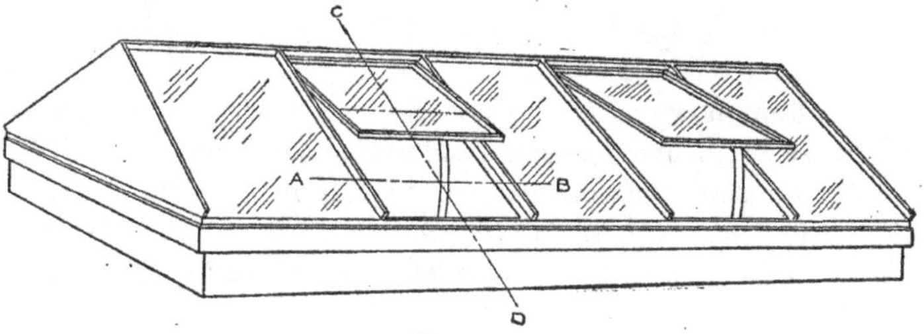

Fig. 155.

D D cover the joint between the glass E E and the stationary half bars. F F are the half caps soldered at *a a* to the bars C C which protect the joints between the glass H H and the bars C C.

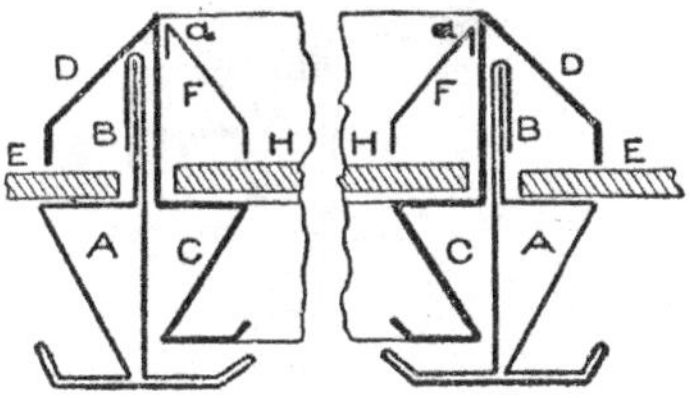

Fig. 156.

VARIOUS SHAPES OF CURBS

In Figs. 157, 158 and 159 are shown a few shapes of curbs which are used in connection with flat skylights. A in Fig. 157 shows the curb for the three sides of a flat skylight, formed in one piece with a joint at B, while C shows the cap, fastened as previously described. "A" shows the height at the lower end of the curb, which is made as high as the glass is thick and allows the water to run over. In Fig. 158, A is another form of skylight formed in one piece and riveted at B; *a* shows the height at the lower end. In the previous figures the frame on which the metal curb rests is of wood, while in Fig. 159 the frame is

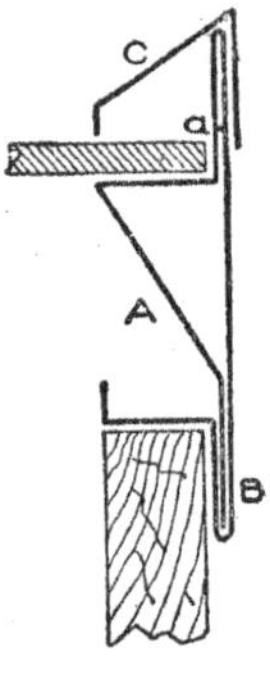

Fig. 157.

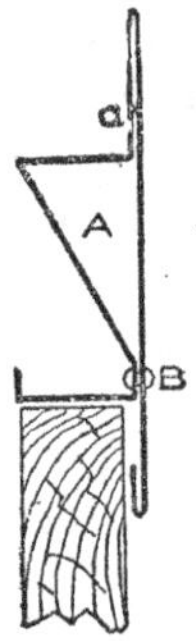

Fig. 158.

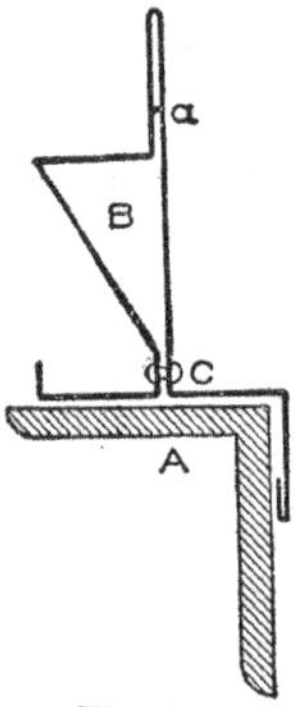

Fig. 159.

of angle iron shown at A. In this case the curb is slightly changed as shown at B; bent in one piece, and riveted at C. In Figs. 160, 161, and 162 are shown various shapes of curbs for pitched skylights in addition to that shown in Fig. 149. A in Fig. 160 shows a curb formed in one piece from *a* to *b* with a condensation hole or tube shown at B.

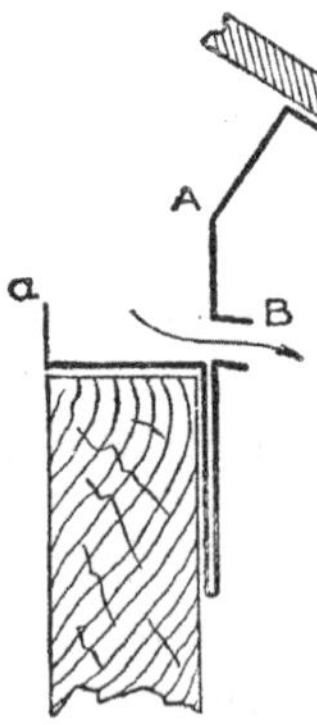

Fig. 160.

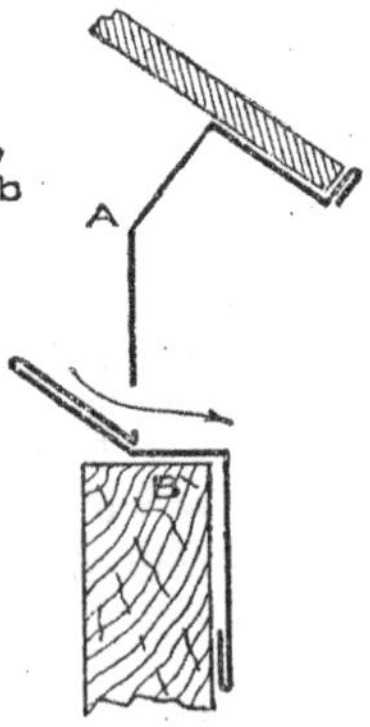

Fig. 161.

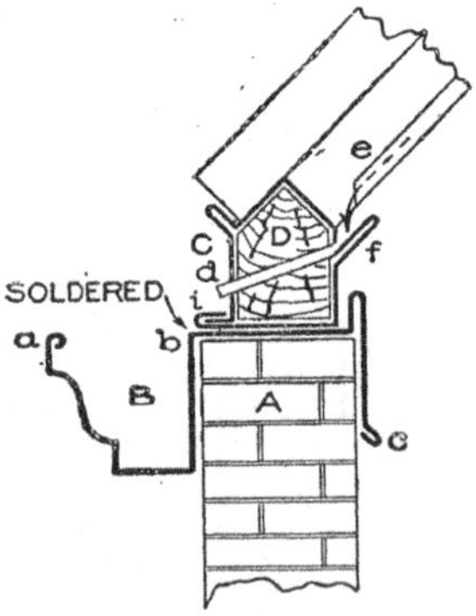

Fig. 162.

In Fig. 161 is shown a slightly modified shape A, with an offset to rest on the curb at B. When a skylight is to be placed over an opening whose walls are brick, a gutter is usually placed around the wall, as shown in Fig. 162, in which A represents a section of the wall on which a gutter, B, is hung, formed from one piece of metal, as shown from *a* to *b* to *c*. On top of this the metal curb C is soldered, which is also formed from one piece with a lock seam at *i*. To stiffen this curb a wooden core is slipped inside as shown at D. From the inside condensation gutter *f* a 14-oz. copper tube runs through the curb, shown at *d*. The condensation from the gutter *e* in the bar, drips into the gutter *f*, out of the tube *d*, into the main gutter B, from which it is conveyed to the outside by a leader.

In Fig. 163 is shown an enlarged section of a raising sash, taken through C D in Fig. 155. A in Fig. 163 shows the ridge bar, B the lower curb and C D the side sections of the bars explained in connection with Fig. 156. E F in Fig. 163 shows the upper frame of the raising sash, fitting onto the half ridge bar A. On each raising sash, at the upper end two hinges H are riveted at E and I, which allow the sash to raise or close by means of a cord, rod, or gearings. J K shows the lower frame of the sash fitting over the curb B. Holes are punched at *a* to allow the condensation to escape into *b*, thence to the outside through C. Over the hinge H a hood or cap is placed which prevents leakage. Fig. 164 shows a section through A B in Fig. 167 and represents a hipped skylight having one-third pitch. By a skylight of one-third pitch is meant a skylight whose altitude or height A B, is equal to one-third of the span C D. If the skylight was to have a pitch of one-fourth or one-fifth, then the altitude A B would equal one-fourth or one-fifth respectively of the span C D.

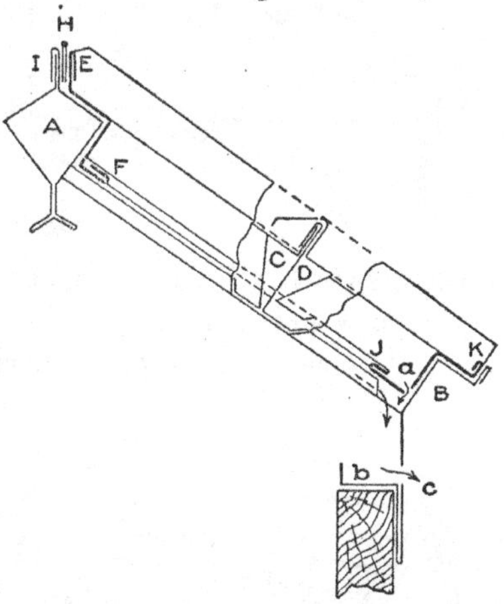

Fig. 163.

The illustration shows the construction of a hipped skylight with ridge ventilator which will be briefly described. C D is the curb; E E the inside ventilator; F F the outside ventilator forming a cap over the glass at *a*. G shows the hood held in position by two cross braces H. J represents a section of the common bar on the rabbets of which the glass K K rests. L shows the condensation gutters on the bar J,

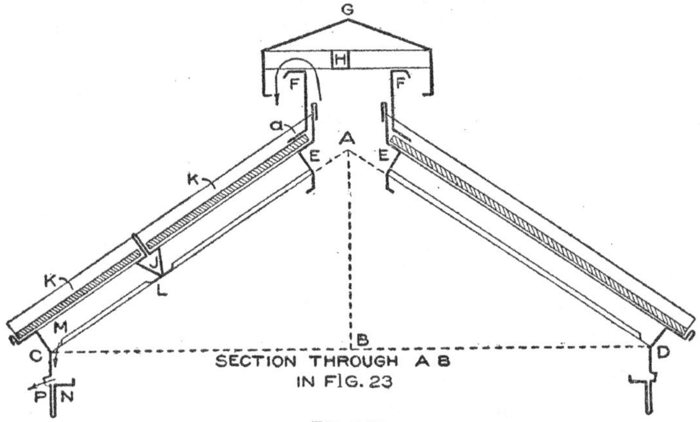

Fig. 164.

which are notched out as shown at M, thus allowing the drip to enter the gutter N and discharge through the tube P. The foul air escapes under the hood G as shown by the arrow.

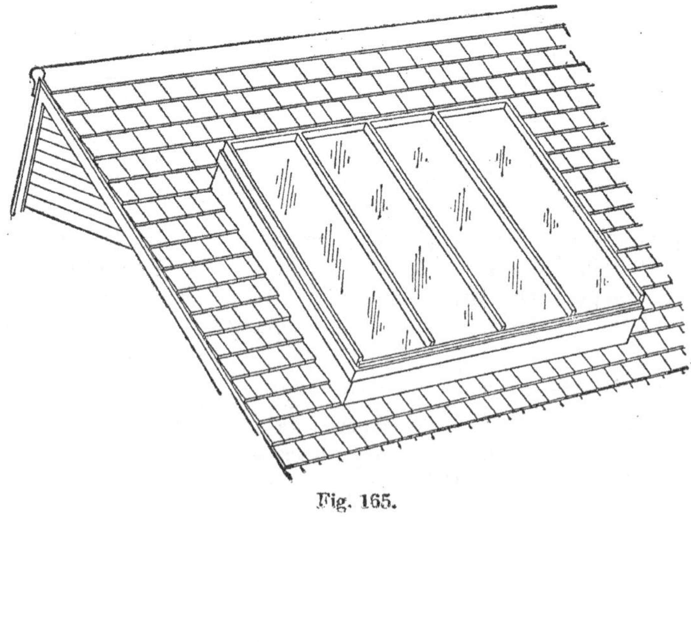

Fig. 165.

VARIOUS STYLES OF SKYLIGHTS

In Fig. 165 is shown what is known as a single-pitch light, and is placed on a curb made by the carpenter which has the desired pitch.

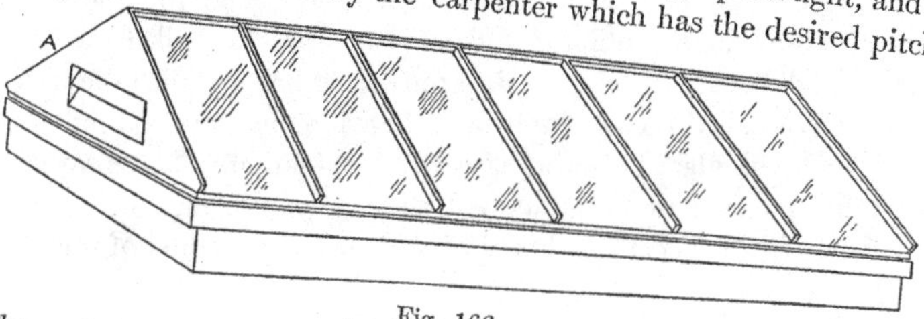

Fig. 166.

These skylights are chiefly used on steep roofs as shown in the illus-tration, and made to set on a wooden curbs pitching the same as the

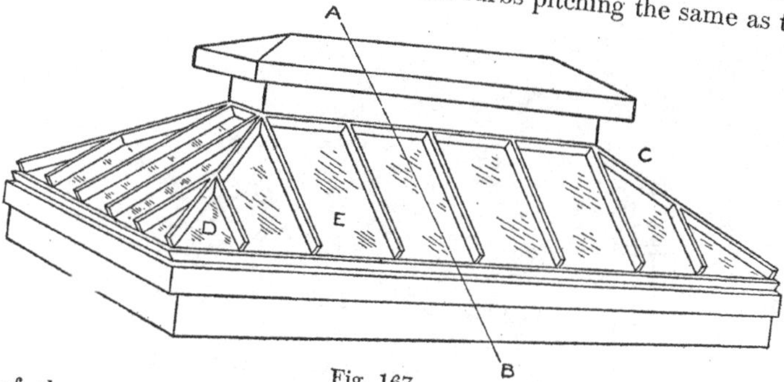

Fig. 167.

roof, the curb first being flashed. Ventilation is obtained by raising one or more lights by means of gearings, as shown in Fig. 155.

Fig. 168.

Fig. 166 shows a double-pitch skylight. Ventilation is obtained by placing louvres at each end as shown at A. Fig. 167 shows a skylight with a ridge ventilator. The corner bar C is called the hip bar; the small bar D, mitering against the corner bar, is called the jack bar, while E is called the common bar. Fig. 168 illustrates a hip monitor skylight with glazed opening sashes for ventilation. These sashes can be opened or closed separately, by means of gearings similar to those shown in Fig. 177 In Fig. 169 is shown the method of raising

Fig. 169.

sashes in conservatories, greenhouses, etc., the same apparatus being applicable to both metal and wooden sashes. Fig. 170 shows a view of a photographer's skylight; if desired, the vertical sashes can be made to open.

In Fig. 171 is shown a flat extension skylight at the rear of a store or building. The upper side and ends are flashed into the brick work and made water-tight with waterproof cement, while the lower side rests on the rear wall to which it is fastened. In some cases the rear gutter is of cast iron, put up by the iron worker, but it is usually made of No. 22 galvanized iron, or 20-oz. cold-rolled copper. To receive the bottom of the gutter and skylight, the wall should be covered by a wooden plate A, Fig. 172, about two inches thick, and another plank set edgeways flush with the inside of the wall, as shown at B. The two planks are not required when a cast iron gutter is used.

Fig. 173 shows a hipped skylight without a ridge ventilator, set on a metal curb in which louvres have been placed. These louvres may be made stationary or movable. When made movable, they are

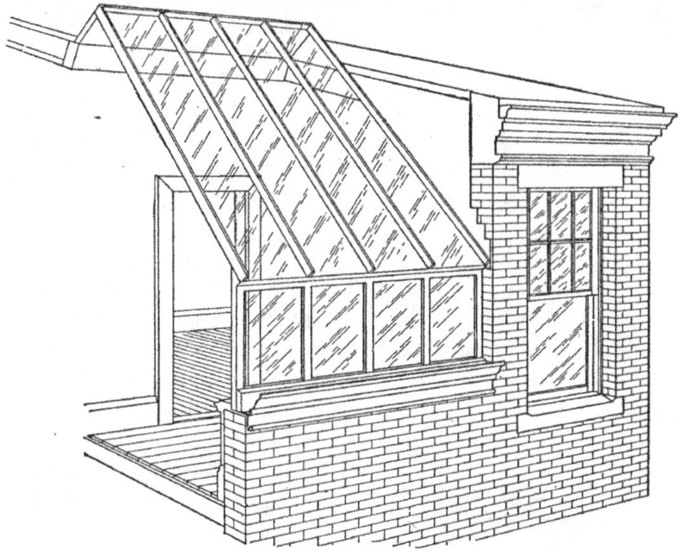

Fig. 170.

constructed as shown in Fig. 174, in which A shows a perspective view, B shows them closed, and C open. They are operated by the quadrants attached to the upright bars *a* and *b*, which in turn are pulled up and down by cords or chains worked from below. When a skylight has a very long span, as in Fig. 175, it is constructed as shown in Fig. 176, in which A represents a T-beam which can be trussed if necessary. This construction allows the water to escape from the bottom of the upper light to the outside of the top of the lower skylight, the curb C of the upper light fitting over the curb B of the lower light.

In Fig. 177 is shown the method of applying the gearings. A shows the side view of the metal or wooden sash partly opened, B the

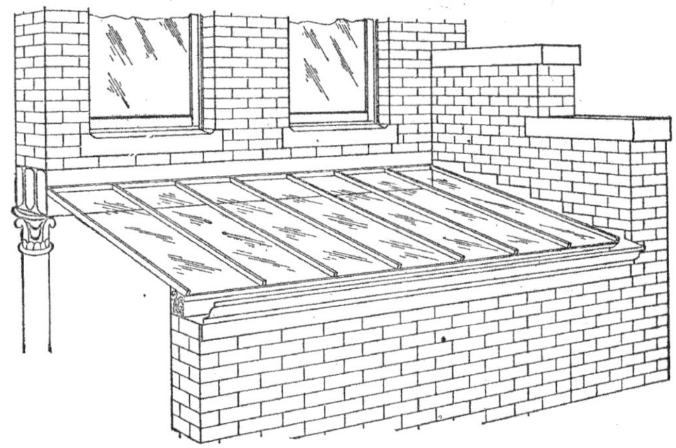

Fig. 171.

end of the main shaft, and C the binder that fastens the main shaft to the upright or rafter. D shows the quadrant wheel attached to main shaft and E is the worm wheel, geared to the quadrant D, communicating motion to the whole shaft. F is a hinged arm fastened to the main shaft B and hinged to the sash. By turning the hand-wheel the sash can be opened at any angle.

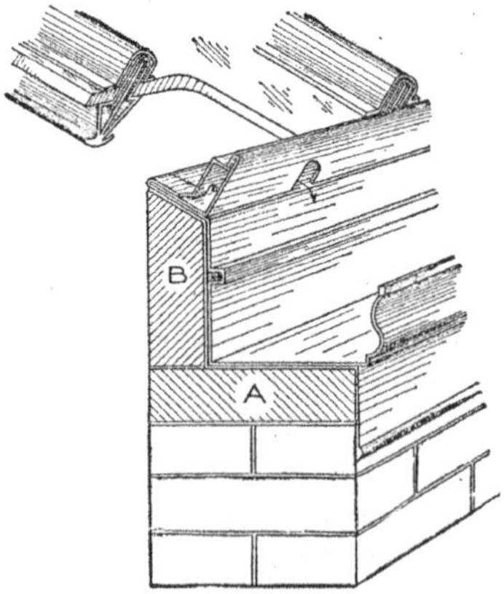

Fig. 172.

DEVELOPMENT OF PATTERNS FOR A HIPPED SKYLIGHT

The following illustrations and text will explain the principles involved in developing the patterns for the ventilator, curb, hip bar, common bar, jack bar, and cross bar or clip, in a hipped skylight. These principles are also applicable to any other form of light, whether flat, double-pitch, single-pitch, etc.

In Fig. 178 is shown a half section, a quarter plan, and a diagonal elevation of a hip bar, including the patterns for the curb, hip, jack, and common bars. The method of making these drawings will be explained in detail, so that the student who pays close attention

Fig. 173

will have no difficulty in laying out any patterns no matter what the pitch of the skylight may be, or what angle its plan may have.

First draw any center line as A B, at right angles to which lay off C 4′, equal to 12 inches. Assuming that the light is to have one-third

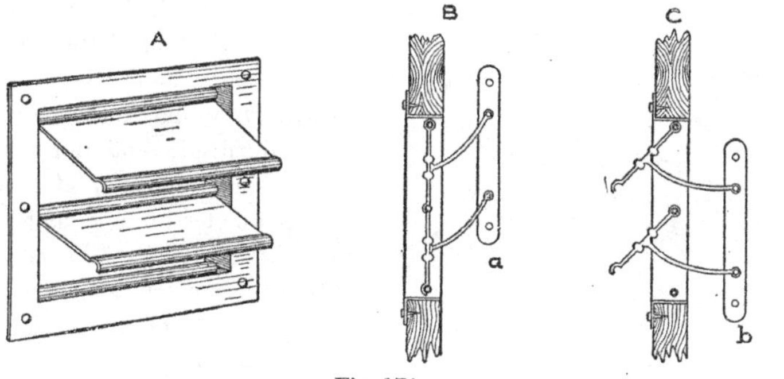

Fig. 174.

pitch, then make the distance C D equal to 8 inches which is one-third of 24 inches, and draw the slant line D 4.′ At right angles to D 4′ place a section of the common bar as shown by E, through which draw lines parallel to D 4′, intersecting the curb shown from *a* to *f* at the bottom and the inside section of the ventilator from F to G at the top. At pleasure draw the section of the outside vent shown from *h* to *l* and the hood shown from *m* to *p*. X represents the section of the brace resting on *i j* to uphold the hood resting on it in the corner *o*. The condensa-

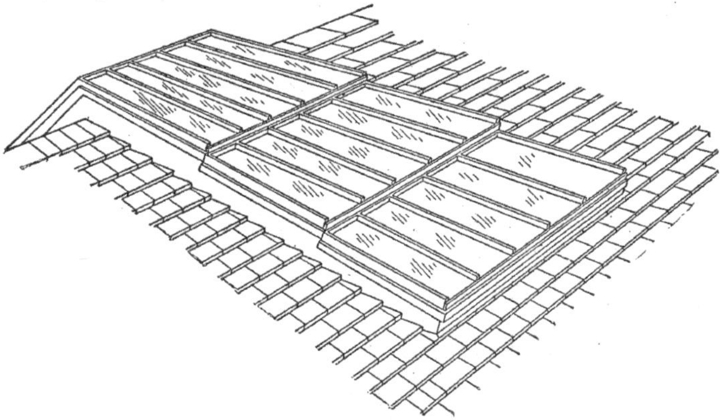

Fig. 175.

tion gutters of the common bar E are cut out at the bottom at 5′ 6′ which allows the drip to go into the gutter *d e f* of the curb and pass out of the opening indicated by the arrow. Number the corners of each half of the common bar section E as shown, from 1 to 6 on each side, through which draw lines parallel to D 4′ until they intersect the curb at the bottom as shown by similar numbers 1′ to 6′, and the inside ventilator at the top by similar figures 1″ to 6″. This completes the one half-section of the skylight. From this section the pattern for the common bar can be obtained without the plan, as follows:

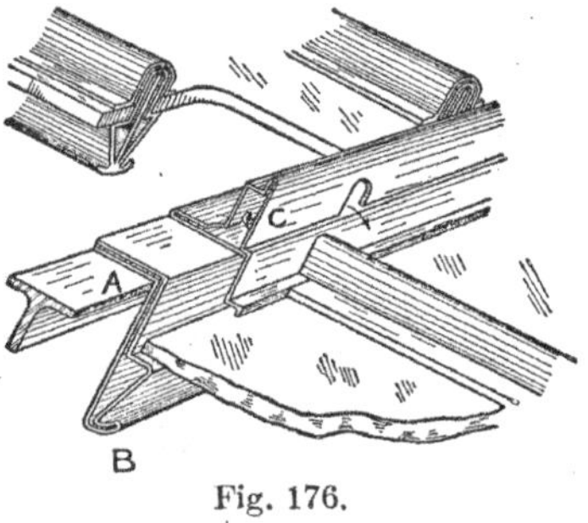

Fig. 176.

At right angles to D 4′ draw the line I J upon which place the stretchout of the section E as shown by similar figures on I J. Through these small figures, and at right angles to I J, draw lines, and intersect them by lines drawn at right angles to D 4′ from similarly numbered intersections 1′ to 6′ on the curb and 1″ to 6″ on the inside ventilator. Trace a line through points thus obtained; then A^1 B^1 C^1 D^1 will be the pattern for the common bar in a hipped skylight. The same method would be employed if a pattern were developed for a flat or a double-pitch light. From this same half section the pattern for the curb is developed by taking the stretchout of the various corners in the curb, *a b* 3′ 4′ *c d e* and *f*, and placing them on the center line A B as shown by similar letters and figures. Through these divisions and at right angles to A B draw lines which intersect with lines drawn at right angles to C 4′ from similar points in the curb section *a f*. Trace a line through points thus obtained; then E^1 F^1 *f a* will be the half pattern for the curb shown in the half section. V represents the condensation hole to be punched into the pattern between each light of glass in the skylight. As the portion *c d* turns up on *c* 4′, use *r* as a center, and with

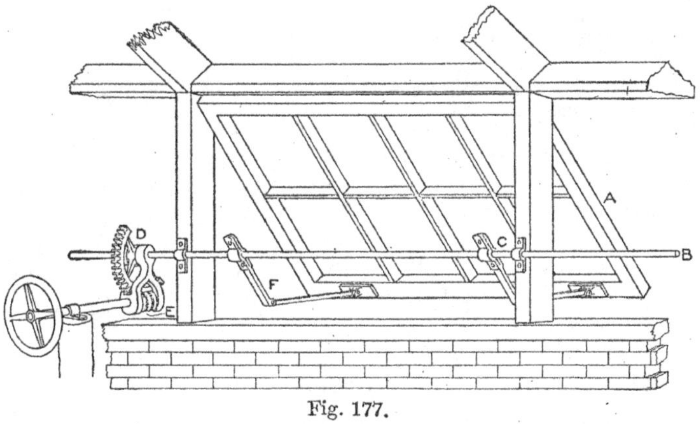

Fig. 177.

the radius *r s′* strike the semicircle shown. Above this semicircle punch the hole V.

Before the patterns can be obtained for the hip and jack bars, a quarter plan view must be constructed which will give the points of intersections between the hip bar and curb, between the hip bar and vent, or ridge bar, and between the hip and jack bar. Therefore, from any point on the center line A B as K, draw K L at right angles to A B. As the skylight forms a right angle in plan, draw from K, at an angle of 45°, the hip or diagonal line K 1°. Take a tracing of the common bar section E with the various figures on same, and place it on the hip line K 1° in plan so that the points 1 4 come directly on the hip as shown by E^1. Through the various figures draw lines parallel to K 1°

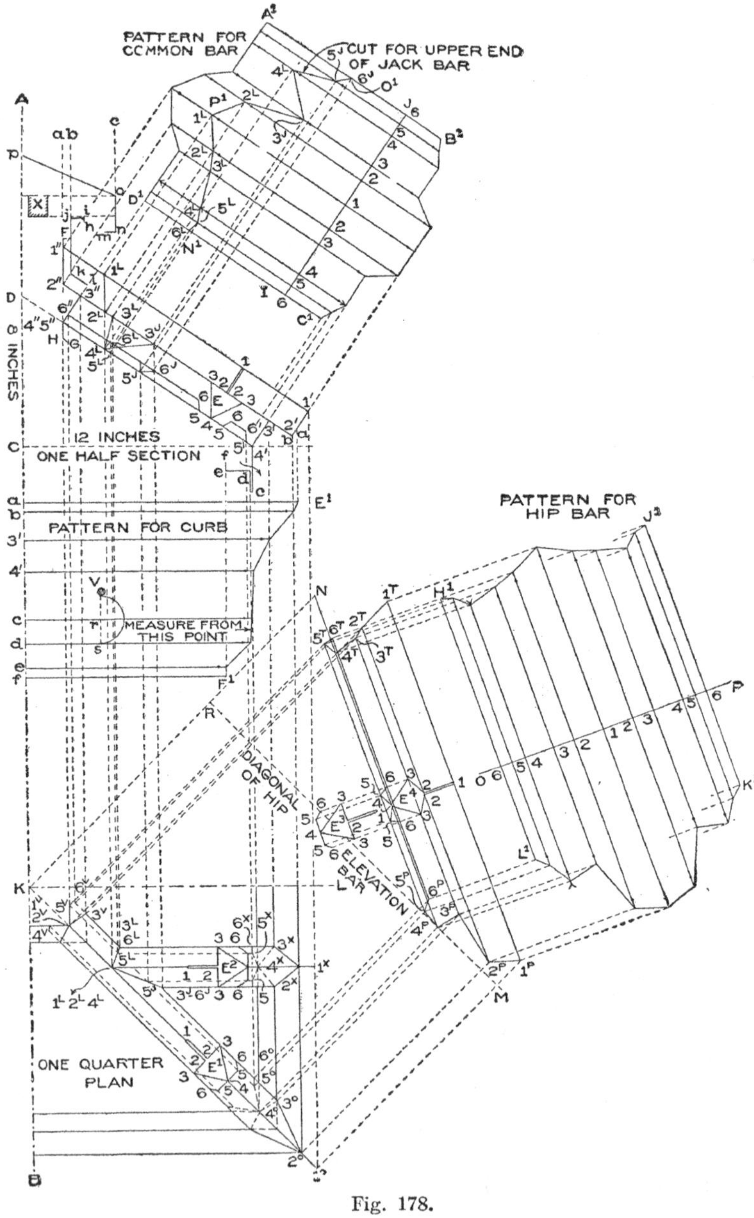
PATTERN FOR
COMMON BAR
CUT FOR UPPER END
OF JACK BAR
8 INCHES
12 INCHES
ONE HALF SECTION
PATTERN FOR CURB
MEASURE FROM
THIS POINT
PATTERN FOR
HIP BAR
DIAGONAL
OF HIP
ELEVATION
BAR
ONE QUARTER
PLAN

Fig. 178.

one-half of which are intersected by vertical lines drawn parallel to A B from similar points of intersection 1′ to 6′ on the curb, and 1″ to 6″ on the ventilator in the half section, as shown respectively in plan by intersections 1° to 6° and 1^V to 6^V. Below the hip line K 1° trace the opposite intersection as shown. It should be understood that the section E^1 in plan does not indicate the true profile of the hip bar (which must be obtained later), but is only placed there to give the horizontal distances in plan. In laying out the work in practice to full size, the upper half intersection of the hip bar in plan is all that is required. It will be noticed that the points of intersections in plan and one half section have similar numbers; and if the student will carefully follow each point the method of these projections will become apparent.

Having obtained the true points of intersections in plan the next step is to obtain a diagonal elevation of the hip bar, from which a true section of the hip bar and pattern are obtained. To do this draw any line as R M parallel to K 1°. This base line R M has the same elevation as the base line C 4′ has in the half section. From the various points 1° to 6° and 1^V to 6^V in plan, erect lines at right angles to K 1° crossing the line R M indefinitely. Now measuring in each and every instance from the line C 4′ in the half section take the various distances to points D 1″ 2″ 3″ 4″ 5″ and 6″ at the top, and to points 1′ 2′ 3′ 4′ 5′ and 6′ at the bottom, and place them in the diagonal elevation measuring in each and every instance from the line R M on the similarly numbered lines drawn from the plan, thus locating respectively the points N 1^T 2^T 3^T 4^T 5^T and 6^T at the top, and 1^P 2^P 3^P 4^P 5^P and 6^P at the bottom. Through the points thus obtained draw the miter lines 1^T to 6^T and 1^P to 6^P and connect the various points by lines as shown, which completes the diagonal elevation of the hip bar intersecting the curb and vent, or ridge. To obtain the true section of the hip bar, take a tracing of the common bar E or E^1 and place it in the position shown by E^3, being careful to place the points 1 4 at right angles to 1^T 1^P as shown. From the various points in the section E^3 at right angles to 1^P 1^T draw lines intersecting similarly numbered lines in the diagonal elevation as shown from 1 to 6 on either side. Connect these points as shown; then E^4 will be the true profile of the hip bar. Note the difference in the two profiles; the normal E^3 and the modified E^4.

Having obtained the true profile E^4 the pattern for the hip bar is obtained by drawing the stretchout line O P at right angles 1^T 1^P.

Take the stretchout of the profile E^4 and place it on O P as shown by similar figures. Through these small figures and at right angles to O P draw lines which intersect by lines drawn at right angles to 1^T 1^P from similarly numbered points at top and bottom, thus obtaining the points of intersections shown. A line traced through the points thus obtained, as shown by H^1 J^1 K^1 L^1 will be the pattern for the hip bar.

For the pattern for the jack bar, take a tracing of the section of the common bar E and place it in the position in plan as shown by E^2 being careful to have the points 1 and 4 at right angles to the line 1^x 1°. It is immaterial how far the section E^2 is placed from the corner 2° as the intersection with the hip bar remains the same no matter how far the section is placed one way or the other. Through the various corners in the section E^2 draw lines at right angles to the line 1° 1^x intersecting one half of the hip bar on similarly numbered lines as shown by the intersections 1^L 2^L 3^L 4^L 5^L 6^L and 1^L 2^J 3^J 4^L 5^J and 6^J; also intersecting the curb in plan at points 1^x to 6^x. The intersection between the jack bar and curb in plan is not necessary in the development of the pattern as the lower cut in the pattern for the common bar is the same as the lower cut in the pattern for the jack bar. However, the intersection is shown in plan to make a complete drawing. At right angles to the line of the jack bar in plan, and from the various intersections with the hip bar, erect lines intersecting similarly numbered lines in the section as shown. Thus from the various intersections shown from 1^L to 6^L in plan, erect vertical lines intersecting the bar in the half section at points shown from 1^L to 6^L. In similar manner from the various points of intersections 3^J, 5^J, and 6^J in plan, erect lines intersecting the bar in the half section at points shown by 3^J 5^J 6^J. Connect these points in the half section, as shown, which represents the line of joint in the section between the hip and jack bars.

For the pattern for the upper cut of the jack bar, the same stretchout can be used as that used for the common bar. Therefore, at right angles to D 4′ and from the various intersections 1^L 2^L 3^L 4^L 5^L and 6^L draw lines intersecting similar numbered lines in the pattern for the common bar as shown by similar figures. In similar manner from the various intersections 3^J 5^J and 6^J in the one half section, draw lines at right angles to D 4′ intersecting similarly numbered lines in the pattern as shown by 3^J 5^J and 6^J. Trace lines from point to point, then the cut shown from N^1 to P^1 will represent the miter for that part shown in plan from 2^L to 6^L, and the cut shown from P^1 to O^1 in the pattern will represent the cut for that part shown in plan from 2^L to 6^J. The lower cut of the jack bar remains the same as that shown in the pattern.

The half pattern for the end of the hood is shown in Fig. 179, and is obtained as follows: Draw any vertical line as A B, upon which place the stretchout of the section of the hood *m n o p* in Fig. 178, as shown by similar letters *m n o p* on A B in Fig. 179. At right angles to A B and through the small letters draw lines, making them equal in length, (measuring from the line A B) to points having similar letters in Fig. 178, also measuring from the center line A B. Connect points shown in Fig. 179, which is the half pattern for the end of the hood. For the half pattern for the end of the outside ventilator, take the

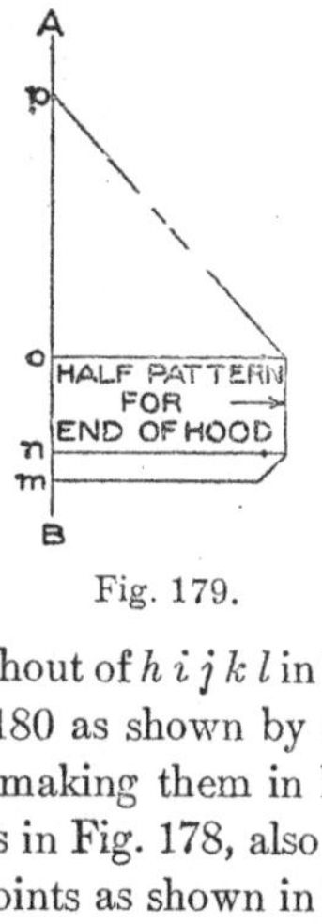

Fig. 179.

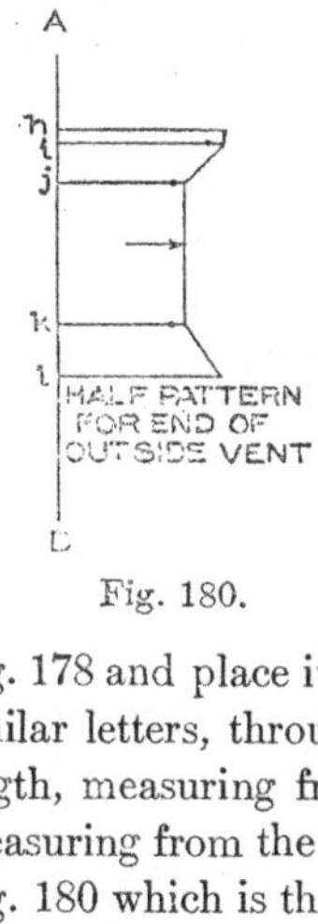

Fig. 180.

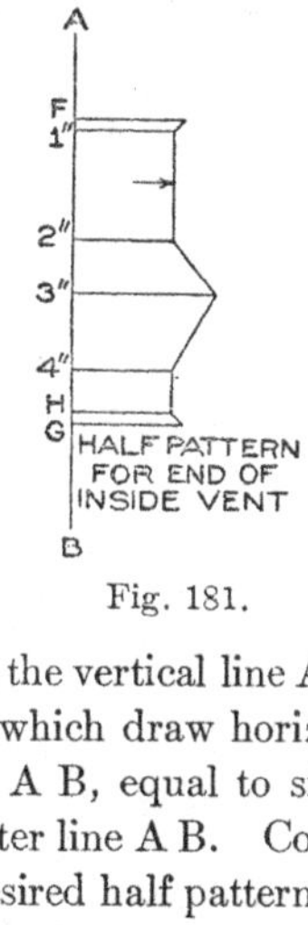

Fig. 181.

stretchout of *h i j k l* in Fig. 178 and place it on the vertical line A B in Fig. 180 as shown by similar letters, through which draw horizontal lines making them in length, measuring from A B, equal to similar letters in Fig. 178, also measuring from the center line A B. Connect the points as shown in Fig. 180 which is the desired half pattern. In Fig. 181 is shown the half pattern for the end of the inside ventilator, the stretchout of which is obtained from F 1″ 2″ 3″ 4″ H G in Fig. 178, the pattern being obtained as explained in connection with Figs. 179 and 180.

When a skylight is to be constructed on which the bars are of such lengths that the glass cannot be obtained in one length, and a cross bar or clip is required as shown by B, in Fig. 150, which miters against the main bar, the pattern for this intersecting cut is obtained as shown in

Fig. 182. Let A represent the section of the main bar, B the elevation of the cross bar, and C its section. Note how this cross bar is bent so that the water follows the direction of the arrow, causing no leaks because the upper glass *a* is bedded in putty, while the lower light *b* is capped by the top flange of the bar C (See Fig. 150). Number all of the corners of the section C as shown, from 1 to 8, from which points draw horizontal lines cutting the main bar A at points 1 to 8 as shown. At right angles to the lines in B draw the vertical line D E upon which

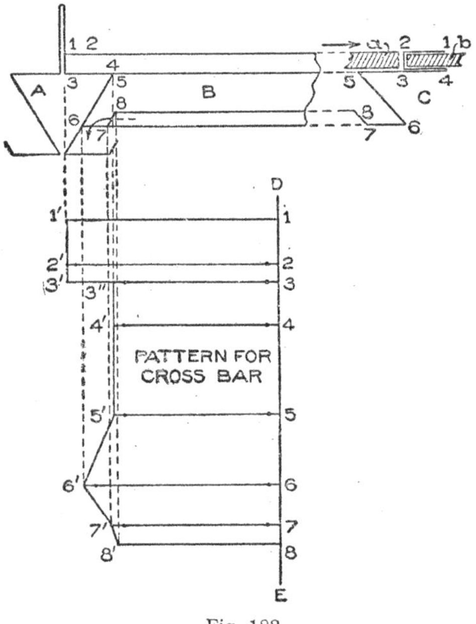

Fig. 182.

place the stretchout of the cross bar C, shown by similar figures, through which draw horizontal lines, intersecting them with lines drawn parallel to D E from similar numbered intersections against the main bar A, thus obtaining the points of intersections 1′ to 8′ in the pattern. Trace a line through points of intersections thus obtained which will be the pattern for the end cut of the cross bar.

In Fig. 183 is shown a carefully drawn working section of the turret sash shown in Fig. 168 at A. These sashes are operated by means of cords, chains or gearings from the inside, the pivot on which they turn being shown by R S in Fig. 183. The method of obtaining the patterns for these sashes will be omitted, as they are only square and butt miters which the student will have no trouble in developing, providing he understands the construction. This will be made clear by the following explanation:

A B represents the upper part of the turret proper with a drip bent on same, as shown at B, against which the sashes close, and a double seam, as shown at A, which makes a tight joint, takes out the twist in bending, and avoids any soldering. This upper part A B is indicated by C in Fig. 168, over which the gutter B is placed as shown by X U Y in Fig. 183. C D represents the lower part of the turret proper or base, which fits over the wooden curb W, and is indicated by D in Fig. 168. E in Fig. 183 represents the mullion made from one piece of metal and double seamed at *a*. This mullion is joined to the top and bottom. The pattern for the top end of the mullion would simply show a square cut, while the pattern for the bottom would represent a butt miter against the slant line *i j*. Before forming up this mullion the holes should be punched in the sides to admit the pivot R S. These mullions are shown in position in Fig. 168 by E E, etc.

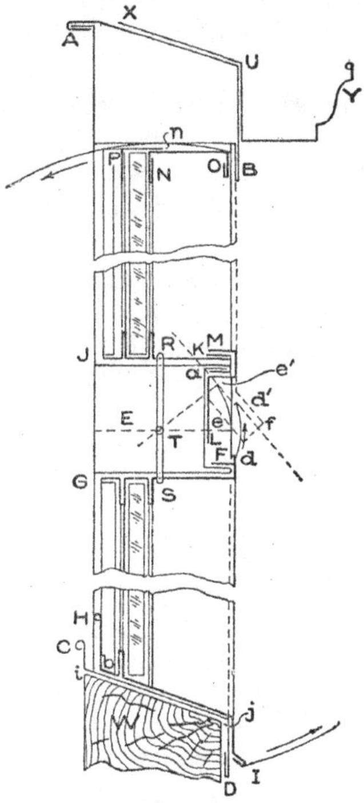

Fig. 183.

F G in Fig. 183 represents the section of the side of the sash below the pivot T. Notice that this lower half of the side of the sash has a lock attachment which hooks into the flange of the mullion E at F. While the side of the sash is bent in one piece, the upper half, above the pivot T, has the lock omitted as shown by J K. Thus when the sash opens, the upper half of the sides turn toward the inside as shown by the arrow at the top, while the lower half swings outward as shown by the arrow at the bottom. When the lower half closes, it locks as shown at F, which makes a water-tight joint; but to obtain a water-tight joint for the upper half, a cap is used, partly shown by L M, into which the upper half of the side of the sash closes as shown at M. This cap is fastened to the upper part of the mullion E with a projecting hood *f* which is placed at the same angle as the sash will have when it is opened as shown by *e e′* and *d d′* or by the dotted lines.

The side of the sash just explained is shown in Fig. 168 at H. The pattern for the side of the sash has a square cut at the top, mitering with H I at the bottom, in Fig. 183, the same as a square miter. H I represents the section of the bottom of the sash. Note where the metal is doubled as at *b*, against which the glass rests in line with the rabbet on the side of the sash. A beaded edge is shown at H which stiffens it. This lower section is shown in Fig. 168 by G and has square cuts on both ends. N O in Fig. 183 shows the section of the top of the sash shown in Fig. 168 by F. The flange N in Fig. 183 is flush with the outside of the glass, thereby allowing the glass to slide into the grooves in the sides of the sash. After the glass is in position the angle P is tacked at *n*. A leader is attached to the gutter Y as shown by B° in Fig. 168. While the method of construction shown in Fig. 183 is generally employed, each shop has different methods; what we have aimed to give is the general construction in use, after knowing which, the student can plan his own construction to suit the conditions which are apt to arise.

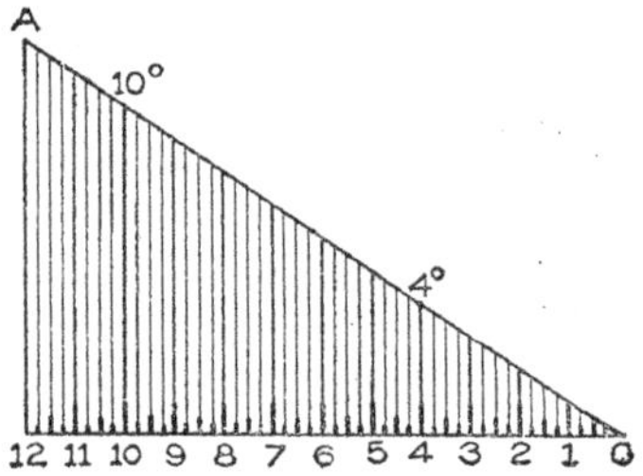

Fig. 184.

In the following illustrations, Figs. 184 to 187, it will be explained how to obtain the true lengths of the ventilator, ridge, hip, jack, and common bars in a hipped skylight, no matter what size the skylight may be. Using this rule only one set of patterns are required, as for example, those developed in connection with Figs. 178, 179, 180, and 181, which in this case has one-third pitch. If, however, a skylight was required whose pitch was different than one-third, a new set of patterns would have to be developed, to which the rule above mentioned would also be applicable for skylights of that particular pitch. Using this rule it should be understood that the size of the curb, or frame, forms the basis for all measurements, and that one of the lines or bends of the bar should meet the line of the curb as shown in Fig. 178, where the bottom of the bar E in the half section meets the line of the curb c 4′ at 4′, and the ridge at the top at 4′. Therefore when laying

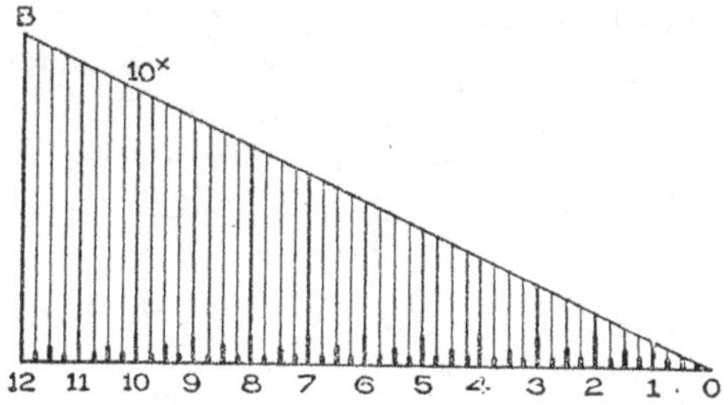

Fig. 185.

out the lengths of the bars, they would have to be measured on the line 4 of the bar E from 4′ to 4″ on the patterns, as will be explained as we proceed.

The first step is to prepare the triangles from which the lengths of the common and jack bars are obtained, also the lengths of the hip bars. After the drawings and patterns have been laid out full size according to the principles explained in Fig. 178, take a tracing of the triangle in the half section D C 4′ and place it as shown by A 12 O, in Fig. 184. Divide O 12, which will be 12 inches in full size, into quarter, half-inches, and inches, the same as on a 2-foot rule, as shown by the figures O to 12. From these divisions erect lines until they intersect the pitch A O which completes the triangle for obtaining the true lengths of jack and common bars for any size skylight. In similar manner take tracing of N R 4^P in the diagonal elevation in Fig. 178 and place it as shown by B 12 O in Fig. 185. The length 12 O then becomes the base of the triangle for the hip bar in a skylight whose base of the triangle for the common and jack bars measures 12 inches

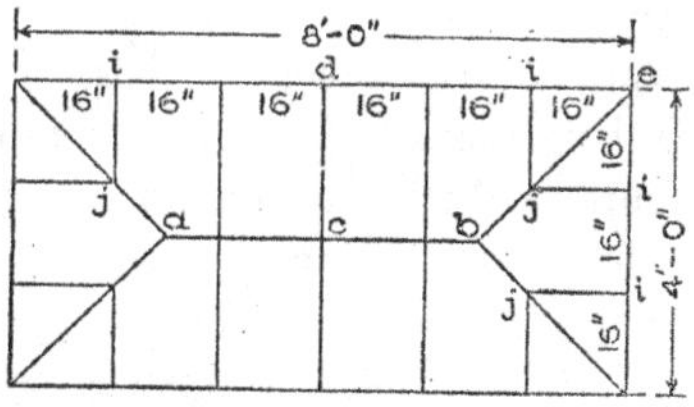

Fig. 186.

as shown in Fig. 184, the heights A 12 in Fig. 184 and B 12 in Fig. 185 being equal. Now divide 12 O in 12 equal spaces which will represent inches when obtaining the measurements for the hip bar. Divide each of the parts into quarter-inches as shown. From these devisions erect lines intersecting the hypothenuse or pitch line B O as shown.

To explain how these triangles are used in practice, Figs. 186 and 187 have been prepared, showing respectively a skylight without and with a ventilator whose curb measures 4 ft. x 8 ft. Three rules are used in connection with the triangles in Figs. 184 and 185, the comprehension of which will make clear all that follows.

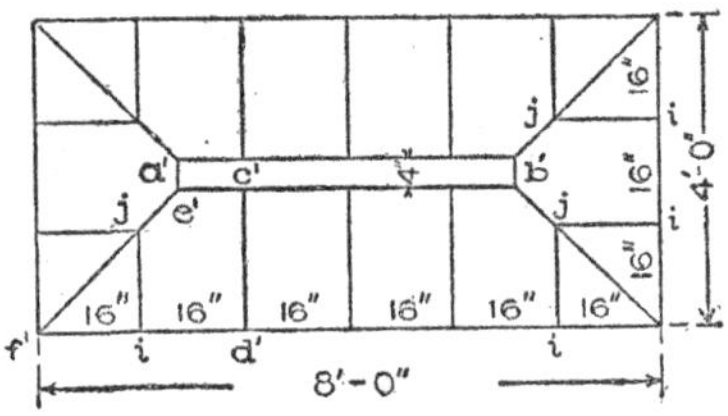

Fig. 187.

Rule 1. To obtain the length of the ridge bar in a skylight without a ventilator, as in Fig. 186, deduct the short side of the frame or curb from the long side.

Example: In Fig. 186, take 8 feet (long side of frame)— 4 feet (short side of frame) = 4 feet (length of ridge bar *a b*).

Rule 2. To find the length of the ventilator in a skylight deduct the short side of the frame from the long side and add the width of the desired ventilator (in this case 4 inches, as shown in Fig. 187).

Example: In Figure 187 take 8 feet (long side of frame) — 4 feet (short side of frame) = 4 feet. 4 feet + 4 inches (width of inside ventilator) = 4 feet 4 inches, (length of inside ventilator *a′ b′*). To find the size of the outside ventilator *h l* and hood *m p* in Fig. 178 simply add twice the distance *a b* and *a c* respectively to the above size, 4 inches, and 4 feet 4 inches, which will give the widths and lengths of the outside vent and hood.

Rule 3. To find the lengths of either common or hip bar (in any size skylight) deduct the width of the ventilator, if any, from the length of the shortest side of frame and divide the remainder by two. Apply the length thus obtained on the base line of its respective triangle for common or hip bars and determine the true lengths of the desired bars, from the hypothenuse.

Example: As no ventilator is shown in Fig. 186, there will be nothing to deduct for it, and the operation is as follows: 4 feet (shortest side of frame) ÷ 2 = 2 feet. We have now the length with which to proceed to the triangle for common and hip bars. Thus the length of the common bar *c d* will be equal to twice the amount of A O in Fig. 184, while the length of the hip bar *b e* in Fig. 186, will be equal to twice the amount of B O in Fig. 185. Referring to Figs. 186 and 187 the jack bars *i j* are spaced 16 inches, therefore, the length of the jack bar for 12 inches will equal A O in Fig. 184, and 4 inches equal to 4° O; both of which are added together for the full length.

The lengths of the common and hip bars will be shorter in Fig. 187 because a ventilator has been used, while in Fig. 186 a ridge bar was employed. To obtain the lengths of the common and hip bars in Fig. 187 use Rule 3: 48 inches (length of short side) − 4 inches (width of inside ventilator) = 44 inches; and 44 inches ÷ 2 = 22 inches or 1 foot 10 inches. Then the length of the common bar *c′ d′* measured with a rule will be equal to A O in Fig. 184 and 10° O added together, and the length of the hip bar *e′ f′* in Fig. 187 will be equal to B O in Fig. 185 and 10^x O added together. Use the same method where fractional parts of an inch occur. In laying out the patterns according to these measurements use the cuts shown in Figs. 178, 179, 180, and 181, being careful to measure from the arrowpoints shown on each pattern.

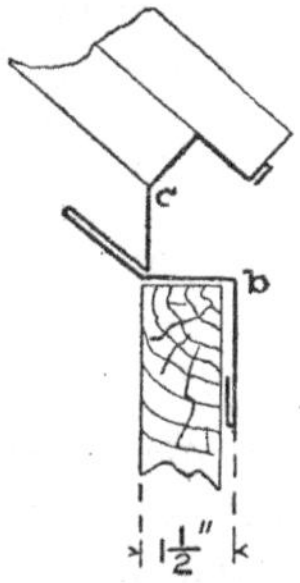

Fig. 188.

It will be noticed in Fig. 178 we always measure on line 4 in the patterns for the hip, common, and jack bars. This is done because the line 4 in the profiles E and E^4 come directly on the slant line of the triangles which were traced to Figs. 184 and 185 and from which the true lengths were obtained. Where a curb might be used, as shown in Fig. 188, which would bring the bottom line of the bar 1½ inches toward the inside of the frame *b*, all around, then instead of using the size of 4 x 8 feet as the basis of measurements deduct 3 inches on each side, making the basis of measurements 3 ft. 9 inches x 7 ft. 9 inches, and proceed as explained above.

ROOFING

A good metal covering on a roof is as important as a good foundation. There are various materials used for this purpose such as terne plate or what is commonly called roofing tin. The rigid body, or the base of roofing tin, consists of thin sheets of steel (black plates) that are coated with an alloy of tin and lead. Where a first-class job is desired soft and cold rolled copper should be used. The soft copper is generally used for cap flashing and allows itself to be dressed down well after the base flashing is in position. The cold-rolled or hard copper is used for the roof coverings. In some cases galvanized sheet iron or steel is employed. No matter whether tin, galvanized iron, or copper is employed the method of construction is the same, and will be explained as we proceed.

Another form of roofing is known as corrugated iron roofing, which consists of black or galvanized sheets, corrugated so as to secure strength and stiffness. Roofs having less than one-third pitch should be covered by what is known as flat-seam roofing, and should be covered (when tin or copper is used) with sheets 10 x 14 inches in size rather than with sheets 14 x 20 inches, because the larger number of seams stiffens the surface and prevents the rattling of the tin in stormy weather. Steep roofs should be covered by what is known as standing-seam roofing made from 14″ x 20″ tin or from 20″ x 28″. Before any metal is placed on a roof the roofer should see that the sheathing boards are well seasoned, dry and free from knots and nailed close together. Before laying the tin plate a good building paper, free from acid, should be laid on the sheathing, or the tin plate should be painted on the underside before laying. Corrugated iron is used for roofs and sides of buildings. It is usually laid directly upon the purlins in roofs, and held in place by means of clips of hoop iron, which encircle the purlins and are riveted to the corrugated iron about 12 inches apart. The method of constructing flat and double-seam roofing, also corrugated iron coverings, will be explained as we proceed.

TABLES

The following tables will prove useful in figuring the quantity of material required to cover a given number of square feet.

FLAT-SEAM ROOFING

Table showing quantity of 14 x 20-inch tin required to cover a given number of square feet with flat seam tin roofing. A sheet of 14 x 20 inches with with ½-inch edges measures, when edged or folded, 13 x 19 inches or 247 square inches. In the following all fractional parts of a sheet are counted a full sheet.

No. of sq. ft.	Sheets required	No. of sq. ft.	Sheets required	No. of sq. ft.	Sheets required	No. of sq. ft.	Sheets required
100	59	330	193	560	327	780	455
110	65	340	199	570	333	790	461
120	70	350	205	580	339	800	467
130	76	360	210	590	344	810	473
140	82	370	216	600	350	820	479
150	88	380	222	610	356	830	484
160	94	390	228	620	362	840	490
170	100	400	234	630	368	850	496
180	105	410	240	640	374	860	502
190	111	420	245	650	379	870	508
200	117	430	251	660	385	880	514
210	123	440	257	670	391	890	519
220	129	450	263	680	397	900	525
230	135	460	269	690	403	910	531
240	140	470	275	700	409	920	537
250	146	480	280	710	414	930	543
260	152	490	286	720	420	940	549
270	158	500	292	730	426	950	554
280	164	510	298	740	432	960	560
290	170	520	304	750	438	970	566
300	175	530	309	760	444	980	572
310	181	540	315	770	449	990	578
320	187	550	321				

1000 square feet, 583 sheets.

A box of 112 sheets 14 x 20 inches will cover approximately 192 square feet.

Example. How much 14 x 20 inch tin with ½-inch edges is required to cover a roof 20 feet x 84 feet? Take 20 × 84 = 1,680 square feet.

Referring to the table for Flat Seam Roofing, 1000 square feet require 583 sheets and 680 square feet require 397 sheets, making a total of 980 sheets.

It should be understood that this amount is figured on the basis of 247 square inches in an edged sheet, which will be a trifle less when the sheets are laid on the roof.

Example. What quantity of 20 x 28-inch tin will be required to lay a standing seam roof, measuring 37 feet long x 45 feet in width? Take 37 × 45 = 1,665 square feet, or 16 squares and 65 feet. Referring to the table for Standing Seam Roofing, 16 squares require 4 boxes and 48 sheets, and 65 feet require 20 sheets, making a total of 4 boxes and 68 sheets.

STANDING-SEAM ROOFING

Table showing the quantity of 20 × 28-inch tin in boxes, and sheets required to lay any given standing-seam roof.

SQ. FEET	SHEETS	SQUARES	SQ. FEET	BOXES	SHEETS	SQUARES	BOXES	SHEETS
1	1	......	68	......	21	35	9	77
2	1	......	69	......	21	36	9	108
3	1	......	70	......	22	37	10	27
4	2	......	71	......	22	38	10	58
5	2	......	72	......	22	39	10	89
6	2	......	73	......	22	40	11	8
7	3	......	74	......	23	41	11	39
8	3	......	75	......	23	42	11	70
9	3	......	76	......	23	43	11	101
10	4	......	77	......	24	44	12	20
11	4	......	78	......	24	45	12	51
12	4	......	79	......	24	46	12	82
13	4	......	80	......	25	47	13	1
14	5	......	81	......	25	48	13	32
15	5	......	82	......	25	49	13	63
16	5	......	83	......	25	50	13	94
17	6	......	84	......	26	51	14	13
18	6	......	85	......	26	52	14	44
19	6	......	86	......	26	53	14	75
20	7	......	87	......	27	54	14	106
21	7	......	88	......	27	55	15	25
22	7	......	89	......	27	56	15	56
23	7	......	90	......	28	57	15	87
24	8	......	91	......	28	58	16	6
25	8	......	92	......	28	59	16	37
26	8	......	93	......	28	60	16	68
27	9	......	94	......	29	61	16	99
28	9	......	95	......	29	62	17	18
29	9	......	96	......	29	63	17	49
30	10	......	97	......	30	64	17	80
31	10	......	98	......	30	65	17	111
32	10	......	99	......	30	66	18	30
33	10	......	100	......	31	67	18	61
34	11	1	......	......	31	68	18	92
35	11	2	......	......	62	69	19	11
36	11	3	......	......	93	70	19	42
37	12	4	......	1	12	71	19	73
38	12	5	......	1	43	72	19	104
39	12	6	......	1	74	73	20	23
40	13	7	......	1	105	74	20	54
41	13	8	......	2	24	75	20	85
42	13	9	......	2	55	76	21	4
43	13	10	......	2	86	77	21	35
44	14	11	......	3	5	78	21	66
45	14	12	......	3	36	79	21	97
46	14	13	......	3	67	80	22	16
47	15	14	......	3	98	81	22	47
48	15	15	......	4	17	82	22	78
49	15	16	......	4	48	83	22	109
50	16	17	......	4	79	84	23	28
51	16	18	......	4	110	85	23	59
52	16	19	......	5	29	86	23	90
53	16	20	......	5	60	87	24	9
54	17	21	......	5	91	88	24	40
55	17	22	......	6	10	89	24	71
56	17	23	......	6	41	90	24	102
57	18	24	......	6	72	91	25	21
58	18	25	......	6	103	92	25	52
59	18	26	......	7	22	93	25	83
60	19	27	......	7	53	94	26	2
61	19	28	......	7	84	95	26	33
62	19	29	......	8	3	96	26	64
63	19	30	......	8	34	97	26	95
64	20	31	......	8	65	98	27	14
65	20	32	......	8	96	99	27	45
66	20	33	......	9	15	100	27	76
67	21	34	......	9	46	-		

Size of sheet before working, 20 × 28 inches. Exposed on roof 27×17¾ inches. Square inches per sheet exposed 479¼ inches. Sheets per box 112.

NET WEIGHT PER BOX TIN PLATES

Basis 14 × 20, 112

Trade term . . .		80-lb.	85-lb.	90-lb.	95-lb.	100-lb.	IC	IXL	IX	IXX	IXXX	IXXXX
Weight per box, lb.		80	85	90	95	100	107	128	135	155	175	195
Size of sheets	Sheets per box											
10 x 14	225	80	85	90	95	100	107	128	135	155	175	195
14 x 20	112	80	85	90	95	100	107	128	135	155	175	195
20 x 28	112	160	170	180	190	200	214	256	270	310	350	390
10 x 20	225	114	121	129	136	143	153	183	193	221	250	279
11 x 22	225	138	147	156	164	172	184	232	234	268	302	337
11½ x 23	225	151	161	170	179	189	202	242	255	293	331	368
12 x 12	225	82	87	93	98	103	110	132	139	159	180	201
12 x 24	112	82	87	93	98	103	110	132	139	159	180	201
13 x 13	225	97	103	109	115	121	129	154	163	187	211	235
13 x 26	112	97	103	109	115	121	129	154	163	187	211	235
14 x 14	225	112	119	126	133	140	150	179	189	217	245	273
14 x 28	112	112	119	126	133	140	160	179	189	217	245	273
15 x 15	225	129	137	145	153	161	172	206	217	249	281	313
16 x 16	225	146	155	165	174	183	196	234	247	283	320	357
17 x 17	225	165	175	186	196	206	221	264	279	320	361	403
18 x 18	112	93	98	104	110	116	124	148	156	179	202	226
19 x 19	112	103	110	116	122	129	138	165	174	200	226	251
20 x 20	112	114	121	129	136	143	153	183	193	221	250	279
21 x 21	112	126	134	142	150	158	169	202	213	244	276	307
22 x 22	112	138	147	156	164	172	184	221	234	268	302	337
23 x 23	112	151	161	170	179	189	202	242	255	293	331	368
24 x 24	112	164	175	185	195	204	220	263	278	319	360	401
26 x 26	112	193	205	217	229	241	258	309	326	374	422	471
16 x 20	112	91	97	103	109	114	122	146	154	177	200	223
14 x 31	112	124	132	140	147	155	166	198	209	240	271	302
11¼ x 22¾	112	73	78	82	87	91	98					
13¼ x 17¾	112	60	71	76	80	84	90					
13¼ x 19¼	112	73	77	82	87	91	97					
13½ x 19½	112	75	80	85	89	94	100					
13½ x 19¾	112	76	81	86	90	95	102					
14 x 18¾	124	83	88	93	98	103	110					
14 x 19¼	120	83	88	93	98	103	110					
14 x 21	112	84	89	95	100	105	112					
14 x 22	112	88	94	99	105	110	118					
14 x 22¼	112	89	95	100	106	111	119					
15½ x 23	112	102	108	115	121	127	136					

STANDARD WEIGHTS AND GAUGES OF TIN PLATES

Trade term	65-lb.	70-lb.	75-lb.	80-lb.	85-lb.	90-lb.	95-lb.	100-lb.
Nearest wire gauge No.	35	35	34	33	32	31	31	30
Weight, square foot, lb.	.298	.322	.345	.367	.390	.413	.436	.459
Weight, box, 14 x 20, lb.	65	70	75	80	85	90	95	100

Trade term	IC	IXL	IX	IXX	IXXX	IXXXX	IXXXXX
Nearest wire gauge No.	30	28	28	27	26	25	24
Weight, square foot, lb.	.491	.588	.619	.712	.803	895	.987
Weight, box, 14 x 20, lb.	107	128	135	155	175	195	215

	IC 14 x 20	IC 20 x 28	IX 14 x 20	IX 20 x 28
Black plates before coating weight per 112 sheets	lb. 95 to 100	lb. 190 to 200	lb. 125 to 130	lb. 250 to 260
When coated the plates weigh per 112 sheets	115 to 120	230 to 240	145 to 150	290 to 300

OTHER FORMS OF METAL ROOFING

There is another form of roofing known as metal slates and shingles, pressed in various geometrical designs with water-tight lock attachments so that no solder is required in laying the roof. Fig. 189 shows the general shape of these metal shingles which are made from tin, galvanized iron, and copper, the dots *a a a a* representing the holes for nailing to the wood sheathing. In Fig. 190, A represents the side lock, showing the first operation in laying the metal slate or shingle on a roof, *a* representing the nail. B, in the same figure, shows the metal slate or shingle in position covering the nail *b*, the valley *c* of the bottom slate allowing the water, if any, to flow over the next lower slate as in A in Fig. 189.

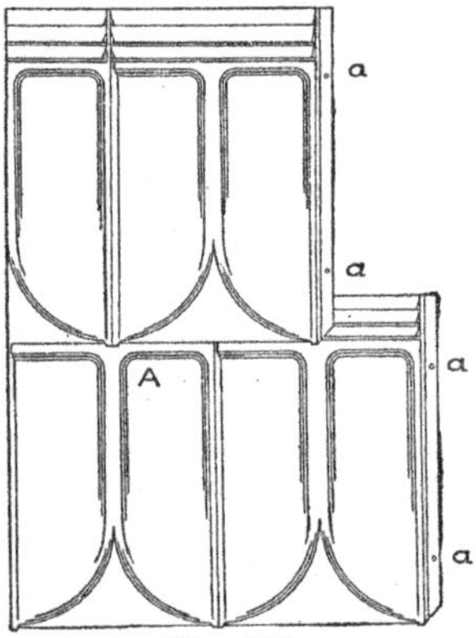

Fig. 189.

In Fig. 191 is shown the bottom slate A covered by the top slate B, the ridges *a a a* keeping the water from backing up. Fig. 192 shows the style of roof on which these shingles are employed, that is, on steep roofs. Note the construction of the ridge roll, A and B in Fig. 192, which is first nailed in position at *a a* etc., after which the shingles B are slipped under the lock *c*. Fig. 193 shows a roll hip covering which is laid from the top downward, the lower end of the hip having a projection piece for nailing at *a*, over which the top end of the next piece is inserted, thus

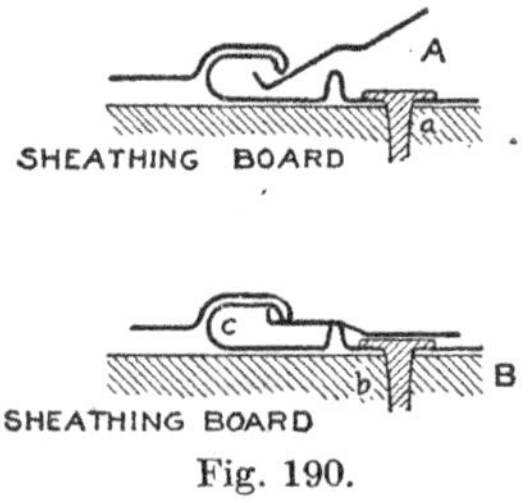

Fig. 190.

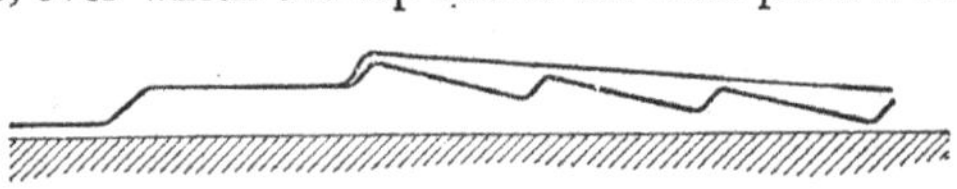
Fig. 191.

covering and concealing the nails. Fig. 194 represents a perspective view of a valley with metal slates, showing how the slates A are locked to the fold in the valley B. There are many other forms of metal shingles, but the shapes shown herewith are known as the Cortright patents.

TOOLS REQUIRED

Fig. 195 shows the various hand tools required by the metal roofer; starting at the left we have the soldering copper, mallet, scraper,

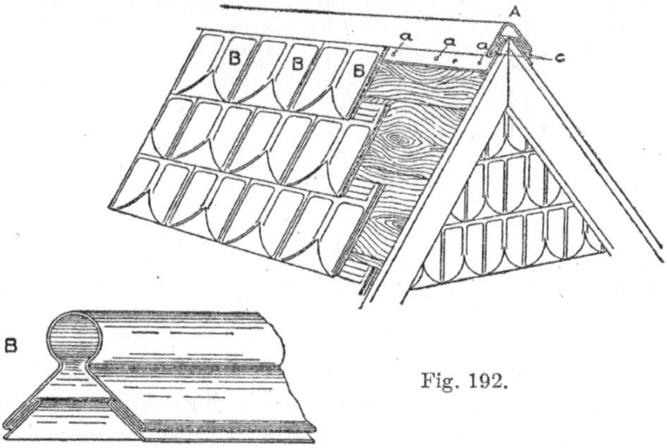

Fig. 192.

stretch-awl, shears, hammer, and dividers. In addition to these hand tools a notching machine is required for cutting off the corners of the

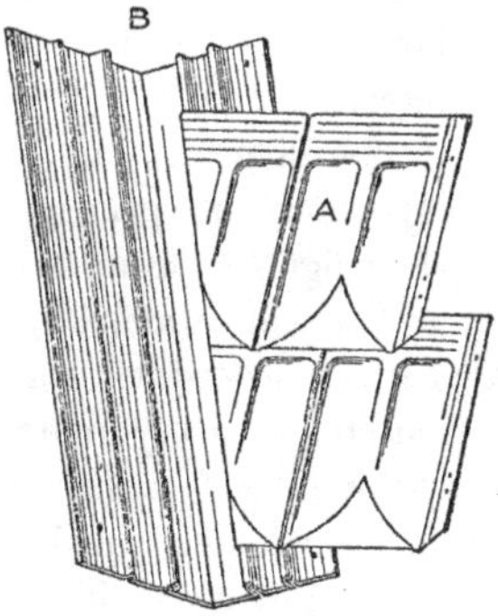

Fig. 194.

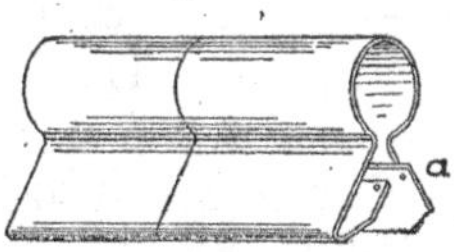

Fig. 193.

sheets, and roofing folders are required for edging the sheets in flat-seam roofing, and hand double seamer and roofing tongs for standing-seam roofing. The roofing double seamer and squeezing tongs can be used for standing-seam roofing (in place of the hand double seamer), which allow the operator to stand in an upright position if the roof is not too steep.

ROOF MENSURATION

While some mechanics understand thoroughly the methods of

laying the various kinds of roofing, there are some, however, who do not understand how to figure from architects' or scale drawings the amount of material required to cover a given surface in a flat, irregular shaped, or hipped roof. The modern house with its gables and va-

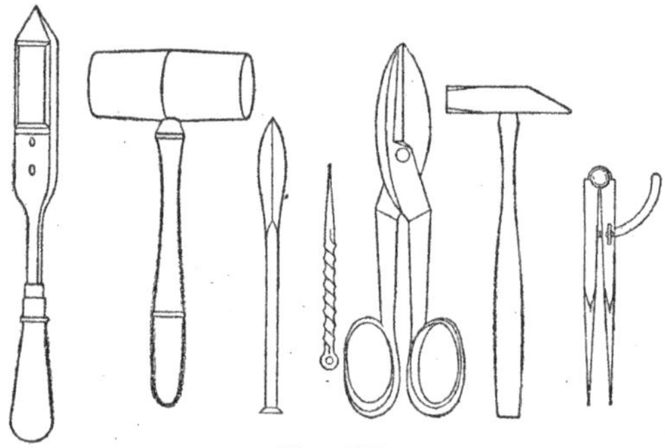

Fig. 195.

rious intersecting roofs, forming hips and valleys, render it necessary to give a short chapter on roof measurement. In Figs. 196 to 198 inclusive are shown respectively the plans with full size measurements for a flat, irregular, and intersected hipped roof, showing how the length of the hips and valleys are obtained direct from the architects' scale drawings.

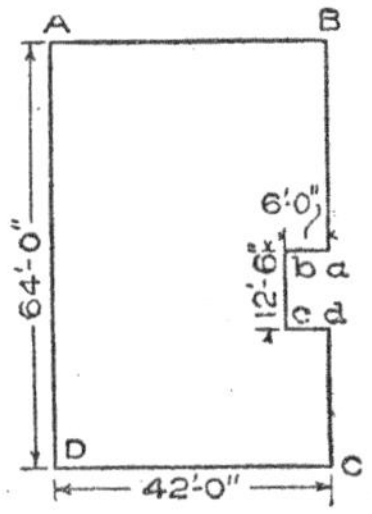

Fig. 196.

The illustrations shown herewith are not drawn to a scale as architects' drawings will be, but the measurements on the diagrams are assumed, which will clearly show the principles which must be applied when figuring from scale drawings. Assuming that the plans from which we are figuring are drawn to a quarter-inch scale, then when measurements are taken, every quarter inch represents one foot. $\frac{1}{8}$ inch = 6 inches, $\frac{1}{16}$ inch = 3 inches, etc. If the drawings were drawn to a half-inch scale, then $\frac{1}{2}$ inch = 12 inches, $\frac{1}{4}$ inch = 6 inches, $\frac{1}{8}$ inch = 3 inches, $\frac{1}{16}$ inch = $1\frac{1}{2}$ inches, etc.

A B C D in Fig. 196 represents a flat roof with a shaft at one side as shown by *a b c d*. In a roof of this kind we will figure it as if there was no air shaft at all. Thus 64 feet × 42 feet = 2,688 square feet. The shaft is 12.5 × 6 feet = 75 square feet; then 2,688 feet − 75 feet =

2,613 square feet of roofing, to which must be added an allowance for the flashing turning up against and into the walls at the sides.

In Fig. 197 is shown a flat roof with a shaft at each side, one shaft being irregular, forming an irregular shaped roof. The rule for obtaining the area is similar to that used for Fig. 196 with the exception that the area of the irregular shaft *x x x x* in Fig. 197 is determined differently to that of the shaft *b c d e*. Thus A B C D = 108 feet × 45 feet = 4,860 square feet. Find the area of *b c d e* which is 9.25 × 39.5 = 365.375 or $365\frac{3}{8}$ square feet. To find the area of the irregular shaft, bisect *xx* and *xx* and obtain *a a*, measure the length of *a a* which is 48 feet, and multiply by 9. Thus 48 × 9 = 412, and 412 + 365.375 = 777.375. The entire roof minus the shafts = 4,860 square feet − 777.375 = 4,082.625 square feet of surface in Fig. 197.

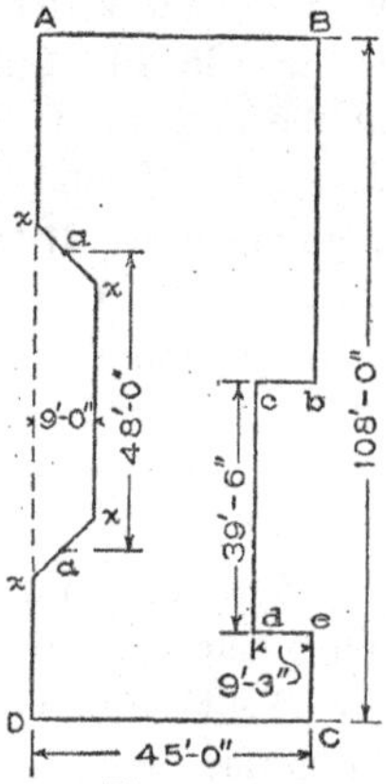

Fig. 197.

In Fig. 198 is shown the plan, front, and side elevations of an intersected hipped roof. A B C D represents the plan of the main build-

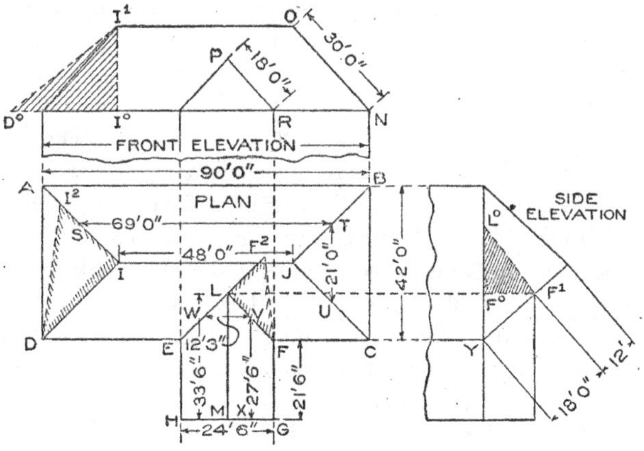

Fig. 198.

ing intersected by the wing E F G H. We will first figure the main roof as if there were no wing attached and then deduct the space taken up by the intersection of the wing. While it may appear difficult to some to figure the quantities in a hipped roof, it is very simple, if the rule is understood. As the pitch of the roof is equal on four sides the length of the rafter shown from O to N in front elevation represents the true length of the pitch on each side. The length of the building at the eave is 90 feet and the length of the ridge 48 feet. Take $90 - 48 = 42$, and $42 \div 2 = 21$. Now either add 21 to the length of the edge or deduct 21 from the length of the eave, which gives 69 feet as shown from S to T. The length of the eave at the end is 42 feet and it runs to an apex at J. Then take 42 feet $\div\ 2 = 21$, as shown from T to U. If desired the hip lines A I, J B and J C can be bisected, obtaining respectively the points S, T, and U, which when measured will be of similar sizes; 69 feet and 21 feet. As the length of the rafter O N is 30 feet, then multiply as follows: $69 \times 30 = 2070$. $21 \times 30 = 630$. Then $630 + 2{,}070 = 2{,}700$, and multiplying by 2 (for opposite sides) gives 5,400 square feet or 54 squares of roofing for the main building. From this amount deduct the intersection E L F in the plan as follows:

The width of the wing is 24 feet 6 inches and it intersects the main roof as shown at E L F. Bisect E L and L F and obtain points W and V, which when measured will be 12 feet 3 inches or one half of HG, 24 feet 6 inches. The wing intersects the main roof from Y to F^1 in the side elevation, a distance of 18 feet. Then take $18 \times 12.25 = 220.5$. Deduct 220.5 from $5400 = 5{,}179.5$. The wing measures 33 feet 6 inches at the ridge L M, and 21 feet 6 inches at the eave F G, thus making the distance from V to X = 27 feet 6 inches. The length of the rafter of the wing is shown in front elevation by P R, and is 18 feet. Then $18 \times 27.5 = 495$, and multiplying by 2 (for opposite side), gives 995 sq. ft. in the wing. We then have a roofing area of 5,179.5 square feet in the main roof and 995 square feet in the wing, making a total of 6,174.5 square feet in the plan shown in Fig. 198.

If it is desired to know the quantity of ridge, hips, and valleys in the roof, the following method is used. The ridge can be taken from the plans by adding $48' + 33'6'' = 81' - 6''$. For the true length of the hip I D in the plan, drop a vertical line from I^1 in the front elevation until it intersects the eave line 1°. On the eave line extended, place the distance I D in the plan as shown from 1° to D° and draw a line from D° to I^1 which will be the true length of the hip I D in the plan. Multiply this length by 4, which will give the amount of ridge capping required. This length of hip can also be obtained from the plan by taking the vertical height of the roof $I^{\circ}\ I'$ in the elevation and placing it at right angles to I D in the plan, as shown, from I to I^2, and draw a line from I^2 to D which is the desired length.

For the length of the valley L F in the plan, drop a vertical line from F^1 in the side elevation until it intersects the eave line at F°. Take the distance F L in the plan and place it as shown from F° to L°, and draw a line from L° to F^1, which is the true length of the valley shown by L F in the plan. Multiply this length by 2, which will give the required number of feet of valley required. This length of valley can also be obtained from the plan by taking the vertical height of the roof of the wing, shown by $F^{\circ}\ F^1$ in the side elevation, and placing it at right angles to F L in the plan, from L to F^2, and draw a line from F^2 to F which is the desired length similar to $F^1\ L^{\circ}$ in the side elevation.

FLAT-SEAM ROOFING

The first step necessary in preparing the plates for flat seam roofing is to notch or cut off the four corners of the plate as shown in Fig. 199 which shows the plate as it is taken from the box, the shaded corners *a a a a* representing the corners which are notched on the notching machine or with the shears. Care must be taken when cutting off these corners not to cut off too little otherwise the sheets will not edge well, and not to cut off too much, otherwise a hole will show at the corners when the sheets are laid. To find the correct amount to be cut off proceed as follows:

Fig. 199.

Assuming that a ½-inch edge is desired, set the dividers at ½ inch and scribe the lines *b a* and *a c* on the sheet shown in Fig. 199, and, where the lines intersect at *a*, draw the line *d e* at an angle of 45 degrees, which represents the true amount and true angle to be cut off on each corner. After all the sheets have been notched, they are edged as shown in Fig. 200, the long sides of the sheet being bent right and left, as shown at *a*, while the short side is bent as shown at *b*, making the notched corner appear as at *e*. In some cases after the sheets are edged the contract requires that the sheets be painted on the underside before laying. This is usually done with a small brush, being careful that the edges of the sheets

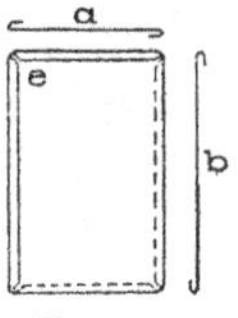

Fig. 200.

are not soiled with paint, which would interfere with soldering. Before laying the sheets the roof boards are sometimes covered with an oil or rosin-sized paper to prevent the moisture or fumes from below from rusting the tin on the underside. As before mentioned, the same method used for laying tin roofing would be applicable for laying copper roofing, with the exception that the copper sheets would have to be tinned about 1½ inches around the edges of the sheets after they are notched, and before they are edged.

In Fig. 201 is shown how a tin roof is started and the sheets laid when a gutter is used at the eaves with a fire wall at the side. A repre-

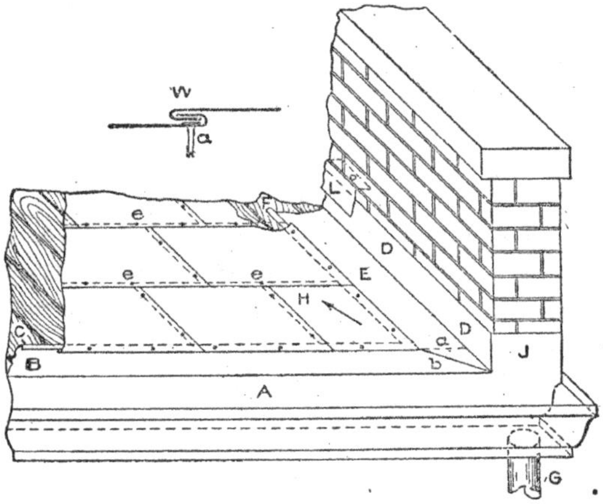

Fig. 201.

sents a galvanized iron gutter with a portion of it lapping on the roof, with a lock at C. In hanging the gutter it is flashed against the fire wall at J; after which the base flashing D D is put in position, flashing out on the roof at E, with a lock at F. Where the base flashing E miters with the flange of the gutter B it is joined as shown at *b*, allowing the flange E of the base flashing as shown by the dotted line *a*. As the water discharges at G, the sheets are laid in the direction of the arrow H, placing the nails at least 6 inches apart, always starting to nail at the butt *e e*, etc. Care should be taken when nailing that the nail heads are well covered by the edges, as shown in W, by *a*. Over the base flashing D D J the cap flashing L is placed, allowing it to go into the wall as at O.

When putting in base flashings there are two methods employed. In Fig. 202 is shown a side flashing between the roof and parapet wall. A shows the flashing turning out on the roof at B, with a lock C, attached and flashed into the wall four courses of brick above the roof line, as shown at D, where wall hooks and paintskins or roofer's cement are used to make a tight joint. Flashings of this kind should always be painted on the underside, and paper should be placed between the brick work and metal, because the moisture in the wall is apt to rust the tin. This method of putting in flashing is not advisable in new work, because when the building is new, the walls and beams are liable to settle and when this occurs the flange D tears out of the wall, and the result is disagreeable leaks that stain the walls. When a new roof is to be placed on an old building where the walls and copings are in place and the brick work and beams have settled, there is not so much danger of leakage.

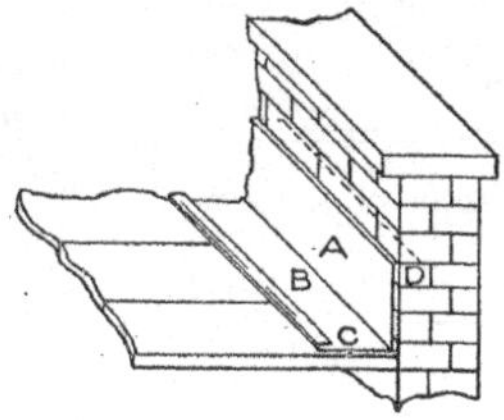

Fig. 202.

The proper method of putting in flashings and one which allows for the expansion and contraction of the metal and the settlement of the building is shown in Fig. 203, in which A shows the cap flashings,

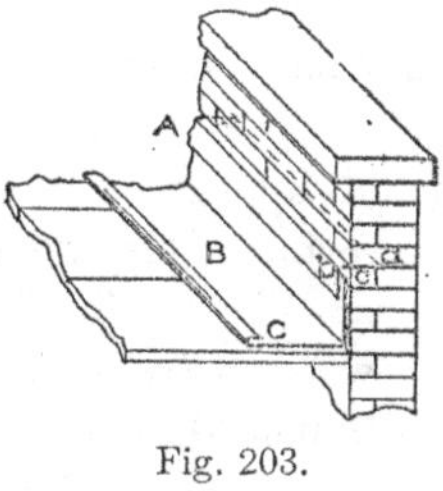

Fig. 203.

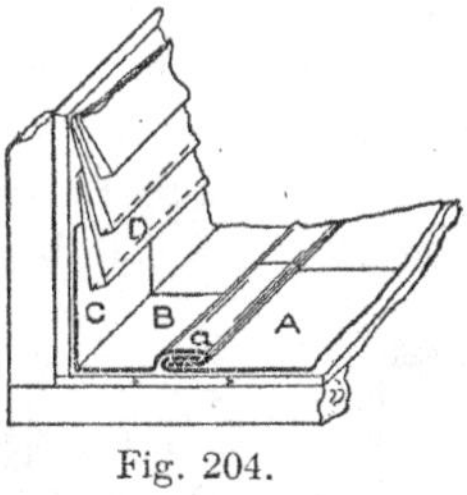

Fig. 204.

painted with two coats of paint before using. When the mason has built his wall up to four courses of brick above the roof line the cap flashing A is placed in position and the wall and coping finished; the base flashing B is then slipped under the cap A. In practice the cap flashing is cut 7 inches, then bent at right angles through the center, making each side *a* and *b* 3½ inches. The base flashing B is then slipped under the cap flashing A as shown at C.

Where the cost is not considered and a good job is desired, it is better to use sheet lead cap flashings in place of tin. They last longer, do not rust, and can be dressed down well to lay tight onto the base flashings. Into the lock C the sheets are attached. After the sheets are laid the seams are flattened down well by means of a heavy mallet, with slightly convex faces, after which the roof is ready for soldering. When a base flashing is required on a roof which abuts against a wall composed of clap boards or shingles as shown in Fig. 204, then, after the last course of tin A has been laid, the flashing B with the lock *a* is locked into the course A and extends the required distance under the boards D. The flashing should always be painted and allowed to dry before it is placed in position. In the previous figures it was shown how the sheets are edged, both sides being edged right and left. In Fig. 205 is shown what is known as a valley sheet, where the short sides are edged both one way, as shown at *a a*, and the long sides right and left as shown at *b b*. Sheets of this kind are used when the water runs together from two directions as shown by A in Fig. 206. By having the locks *a* and *a* turned one way the roof is laid in both directions.

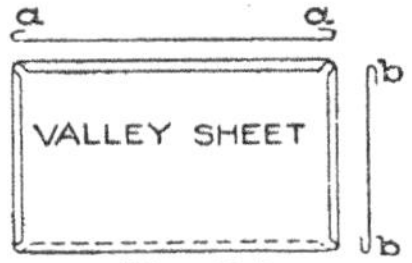

Fig. 205.

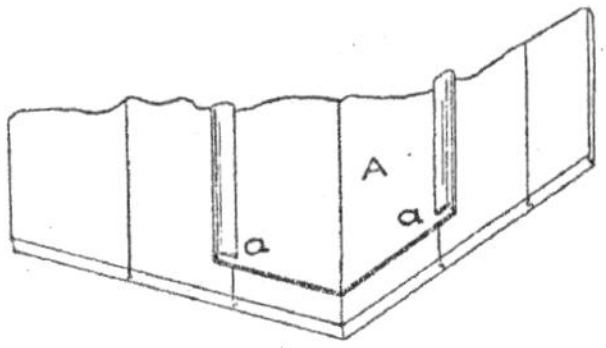

Fig. 206.

Fig. 207 shows a part plan of a roof and chimney A, around which the flashing B C D E is to be placed, and explains how the corners C and D are double seamed, whether on a chimney, bulkhead, or any other object on a roof when the water flows in the direction of the arrow F. The first operation is shown at *a* and the final operation at *b*. Thus it will be seen that the water flows past the seam and not against it. In laying flat seam roofing especially when copper is used, allowance must be made for the expansion and contraction of the sheets.

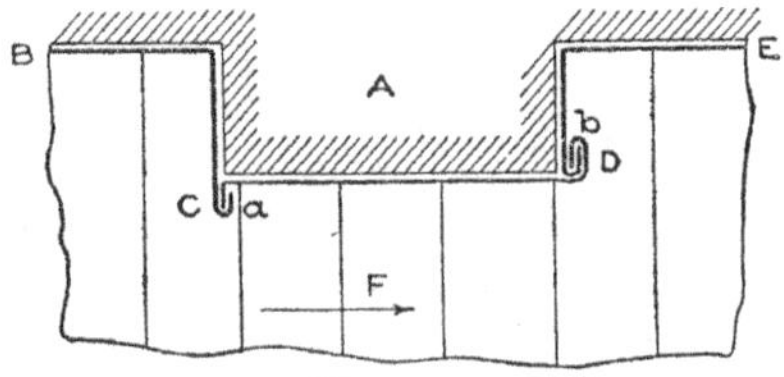

Fig. 207.

Care should be taken not to nail directly through the sheet as is shown in W, Fig. 201. While this method is generally employed in tin roofing, on a good job, as well as on copper roofing, cleats as shown at D in Fig. 208 should be used.

To show how they are used, A and B represent two locked-edged sheets. The lock on the cleat D is locked into the edge of the sheets and nailed into the roof boards at *a b c* and *d*, ar as often as required.

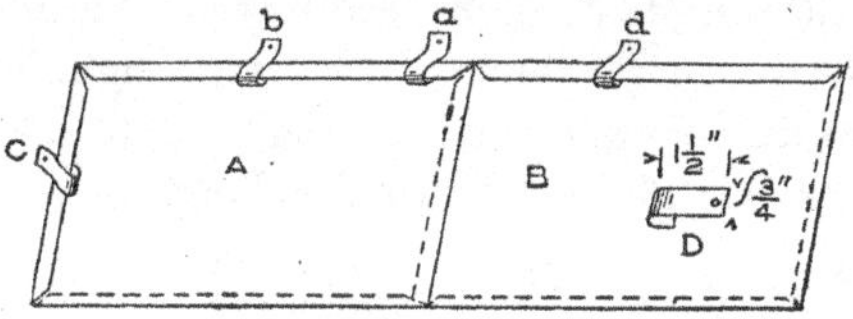

Fig. 208.

In this manner the entire roof can be fastened with cleats without having a nail driven into the sheets, thereby allowing for expansion and contraction of the metal. The closer these cleats are placed, the firmer the roof will be and the better the seams will hold. By using fewer cleats, time may be saved in laying the roof, but double this time is lost when soldering the seams, for the heat of the soldering copper

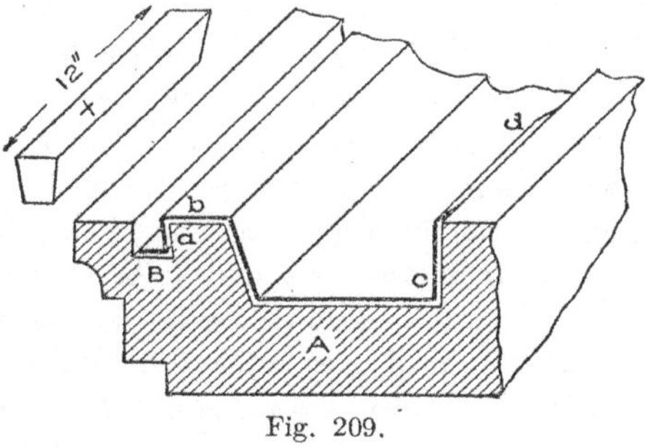

Fig. 209.

will raise the seams, causing a succession of buckles, which retard soldering and require 10 per cent more solder. When the seams are nailed or cleated close it lays flat and smooth and the soldering is done with ease and less solder.

When a connection is to be made between metal and stone or terra cotta, the method shown in Fig. 209 is employed. This illustration shows a stone or terra-cotta cornice A. The heavy line *a b c d* represents the gutter lining, which is usually made from 20-oz. cold-rolled copper. If the cornice A is of stone, the stone cutter cuts a raggle into the top of the cornice A as at B, dove-tail in shape, after which the lining *a b c d* is put in position as shown. Then, being careful that there is no water or moisture in the raggle B, molten lead is poured into the raggle and after it is cooled it is dressed down well with the caulking chisel and hammer.

By having the dove-tail cut, the lead is secured firmly in position, holding down the edge of the lining and making a tight joint. Should the cornice be of terra cotta this raggle is cut into the clay before it is baked in the ovens. This method of making connection between

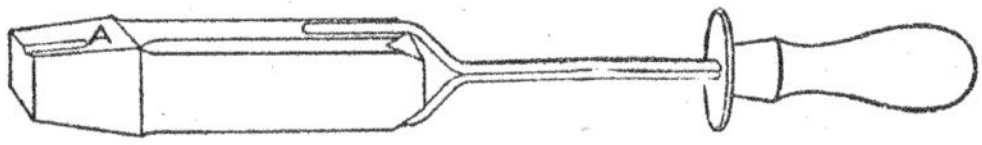

Fig. 210.

metal and stone is the same no matter whether a gutter or upright wall is to be flashed. When a flashing between a stone wall and roof is to be made tight, then instead of using molten lead, cakes of lead are cast in molds made for this purpose, about 12 inches long, and these are driven into the raggle B as shown in Fig. 209 at X.

The most important step in roofing is the soldering. The style of soldering copper employed is shown in Fig. 210 and weighs at least 8 pounds to the pair. When rosin is used as a flux, it is also employed in tinning the coppers, but when acid is used as a flux for soldering zinc or galvanized iron, salammoniac is used for tinning the coppers. It will be noticed that the soldering coppers are forged square at the ends, and have a groove filed in one side as shown at A. When the copper is turned upward the groove should be filed toward the lower side within $\frac{1}{4}$ inch from the corner, so that when the groove is placed upon the seam, as shown in Fig. 211, it acts as a guide to the copper as the latter is drawn along the seam. The groove *a* being in the position shown, the largest heated surface *b* rests directly on the seam, "soaking" it thoroughly with solder. As the heat draws the solder between the locks, about 6 pounds of $\frac{1}{2}$ and $\frac{1}{2}$ solder are required for 100 square feet of surface using 14 x 20-inch tin. The use of acid in soldering seams in a tin roof is to be avoided as acid coming in contact with the

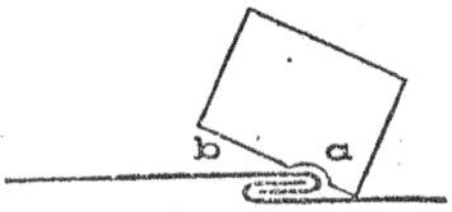

Fig. 211.

bare edges and corners, where the sheets are folded and seamed together, will cause rusting. No other soldering flux but good clean rosin should be employed. The same flux (rosin) should be used when soldering copper roofing whose edges have previously been tinned with rosin.

We will now consider the soldering of upright seams. The soldering copper to be employed for this purpose is shaped as shown in Fig. 212. It is forged to a wedge shape, about 1 inch wide and ¼ inch

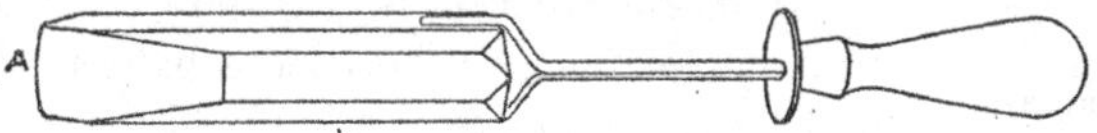

Fig. 212.

thick at the end, and is tinned on one side and the end only; if tinned otherwise, the solder, instead of remaining on the tinned side when soldering, would flow downward; by having the soldering copper tinned on one side only, the remaining sides are black and do not tend to draw the solder downward. The soldering copper being thus prepared, the upright seam, shown in Fig. 213, where the sheet B overlaps the sheet A 1″, is soldered by first tacking the seam to make it lay close, then thoroughly soaking the seam, and then placing ridges of solder across it to strengthen the same. In using the soldering copper it should be held in the position shown by C, which allows the solder to flow forward and into the seam, while if the copper were held as shown by D, the solder would flow backward and away from the seam. In "soaking" the seam with solder the copper should be placed directly over the lapped part, so that the metal gets thoroughly heated and draws the solder between the joint. It makes no difference where this cross joint occurs; the same methods are used.

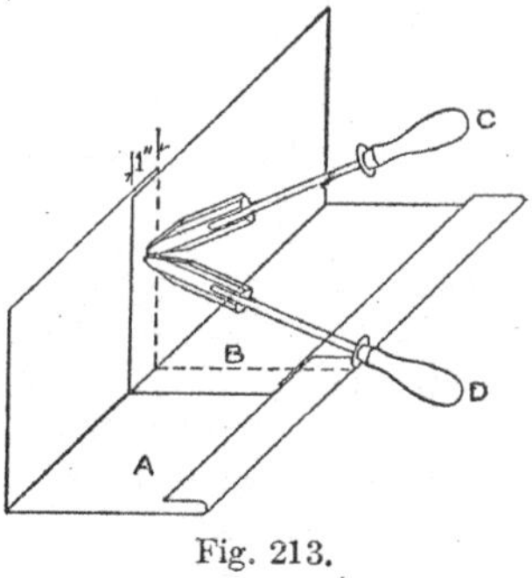

Fig. 213.

The roof being completed, the rosin is scraped off the seams and the roof cleaned and painted with good iron oxide and linseed oil paint. Some roofers omit the scraping of rosin and paint directly over it. This is the cause of rusting of seams which sometimes occurs. If the paint is applied to the rosin, the latter, with time, will crack, and the rain will soak under the cracked rosin to the tin surface. Even when the surface of the roof is dry, by raising the cracked rosin, moisture will often be found underneath, which naturally tends to rust the plate more and more with each storm. If the rosin is removed, the entire tin surface is protected by paint.

One of the most difficult jobs in flat-seams roofing is that of covering a conical tower. As the roof in question is round in plan and tapering in elevation, it is necessary to know the method of cutting the various patterns for the sheets. In Fig. 214 A B C shows the elevation of a tower to be covered with flat seam roofing, using 10 × 14-inch tin at the base. Assuming that the tower through B C is 10 feet 6 inches, or 126 inches, in diameter, the circumference is obtained by multiplying 126 by 3.1416 which equals 395.8416, or say 396 inches. As 10 x 14-inch plate is to be used at the base of the tower the nearest width which can be employed, and which will divide the space into equal spaces, is $13\frac{1}{5}$ inches without edges, thus dividing the circumference in 30 equal spaces. This width of $13\frac{1}{5}$ inches together with the length of the rafter A B or B C in elevation, will be the basis from which all the patterns for the various courses will be laid off.

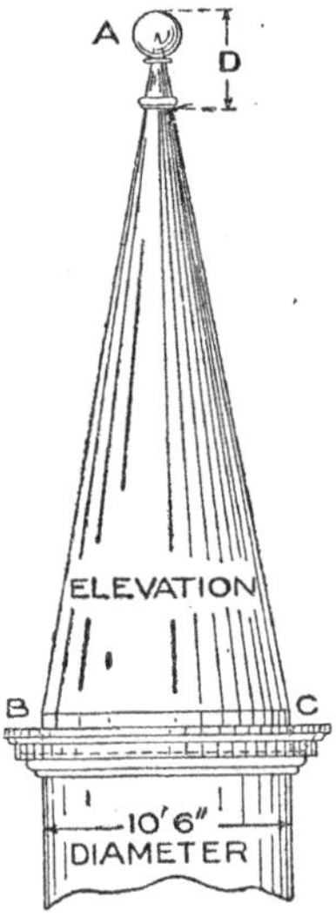

Fig. 214.

At any convenient place in the shop or at the building, stretch a piece of tar felting of the required length, tacking it at the four corners with nails to keep the paper from moving. Upon the center of the felting strike a chalk line as A B in Fig. 215, making it equal to the length of the rafter A B or A C in Fig. 214. At right angles to A B in Fig. 215 at either side, draw the lines B D and B C each equal to $6\frac{3}{5}$ inches, being one half of the $13\frac{1}{5}$ above referred to. From the points C and D draw lines to the apex A (shown broken). As the width of the sheet used is 10 inches and as we assume an edge of $\frac{3}{8}$ inch for each side, thus leaving $9\frac{1}{4}$ inches, measure on the vertical line A B lengths of $9\frac{1}{4}$ inches in succession, until the apex A is reached, leaving the last sheet at the top to come as it may. Through the points thus obtained on A B draw lines parallel to C D intersecting the lines A C and A D as shown. Then the various shapes marked 1 2 3 etc. will be the net patterns for similarly numbered courses. Take the shears and cut out the patterns on the felting and number them as required.

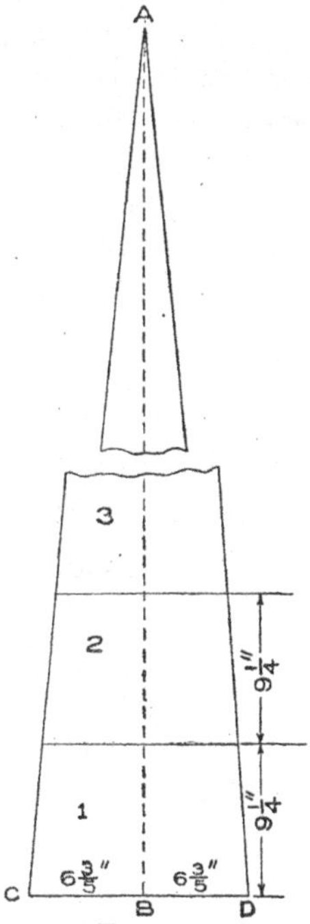

Fig. 215.

For example, take the paper pattern No. 1, place it on a sheet of tin as shown in Fig. 216, and allow $\frac{3}{8}$-inch edges all around, and notch the corners A B C and D. Mark on the tin pattern "No. 1, 29 more", as 30 sheets are required to go around the tower, and cut 29 more for course No. 1. Treat all of the paper patterns from No. 1 to the apex in similar manner. Of course where the patterns become smaller in size at the top, the waste from other patterns can be used.

In Fig. 217 is shown how the sheets should be edged, always being careful to have the narrow side towards the top with the edge toward the outside, the same as in flat seam roofing. Lay the sheets in the usual manner, breaking joints as in general practice. As the seams are not soldered care must be taken to lock the edges well. After the entire roof is laid and before closing the seams with the mallet take a small brush and paint the locks with thick white lead, then close with the mallet. This will make a water-tight job. After the roof is completed the finial D in Fig. 214 is put in position.

Fig. 216.

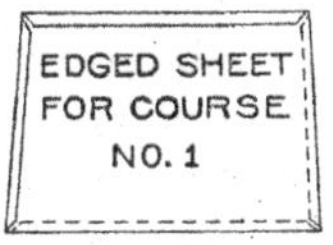

Fig. 217.

As the method used for obtaining the patterns for the various sheets in Fig. 215 is based upon the principle used in obtaining the envelope of a right cone, some student may say that in accurate patterns the line from C to D and all following lines should be curved, as if struck with a radius from the center A, and not straight as shown. To those the writer would say that the curve would be so little on a small pattern, where the radius is so long, that a straight line answers the purpose just as well in all practical work; for it would amount to considerable labor to turn edges on the curved cut of the sheet, and there is certainly no necessity for it.

When different metals are to be connected together, as for instance tin roofing to copper flashing, or copper tubes to galvanized iron gutters, or zinc flashings in connection with copper linings, care must be taken to have the copper sheets thoroughly tinned on both sides where it joins to the galvanized iron, zinc, or other metal, to avoid any electrolysis between the two metals. It is a fact not well known to roofers that if we take a glass jar and fill it with water and place it in separately, two clean strips, one of zinc and the other of copper, and connect the two with a thin copper wire, an electrical action is the result, and if the connection remains for a long time (as the action is very faint) the zinc would be destroyed, because, it may be said, the zinc furnishes the fuel for the electrical action, the same as wood furnishes the fuel for the fire. Therefore, if the copper was not tinned, before locking into the other metal, and the joint became wet with rain, the coating of the metal would be destroyed by the electrical action between the two metals, and the iron would rust through.

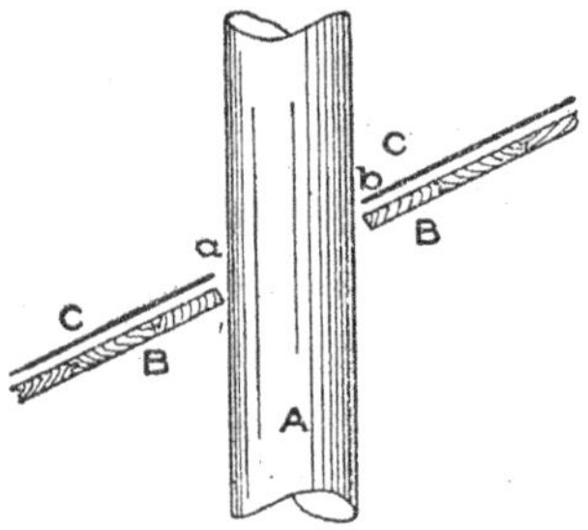

Fig. 218.

While the roofer is seldom called upon to lay out patterns for any roofing work occasion may arise that a roof flashing is required around a pipe passing through a roof of any pitch, as shown in Fig. 218, in which A represents a smoke or vent pipe passing through the roof B B, the metal roof flashing being indicated by C C. If the roof B B were level the opening to be cut into the flashing C C would simply be a true circle the same diameter as the pipe A. But where the roof pitches the opening in the flashing becomes an ellipse, whose minor axis is the same as the diameter of the pipe, and whose major axis is equal to the pitch *a b*. In Fig. 219 is shown how this opening is obtained by the use of a few nails, a string, and a pencil, which the roofer will always have handy.

First draw the line A B representing the slant of the roof, and then make the pipe of the desired size passing through this line at its proper angle to the roof line. Next draw the center line R S of the pipe, as shown. Call the point where this line intersects the roof line, I, and the points where D E and C F intersect A B, G and H respectively. Through I draw K L at right angles to A B, making K I and I L each equal to the half diameter of the pipe. Having established the minor axis K L and the major axis G H, the ellipse is made by taking I H, or half the major axis, as a radius, and with L as a center strike arcs intersecting the major axis, at points M and N. Drive a small nail in each of these two points and attach a string to the nails as shown by the dotted lines K M N, in such a way that when a pencil point is placed in the string it will reach K. Move the pencil along the string, keeping it taut all the time until the ellipse K H L G is obtained. Note how the position of the string changes when it reaches *a*, then *b*, etc.

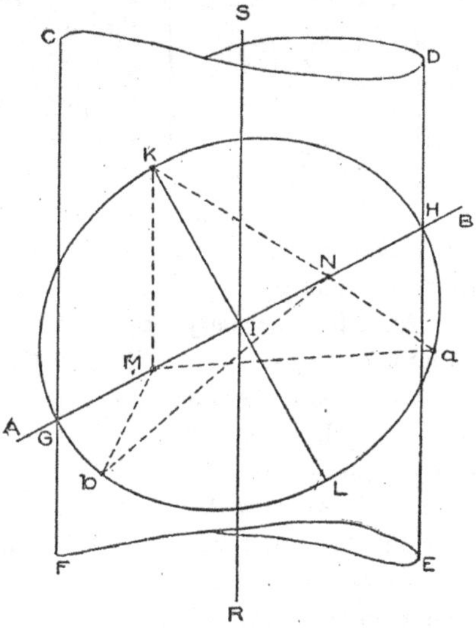

Fig. 219.

STANDING-SEAM ROOFING

Another form of metal roofing is that known as standing seam, which is used on steep roofs not less than $\frac{1}{6}$ pitch, or $\frac{1}{6}$ the width of the building. It consists of metal sheets whose cross or horizontal seams are locked as in flat seam roofing, and whose vertical seams are standing locked seams, as will be described in connection with Figs.

220 to 229 inclusive. Assume that 14 x 20-inch sheets are used and the sheets are edged on the 20-inch sides only, as shown by A in Fig. 220, making the sheet 13 x 20 inches. After the required number of sheets have been edged, and assuming that the length of the pitched roof is 30 feet, then as many sheets are locked together as will be required, and the seams are closed with the mallet and soldered. In practice these strips are prepared of the required length in the shop, painted on the underside, and when dry are rolled up and sent to the building. If desired they can be laid out at the building, which avoids the buckling caused by rolling and transportation from the shop to the job.

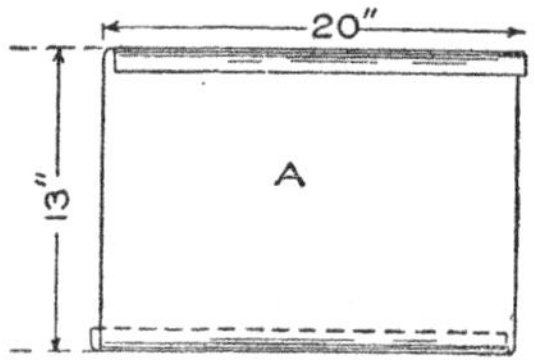

Fig. 220.

After the necessary strips have been prepared they are bent up with the roofing tongs, or, what is better and quicker, the roofing edger for standing-seam roofing. This is a machine into which the strips of tin are fed, being discharged in the required bent form shown at A or B in Fig. 221, bent up 1 inch on one side and $1\frac{1}{4}$ inches on the other side. Or the machine will, if desired, bend up $1\frac{1}{4}$ inches and $1\frac{1}{2}$ inches, giving a $\frac{3}{4}$-inch finished doubled seam in the first case and a 1-inch seam in the second. When laying standing-seam roofing, in no case should any nails be driven into the sheets. This applies to tin, copper or galvanized iron sheets. A cleat should be used, as shown in Fig. 222, which also shows the full size for laying the sheets given in Fig. 221. Thus it will be seen in Fig. 222 that $\frac{1}{4}$ inch has been added over the measurements in Fig. 221, thus allowing edges.

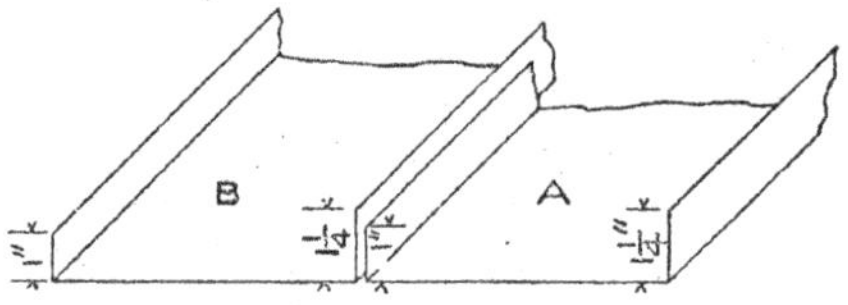

Fig. 221.

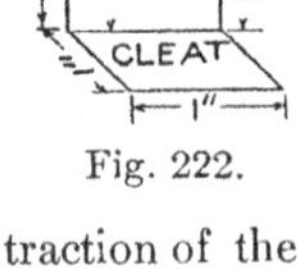

Fig. 222.

These cleats shown in Fig. 222 are made from scrap metal; they allow for the expansion and contraction of the roofing and are used in practice as shown in Fig. 223, which represents the first operation in laying a standing-seam roof, and in which A represents the gutter with a lock attached at B. The gutter being fastened in position by means of cleats under the lock B—the same as in flat seam roofing—the standing seam strips are laid as follows: Take the strip C and lock it well into the lock B of the gutter A as shown, and place the cleat shown in Fig. 222 tightly against the upright bend of the strip C in Fig. 223 as shown at D, and fasten it to the roof by means of a 1-inch roofing nail *a*.

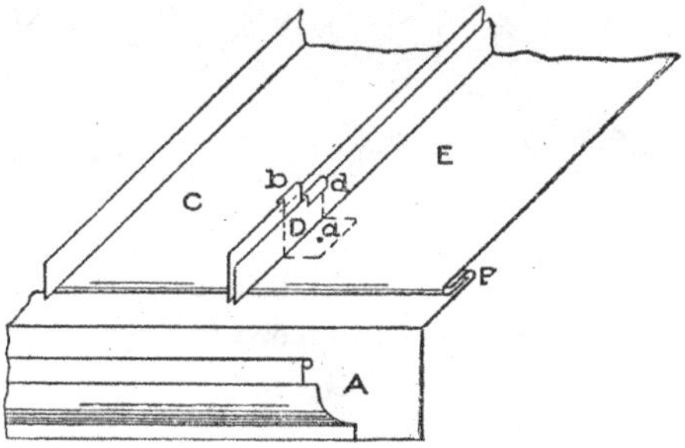

Fig. 223.

Press the strip C firmly onto the roof and turn over edge *b* of the cleat D. This holds the sheet C in position. Now take the next sheet E, press it down and against the cleat D and turn over the edge *d*, which holds E in position. These cleats should be placed about 18 inches

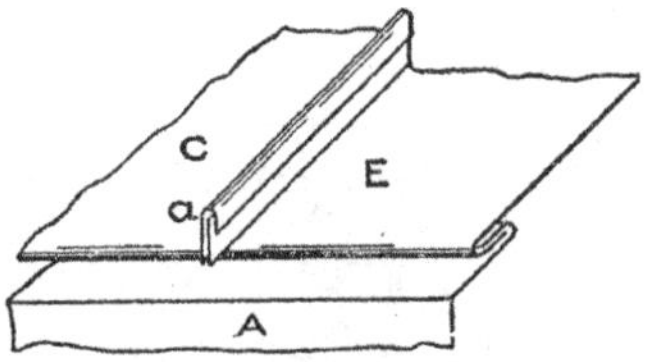

Fig. 224.

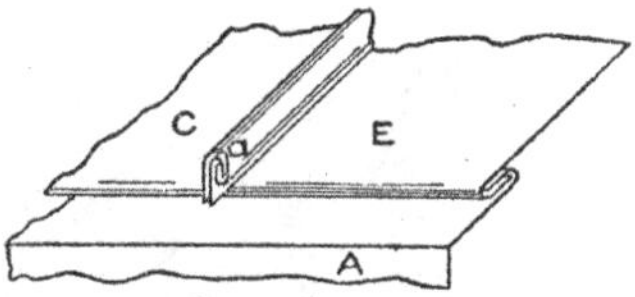

Fig. 225.

apart and by using them it will be seen that no nails have been driven through the sheets, the entire roof being held in position by means of the cleats only.

The second operation is shown in Fig. 224. By means of the hand double seamer and mallet or with the roofing double seamers and squeezing tongs, the single seam is made as shown at *a*. The third and last operation is shown in Fig. 225 where by the use of the same tools the doubled seam *a* is obtained. In Fig. 226 is shown how the finish is made with a comb ridge at the top. The sheets A A A have on the one side the single edge as shown, while the opposite side B has a double edge turned over as shown at *a*. Then, standing seams *b b b* are soldered down to *e*.

In Fig. 227 is shown how the side of a wall is flashed and counter

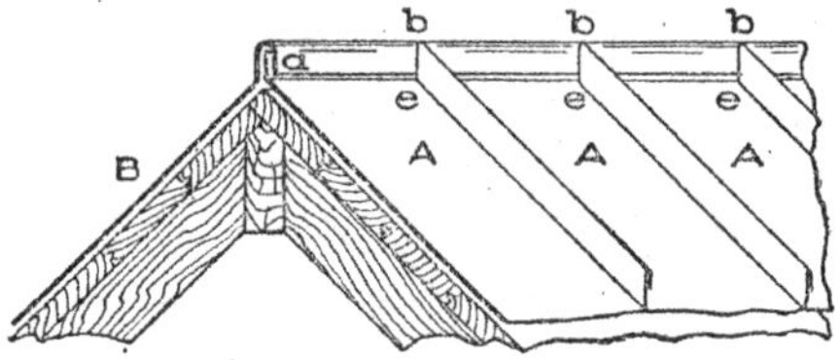

Fig. 226.

flashed. A shows the gutter, B the leader or rain water conductor, and C the lock on the gutter A, fastened to the roof boards by cleats

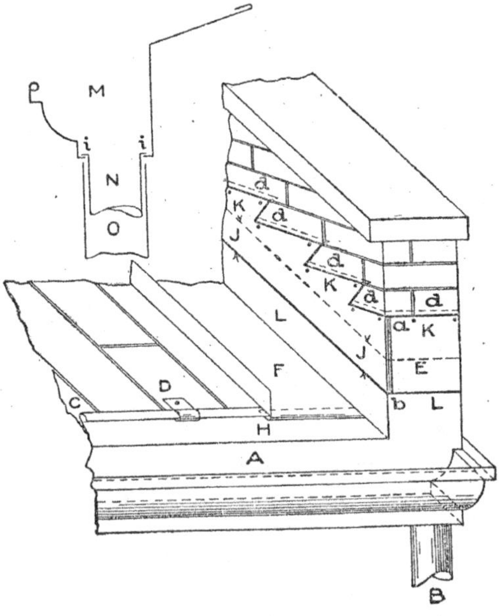

Fig. 227.

as shown at D. The back of the gutter is flashed up against the wall as high as shown by the dotted line E. F represents a standing-seam strip locked into the gutter at H and flashed up against the wall as high as shown by the dotted line J J. As the flashing J J E is not fastened at any part to the wall the beams or wall can settle without disturbing the flashing. The counter or cap flashing K K K is now stepped as shown by the heavy lines, the joints of the brick work being cut out to allow a one-inch flange *d d d* etc. to enter. This is well fastened with flashing hooks, as indicated by the small dots, and then made water-tight with roofer's cement. As will be seen the cap flashing overlaps the base flashing a distance indicated by J J and covers to L L; the corner is double seamed at *a b*. M shows a sectional view through the gutter showing how the tubes and leaders are joined. The tube N is flanged out as shown at *i i*, and soldered to the gutter; the leader O is then slipped over the tube N as shown, and fastened.

In the section on Flat-Seam Roofing it was explained how a conical tower, Fig. 214, would be covered. It will be shown now how this tower would be covered with standing-seam roofing. As the circumference of the tower at the base is 396 inches, and assuming that 14 x 20-inch tin plate is to be used at the base of the tower, the nearest width which can be employed and which will divide the base into equal spaces is $17\frac{5}{23}$ inches, without edges, thus dividing the circumference into 23 equal parts. Then the width of $17\frac{5}{23}$ inches and the length of the rafter A B or A C in elevation will be the basis from which to construct the pattern for the standing seam strip, for which proceed as follows:

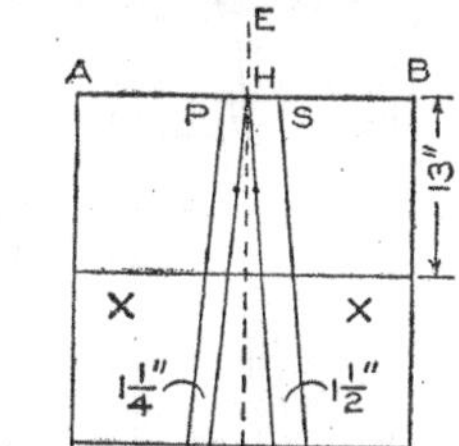

Fig. 228.

Let A B C D in Fig. 228 represent a 20-inch wide strip locked and soldered to the required length. Through the center of the strip draw the line E F. Now measure the length of the rafter A B or A C in Fig. 214 and place it on the line E F in Fig. 228 as shown from H to F. At right angles to H F on either side draw F O and F L making each equal to $8\frac{14}{23}$ inches, being one half of the $17\frac{5}{23}$ above referred to.

From points L and O draw lines to the apex H (shown broken). At right angles to H L and H O draw lines H P equal to 1¼ inches and H S equal to 1½ inches respectively. In similar manner draw L D and O C and connect by lines the points P D and S C. Then will P S C D be the pattern for the standing seam strip, of which 22 more will be required. When the strips are all cut out, use the roofing tongs and bend up the sides, after which they are laid on the tower, fastened with cleats, and double seamed with the hand seamer and mallet in the usual manner.

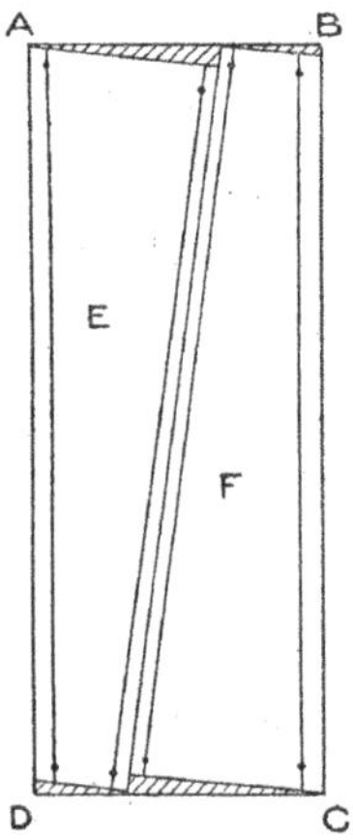

Fig. 229.

If the tower was done in copper or galvanized sheet iron or steel, where 8-foot sheets could be used, as many sheets would be cross-locked together as required; then metal could be saved, and waste avoided, by cutting the sheets as shown in Fig. 229 in which A B C D shows the sheets of metal locked together, and E and F the pattern sheets, the only waste being that shown by the shaded portion. Where the finial D in Fig. 214 sets over the tower, the standing seams are turned over flat as much as is required to receive the finial, or small notches would be cut into the base of the finial, to allow it to slip over the standing seams. Before closing the seams, they are painted with white lead with a tool brush, then closed up tight, which makes a good tight job.

CORRUGATED IRON ROOFING AND SIDING

Corrugated iron is used for roofs and sides of buildings. It is usually laid directly upon the purlins in roofs constructed as shown in Figs. 230 and 231, the former being constructed to receive sidings of corrugated iron, while in the latter figure the side walls of the building are brick. Special care must be taken that the projecting edges of the corrugated iron at the eaves and gable ends of the roof are well secured, otherwise the wind will loosen the sheets and fold them up. The corrugations are made of various sizes such as 5-inch, 2½-inch, 1¼-inch and ¾-inch, the measurements always being from A to B in Fig. 232, and the depth being shown by C. The smaller corrugations give a more pleasing appearance, but the larger corrugations are stiffer and will span a greater distance, thereby permitting the purlins to be further apart.

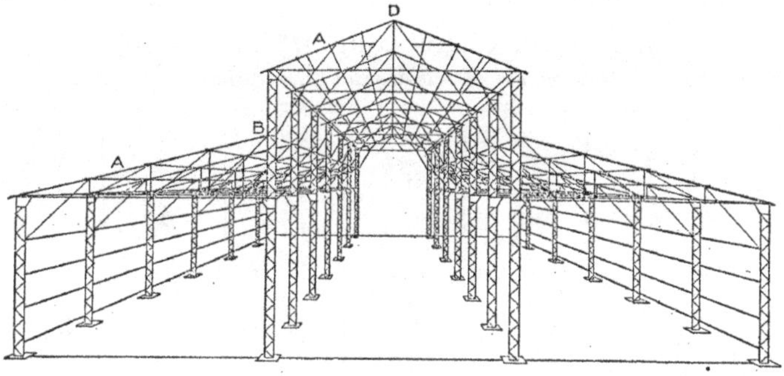

Fig. 230.

The thickness of the metal generally used for roofing and siding varies from No. 24 to No. 16 gauge. By actual trial made by The

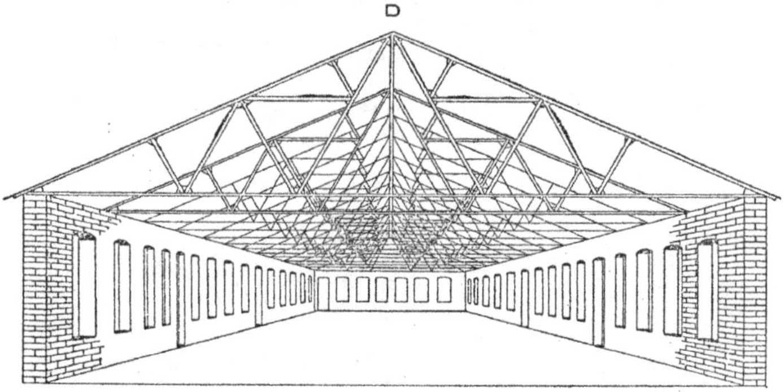

Fig. 231.

Keystone Bridge Company it was found that corrugated iron No. 20, spanning 6 feet, began to give permanent deflection at a load of 30 lb. per square foot, and that it collapsed with a load of 60 lb. per square foot. The distance between centers of purlins should, therefore, not exceed 6 feet, and preferably be less than this.

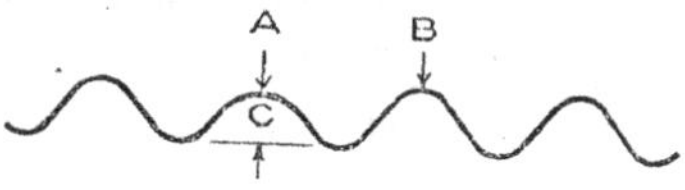

Fig. 232.

TABLES

The following tables will prove of value when desiring any information to which they appertain.

MEASUREMENTS OF CORRUGATED SHEETS

Dimensions of Sheets and Corrugations.

Kind of corrugation	Width of corrugation	Depth of corrugation	No. of corrugations to the sheet	Covering width after lapping one corrugation	Width of sheet after corrugated	Length of the longest sheets furnished
5 inch.	5 inch.	1 inch.	6	24 inch.	27 inch.	10 feet.
2½ inch.	2½ inch.	½ to ⅝ inch.	10	24 inch.	26 inch.	10 feet.
1¼ inch.	1¼ inch.	⅜ to ½ inch.	19½	24 inch.	26 inch.	10 feet.
¾ inch.	¾ inch.	¼ inch.	34½	25 inch.	26 inch.	8 feet.

RESULTS OF TEST

of a corrugated sheet No. 20, 2 feet wide, 6 feet long between supports, loaded uniformly with fire clay.

Load per square foot. lb.	Deflection at center under load. Inches.	Permanent Deflection, load removed.
5	½	0
10	¾	0
15	1	0
20	1¼	0
25	1½	0
30	1⅞	⅛
35	2¼	½
40	2⅝	¾
45	3½	1⅛
50	4	1½
55	6½	Not noted.
60	Broke down.	" "

The following table shows the distance apart the supports should be for different gauges of corrugated sheets:

Nos. 16 and 18 .... 6 to 7 feet apart.
Nos. 20 and 22 .... 4 to 5 feet apart.
No. 24 .... 2 to 4 feet apart.
No. 28 .... 2 feet apart.

The following table is calculated for sheets 30½ inches wide before corrugating.

No. by Birmingham gauge	Thickness Inch	Weight per square ft., flat Lb.	Weight per square ft., corrugated Lb.	Weight per square of 100 square feet, when laid, allowing 6 inches lap in length and 2½ inches or one corrugation in width of sheet, for sheet lengths of:						Weight per square ft., flat, galvanized Lb.
				5 feet	6 feet	7 feet	8 feet	9 feet	10 feet	
16	.065	2.61	3.28	365	358	353	350	348	346	2.95
18	.049	1,97	2.48	275	270	267	264	262	261	2.31
20	.035	1.40	1.76	196	192	190	188	186	185	1.74
22	.028	1.12	1.41	156	154	152	150	149	148	1.46
24	.022	.88	1.11	123	121	119	118	117	117	1.22
26	.018	.72	.91	101	99	97	97	96	95	1.06

LAYING CORRUGATED ROOFING

When laying corrugated iron on wood sheathing use galvanized iron nails and lead washers. The advantage in using lead washers is that they make a tight joint and prevent leaking and rusting at the nail hole; the washer being soft it easily shapes itself to any curve. In Fig. 233 is shown how these washers are used; A shows the full size nail

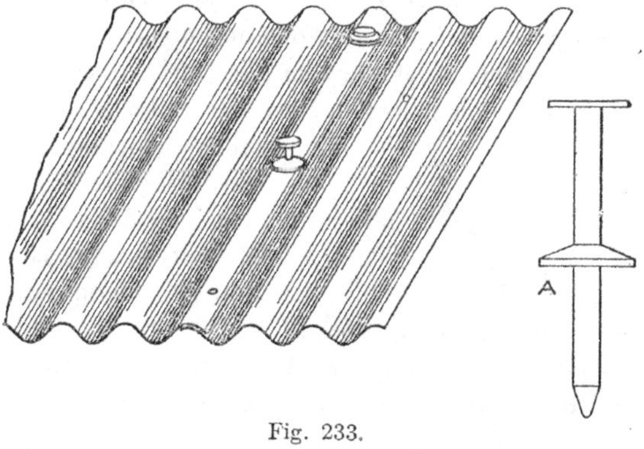

Fig. 233.

and washer. When laying, commence at the left hand corner of the eave and end of the building. Continue laying to the ridge by lapping the second sheet over the first 4 inches,the left-hand edge being finished by means of a gable band A, formed as shown in Fig. 234, into which the corrugated sheet B is well bedded in roofer's cement C. When it is not desired to use this gable band the sheet must be well secured at the edge to keep the wind from raising the sheets from the roof in a storm, as at A in Fig. 230.

Should the gable have a fire wall, then let the sheets A butt against the wall and flash with corrugated flashing as shown in Fig. 235, over which the regular cap or counter flashing is placed as explained in connection with Fig. 227. Should the ridge of the roof A butt against a wall, as shown at B in Fig. 230, then an end-wall flashing is used as is shown in Fig. 236 which must also be capped, by either using cap flashing or allowing the corrugated siding to overlap this end-wall flashing

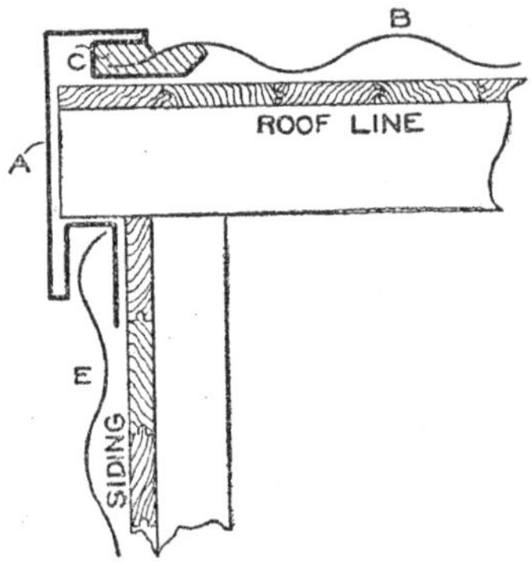

Fig. 234.

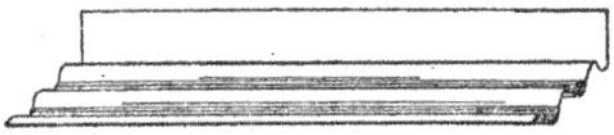

Fig. 235.

as would be the case at B in Fig. 230. Now commence the second course at the eaves, giving one and one half corrugations for side lap, being careful that the side corrugations center each other exactly and nail with washers as shown in Fig. 237. Nail at every other corrugation at end laps, and at about every 6 inches at side laps, nailing through top of corrugation as shown in Fig. 237. Continue laying in this manner until the roof is covered.

Fig. 236.

The same rule is to be observed in regard to laps and flashing if the corrugated iron were to be fastened to iron purlins, and the method of fastening to the iron frames would be accomplished as shown in Figs. 238 to 240 inclusive. Assuming that steel structures are to be covered, as shown in Figs. 230 and 231, then let A in Fig. 238 be the iron rafter, B the cross angles on which the sheets D are laid, then by means of the clip or clamp C, which is made from hoop iron and bent around the angle B, the sheets are riveted in position. In Fig. 239 is shown another form of clamp, which is bent over the bottom of the angle iron.

Fig. 237.

Fig. 240 shows still another method, where the clamp F is riveted to the sheet B at E, then turned around the angle A at D. To avoid having the storm drive in between the corrugated opening at the eaves, corrugated wood filler is used as shown in Fig. 241. This keeps out the

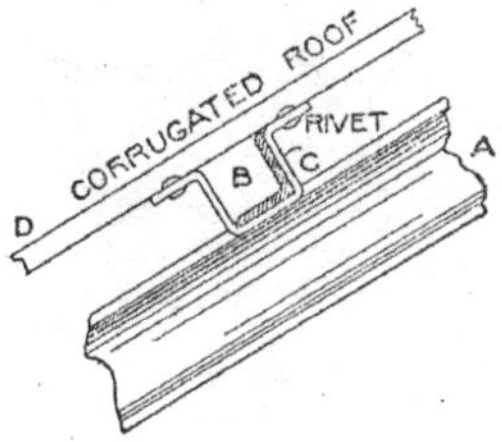

Fig. 238.

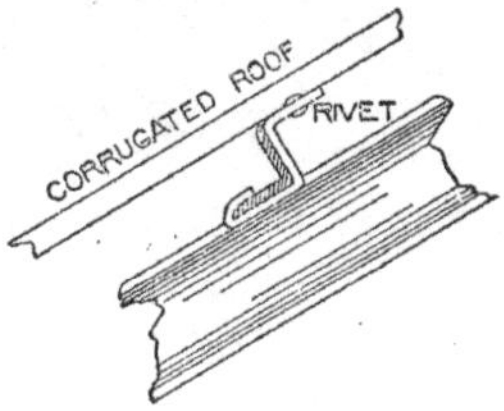

Fig. 239.

snow and sleet. On iron framing this is made of pressed metal. Another form of corrugated iron roofing is shown in Fig. 242. This is put down with cleats in a manner similar to standing-seam roofing.

If there are hips on the roof, the corrugated iron should be carefully cut and the hip covered with sheet lead. This is best done by having a wooden cove or filler placed on the hip, against which the roofing butts. Sheet lead is then formed over this wooden core and into the corrugations, and fastened by means of wood screws through the lead cap into the wooden core. The lead being soft, it can be worked into any desired shape. When a valley occurs in a hipped roof, form from plain sheet iron a valley as shown in Fig. 243, being sure to give it two coats of paint before laying, and make it from 24-inch wide sheets, bending up 12 inches on each side. Fit it in the valley, and cut the corrugated iron to fit the required angle. Then lap the corrugated iron over the valley from 6 to 8 inches.

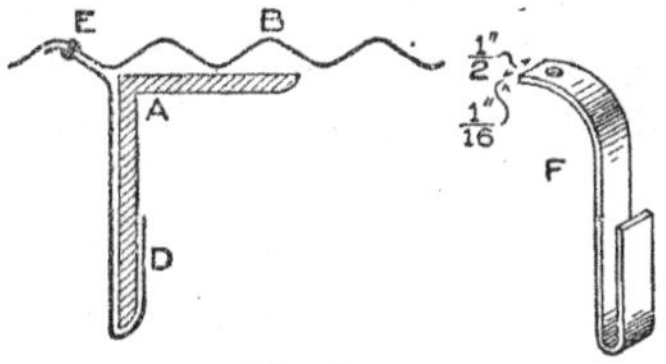

Fig. 240.

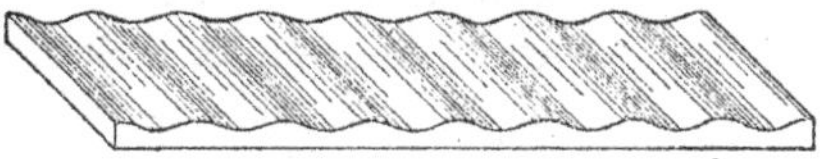

Fig. 241.

When a chimney is to be flashed, as shown in Fig. 244, use plain iron, bending up and flashing into the chimney joints, and allowing the flashing to turn up under the corrugated iron at the top about 12 inches and over the corrugated iron at the bottom about the same distance. At the side the flashing should have the shape of the corrugated iron and receive a lap of about 8 inches, the entire flashing

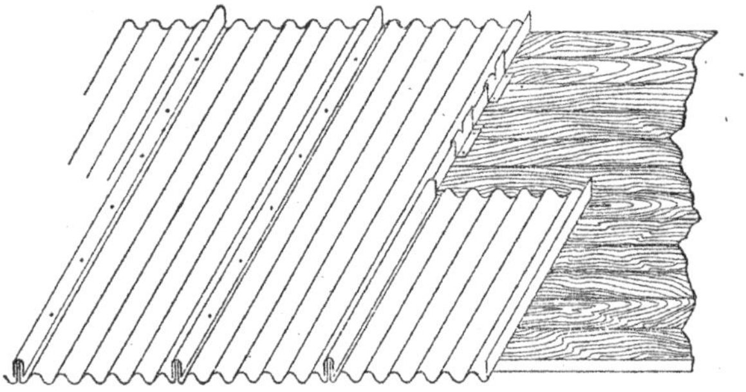

Fig. 242.

being well bedded in roofer's cement. When a water-tight joint is required around a smoke stack, as shown in Fig. 245, the corrugated iron is first cut out as shown, then a flashing built around one half the upper part of the stack to keep the water from entering inside. This is best done by using heavy sheet lead and riveting it to the sheets, using strips of similar corrugated iron as a washer to avoid damaging the lead. Before riveting, the flashing must be well bedded in roofer's cement and then make a beveled angle of cement to make a good joint. After this upright flashing is in position a collar is set over the same and fastened to the stack by means of an iron ring bolted and made tight as shown. Cement is used to make a water-tight joint around the stack. This construction gives room for the stack to sway and allows the heat to escape.

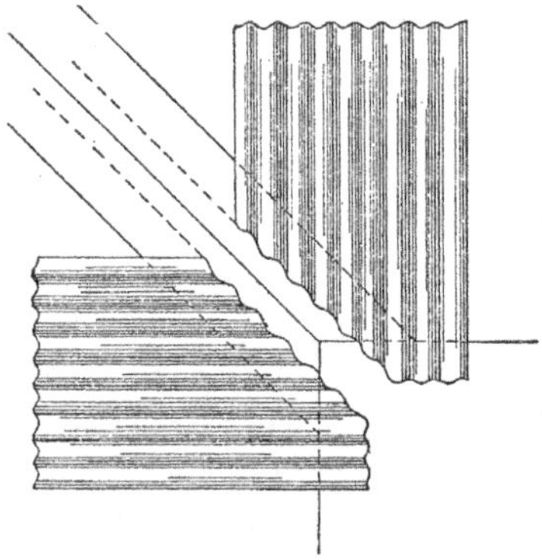

Fig. 243.

Sometimes the end-wall flashing shown in Fig. 236 can be used to good advantage in building the upright flashing in Fig. 245. Where the corrugated iron meets at the ridge, as at D and D in Figs. 230 and

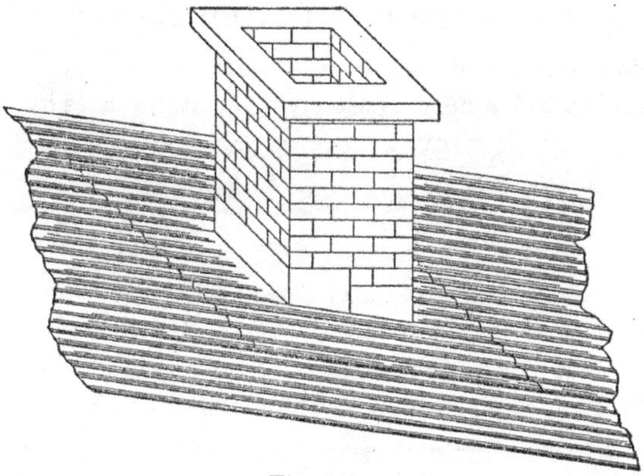

Fig. 244.

231, a wooden core is placed in position as explained in connection with the hip ridge, and an angle ridge, pressed by dealers who furnish the

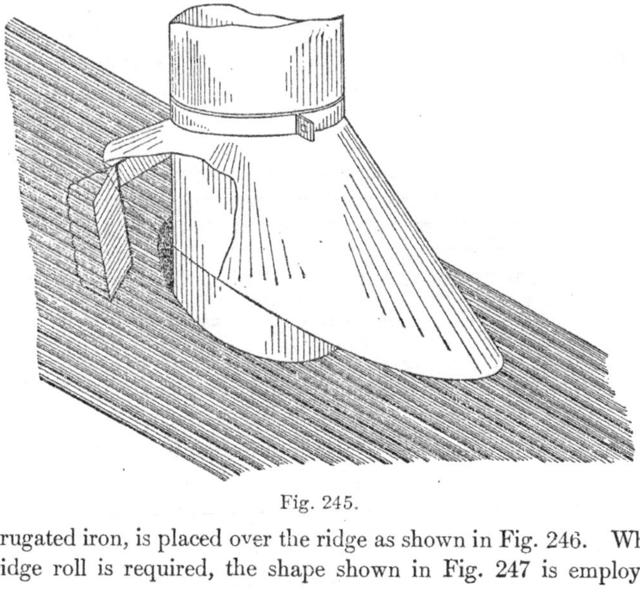

Fig. 245.

corrugated iron, is placed over the ridge as shown in Fig. 246. When a ridge roll is required, the shape shown in Fig. 247 is employed.

These ridges are fastened direct to the roof sheets by means of riveting or bolting.

LAYING CORRUGATED SIDING

Before putting on any corrugated siding or clapboarding, as shown in Fig. 248, a finish is usually made at the eaves by means of a

Fig. 246.

hanging gutter or a plain cornice, shown in Fig. 249, which is fastened to the projecting wooden or iron rafters. This method is generally used on elevators, mills, factories, barns, etc., where corrugated iron, crimped iron or clapboards are used for either roofing or siding. This

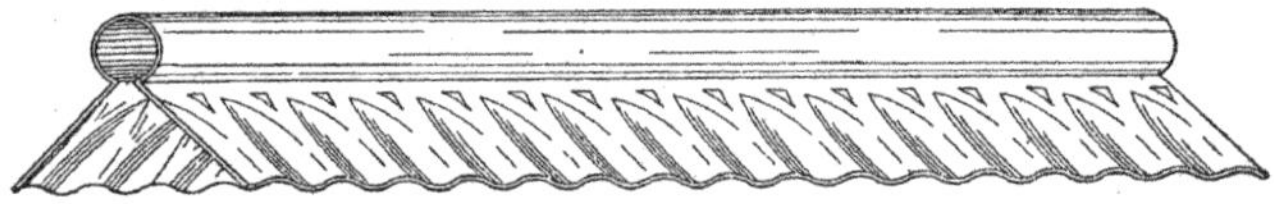

Fig. 247.

style of cornice covers the eaves and gable projections, so as to make the building entirely ironclad. When laying the siding commence at the left hand corner, laying the courses from base to cornice, giving the sheets a lap of two inches as the ends and one and one half corruga-

Fig. 248.

tions at the sides. Nail side laps every 6 inches and end laps at every other corrugation, driving the nails as shown in Fig. 250.

Where the sheets must be fastened to iron framing use the same method as explained in connection with Figs. 238, 239 and 240. In this case, instead of nailing the sheets, they would be riveted. If siding is put on the wooden studding care should be taken to space the studding the same distance apart as the laying width of the iron used. In this case pieces of studding should be placed between the uprights at the end of each sheet to nail the laps. When covering grain elevators

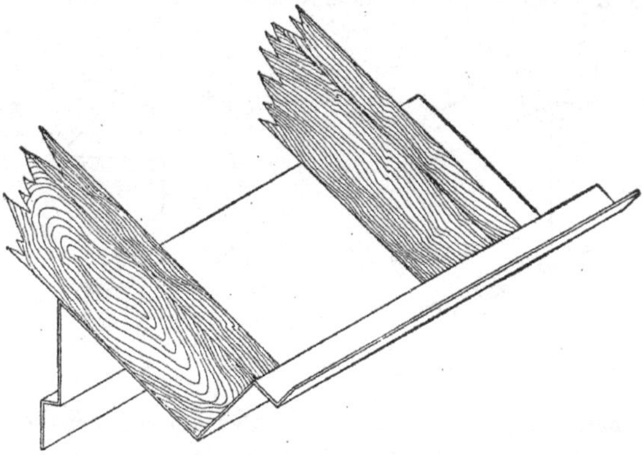

Fig. 249.

it is necessary to use swinging scaffolds. Commence at the base and carry up the course to the eave, the length of the scaffold. Commence at the left hand and give the sheets a lap of one corrugation on the side and a two-inch lap at the end. Nail or rivet in every corrugation 3 inches from the lower end of the sheet; this allows for settling of the building.

Fig. 250.

When any structure is to be covered on two or more sides, corner casings made of flat iron are employed, of a shape similar to that shown at B, Fig. 251. It will be seen that a rabbet is bent on both sides *a* and *b* to admit the siding. This makes a neat finish on the outside and hides the rough edges of the siding. If a window opening is to have casings a jamb is used as shown at A, Fig. 251, which has a similar rabbet at *a* to receive the siding, and a square bend at *b* to nail against the frame. In Fig. 252 is shown the cap of a window or opening. It is

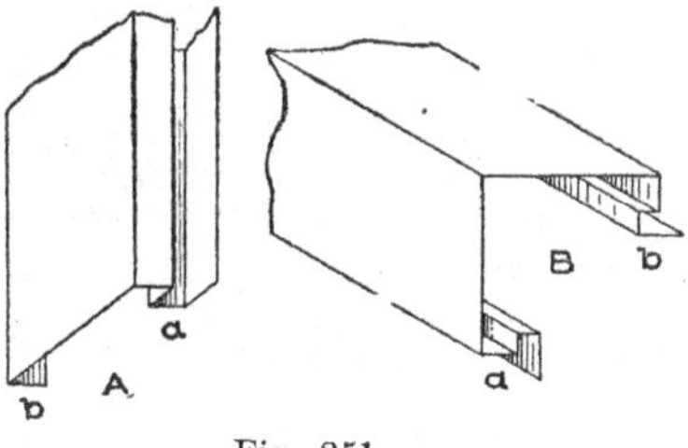

Fig. 251.

bent so that *a* is nailed to the window or other frame at the bottom, while *b* forms a flashing over which the siding will set. Fig. 253 shows the sill of a window, which has a rabbet at *a*, in which the siding is

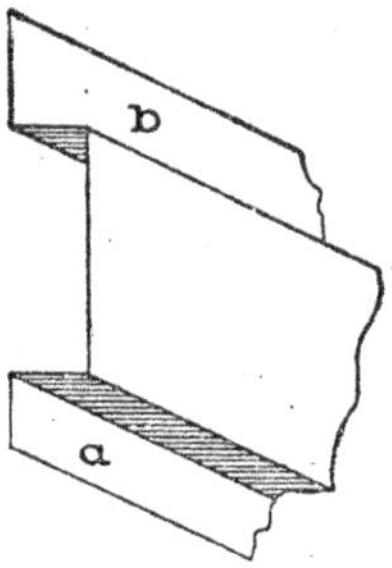

Fig. 252.

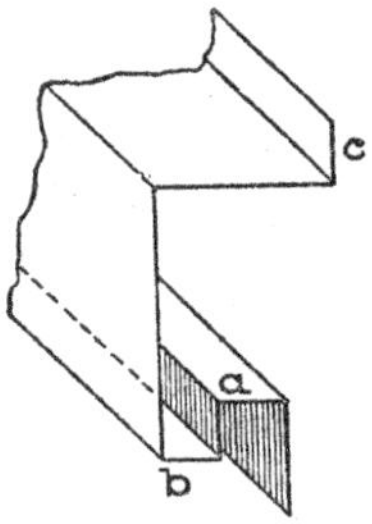

Fig. 253.

slipped; then *b* forms a drip, and any water coming over the sill passes over the siding without danger of leaks; *c* is nailed in white lead to the window frame.

Another use to which corrugated iron is put is to cover sheds and awnings. Sheets laid on wood are nailed in the usual manner, while sheets laid on angle iron construction are fastened as explained in the

Fig. 254.

preceding sections. In Fig. 254 is shown an awning over a store laid on angle iron supports. In work of this kind, to make a neat appearance, the sheets are curved to conform to the iron bracket A.

CORNICE OVER BRICK BAY *

An elevation and plan of a brick bay are shown in the illustration, the sides of which are 8 inches, 3 feet 2 inches and 5 feet 10 inches wide. Laps or flanges for soldering are to be allowed on the 3 feet 2 inch pieces and no laps on the 8 inch and 5 feet 10 inch pieces. The lookouts or iron braces are indicated in the plan by the heavy dashes making a total of 9 required.

After the detail section is drawn and knowing the angle of the bay in plan, the angle is placed as shown by ABC, being careful to place CB on a line drawn vertically from 3–4 in the section. The miter line is then drawn as shown by BD, the section divided into equal spaces, and vertical lines dropped to the miter line BD as shown. At right angles to BC the girth of the section is drawn as shown by similar figures from 1 to 26, through which points at right angles to 1–26, lines are drawn and intersected by similar numbered lines drawn from the miter line BD at right angles to BC, thus obtaining the upper miter cut shown. Now using this miter cut in practice, make the distance from either points 25 or 24 (which represents the line of the wall) equal to 8 inches, 3 feet 2 inches and 5 feet 10 inches. The 3 feet 2 inches and 5 feet 10 inches have opposite miter cuts as shown.

As will be seen by the plan, two eight inch pieces will be required, one right and one left and two 3 feet 2 inch and one 5 feet 10 inch pieces. Nine iron lookouts will be required formed to the shape shown in the detail section, where holes are punched for bolting as there indicated.

* The illustration referred to will be found on the back of this page.

PATTERNS AND LAY-OUT FOR CORNICE OVER BRICK BAY

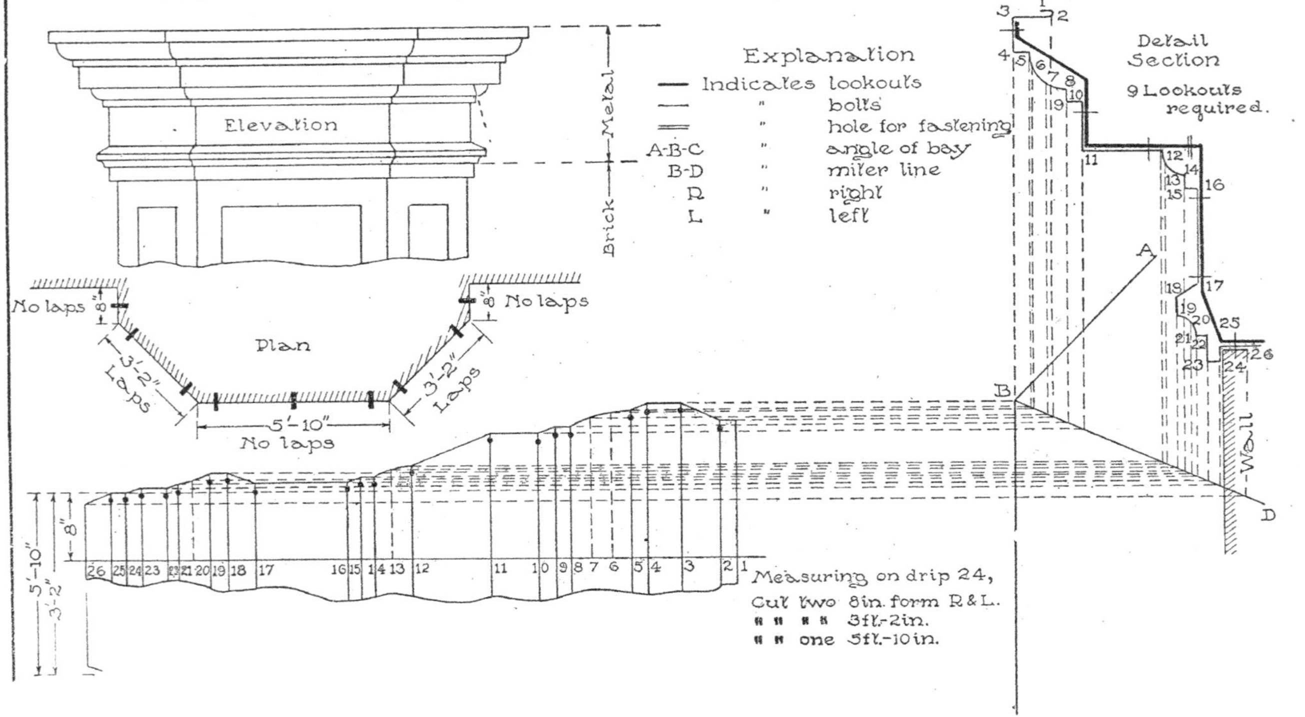

SHEET METAL WORK

PART IV

CORNICE WORK

There is no trade in the building line to-day which has made such rapid progress as that of Sheet-Metal Cornice, or Architectural Sheet-Metal Work. It is not very long since the general scope of this branch of craftsmanship merely represented a tin-shop business on a large scale. But as things are to-day, this is changed. From an enlarged tin-shop business, sheet-metal cornice work, including under that title every branch of architectural sheet-metal work, has become one of the substantial industries of the country, comparing favorably with almost any other mechanical branch in the building trades. Nor is this work confined to the larger cities. In the smaller towns is shown the progress of architectural sheet-metal work in the erection of entire building fronts constructed from sheet metal.

CONSTRUCTION

Sheet-metal cornices have heretofore, in a great measure, been duplications of the designs commonly employed in wood, which, in turn, with minor modifications, were imitations of stone.

With the marked advancement of this industry, however, this need no longer be the case. A sheet-metal cornice is not now imitative. It possesses a variety and beauty peculiarly its own. No pattern is too complex or too difficult. Designs are satisfactorily executed in sheet metal which are impossible to produce in any other material. By the free and judicious application of pressed metal ornaments, a product is obtained that equals carved work. For boldness of figure, sharp and clean-cut lines, sheet-metal work takes the lead of all competitors.

In order that there may be no misunderstanding as to the various parts contained in what the sheet-metal worker calls a "cornice," Fig. 255 has been prepared, which gives the names of all the members in the "entablature"—the architectural name for what in the shop is known as the cornice. The term "entablature" is seldom heard among mechanics, a very general use of the word "cornice" having supplanted it in the common language of business.

An entablature consists of three principal parts—the *cornice,* the *frieze,* and the *architrave.* A glance at the illustration will serve to show the relation that each bears to the others. Among mechanics the shop term for architrave is *foot-moulding*; for frieze, *panel*; and for

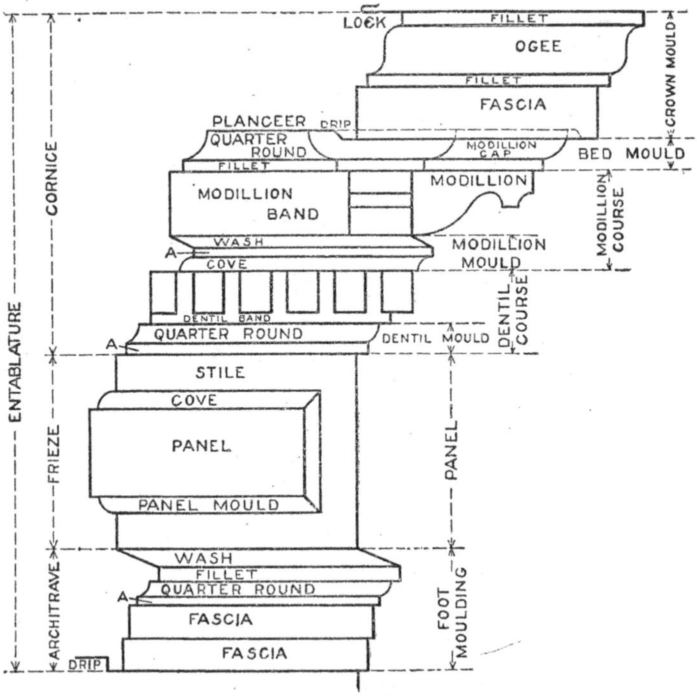

Fig. 255.

the subdivisions of the cornice, *dentil course, modillion course, bed-mould,* and *crown-mould.* In the modillion course, are the *modillion-band* and *modillion-mould*; while in the dentil course are the *dentil-band* and *dentil-mould. Drips* are shown at the bottom of the crown- and foot-mould fascias, and the ceiling under the crown mould is called the *planceer.* The edge at the top of the cornice is called a *lock,* and is used to lock the metal roofing into, when covering the top of the cornice. In the panel, there are the *panel* proper, the *panel-mould,* and the *stile.* The side and front of the modillion are also shown.

Fig. 256 shows the side and front view of what is known as a bracket. Large terminal brackets in cornices, which project beyond the mouldings, and against which the mouldings end, are called trusses, a front and a side view of which are shown in Fig. 257. A block placed above a common bracket against which the moulding ends, is called a *stop block,* a front and a side view of which are shown in Fig. 258.

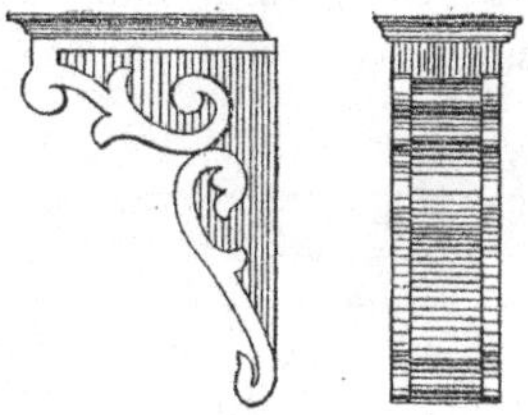

Fig. 256.

Fig. 257.

Fig. 259 is the front elevation of a cornice, in which are shown the truss, the bracket, the modillion, the dentil, and the panel. It is sometimes the case, in the construction of a cornice, that a bracket or modillion is called for, whose front and sides are carved as shown in the front and side views in Fig. 260. In that case, the brackets are obtained from dealers in pressed ornaments, who make a specialty of this kind of work. The same applies to capitals which would be required for pilasters or columns, such as those shown in Figs. 261 and 262. The pilaster or column would be formed up in sheet metal, and the capital purchased and soldered in position. In Fig. 263, A shows an inclined moulding, which, as far as general position is concerned, would be the same as a gable moulding.

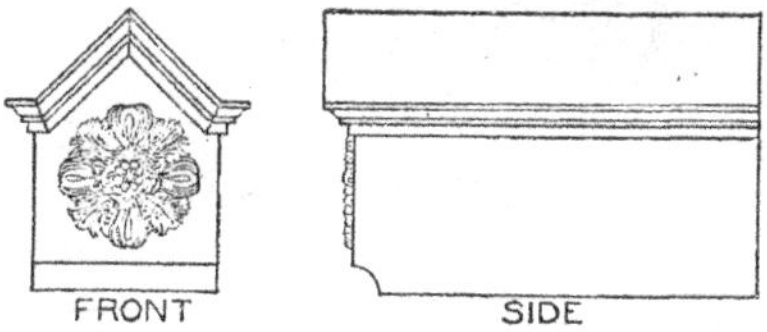

Fig. 258.

Raking mouldings are those which are inclined as in a gable or pediment; but, inasmuch as to miter an inclined moulding (as A) into a horizontal moulding (as B and C), under certain conditions, necessitates a change of profile, the term "to rake," among sheet-metal workers, has come to mean "to change profiles" for the accomplishment of

FRONT ELEVATION

Fig. 259

such a miter. Hence the term "raked moulding" means one whose profile has been changed to admit of mitering.

The term *miter*, in common usage, designates a joint in a moulding at any angle.

Drawings form a very important part in sheet-metal architectural

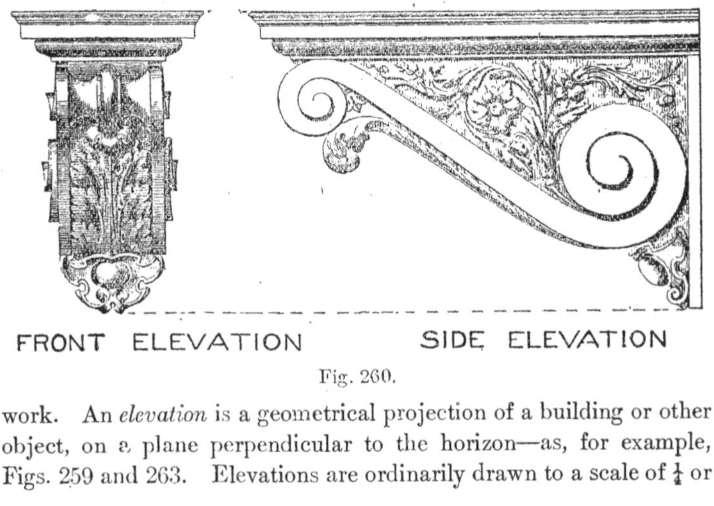

FRONT ELEVATION SIDE ELEVATION

Fig. 260.

work. An *elevation* is a geometrical projection of a building or other object, on a plane perpendicular to the horizon—as, for example, Figs. 259 and 263. Elevations are ordinarily drawn to a scale of ¼ or

$\frac{1}{2}$ inch to the foot. A *sectional drawing* shows a view of a building or other object as it would appear if cut in two at a given vertical line—as, for example, Fig. 255. *Detail drawings* are ordinarily full size, and

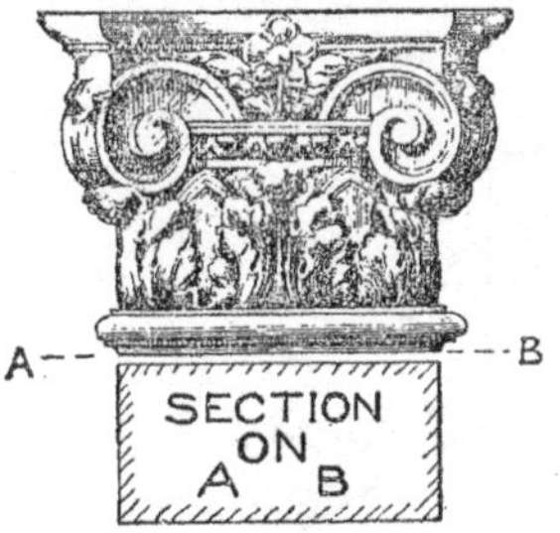

Fig. 261.

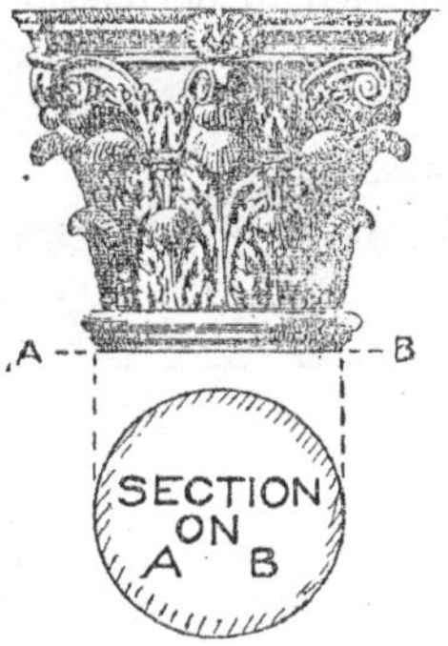

Fig. 262.

are often called *working drawings. Tracings* are duplicate drawings, made by tracing upon transparent cloth or paper placed over the orig-

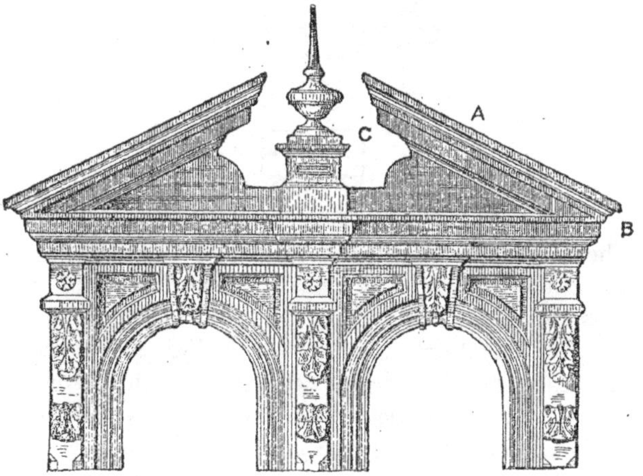

Fig. 263.

inal drawing. Many other terms might be introduced here; but enough, we believe, have been presented to give the student the leading general points.

A few words are necessary on the subject of *fastening the cornice to the wall.*

Sheet-metal cornices are made of such a wide range of sizes, and are required to be placed in so many different locations, that the methods of construction, when wooden lookouts are employed and

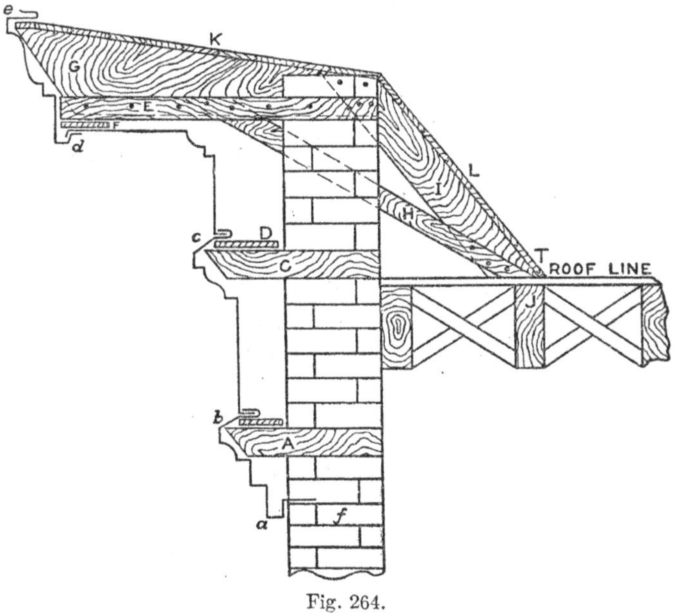

Fig. 264.

when the cornice is put together at the building in parts, are worthy of the most careful study. The general order of procedure in putting up, is as follows:

The foot-moulding or architrave *a b* (Fig. 264) is set upon the wall finished up to *f*, the drip *a* being drawn tight against the wall. The brickwork is then carried up, and the lookout A placed in position, the wall being carried up a few courses higher to hold the lookout in position. A board B is then nailed on top of the lookouts (which should be placed about three feet apart); and on this the flange of the foot-mould *b* is fastened. The frieze or panel *b c* is now placed into the lock B, which is closed and soldered; when the lookout C and the board D are placed in their proper positions, as before described.

The planceer and bed-mould *c d* are now locked and soldered at D, and the lookout E placed in position, with a board F placed under the lookouts the entire length of the cornice; onto this board the planceer is fastened. Having the proper measurements, the framer now constructs his lookouts or brackets G H I E, fastening to the beam at T, when the crown-mould *d e* is fastened to the planceer, through the flange of the drip at *d*, and at the top at *e*. The joints between lengths of mouldings, are made by lapping, riveting, or bolting, care being taken that they are joined so neatly as to hide all indications of a seam when finished and viewed from a short distance.

If brackets or modillions are to be placed in position, they are riveted or bolted in position; or sometimes the back of the cornice is blocked out with wood, and the brackets screwed in position through their flanges.

While a galvanized-iron cornice thus constructed on wooden lookouts will resist fire for a long time, a strictly fireproof cornice is obtained only by the use of metal for supports and fastenings, to the entire exclusion of wood. This fireproof method of construction is shown in Fig. 265. Instead of putting up in parts on the building, the cornice is constructed in one piece in the shop or upon the ground, and hoisted to the top of the wall in long lengths easily handled. A drip *a* is used at the bottom of the foot-mould, and the joints made in the way indicated at *b* and *c*, with a lock at *d*. Band iron supports and braces are used, formed to the general contour of the parts as shown by A B C, and bolted direct to the cornice, as shown, before hoisting.

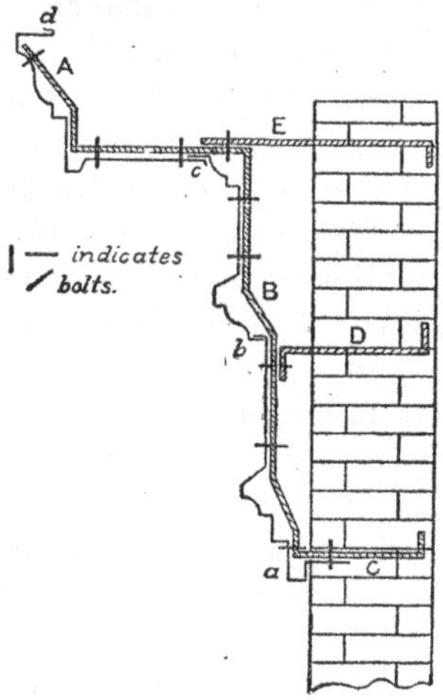

Fig. 265.

When the cornice sets on the wall as at C, anchors are fastened to the main brace, as at D and E, with an end bent up or down for fastening. If the cornice sets perfectly plumb, the mason carries up his wall, which holds the cornice in a firm position. The top and back are then framed in the usual manner and covered by the metal roofer. In constructing cornices in this manner, the mouldings are run through solid, behind all brackets and modillions. The brackets and modillions are attached by means of riveting through outside flanges.

SHOP TOOLS

One of the most important tools in cornice or architectural sheet-metal working shop is the *brake*. On those operated by hand, sheets are bent up to 8 feet in one continuous length. In the larger shops, power presses or brakes are used, in which sheets are formed up to 10 feet in length, the press being so constructed that they will form ogees, squares, or acute bends in one operation.

Large 8- or 10-feet *squaring shears* also form an important addition to the shop, and are operated by foot or power.

When cornices are constructed where the planceer or frieze is very wide, it is usual to put crimped metal in, to avoid the waves and buckles showing in the flat surface; for this purpose the *crimping machine* is used.

In preparing the iron braces for use in the construction of fireproof cornices, a *punching machine* and *slitting shears* are used for cutting the band iron and punching holes in it to admit the bolts. While braces are sometimes bent in a vise, a small machine known as a *brace bender* is of great value m the shop. In large fireproof building constructions, it is necessary that all doors, window frames, and even sashes be covered with metal, and made in so neat a manner that, when painted and grained, no differences will be apparent to indicate whether the material is wood or metal, the smallest bends down to $\frac{1}{8}$ inch being obtained. This, of course, cannot be done on the brakes just mentioned, but is done by means of the *draw-bench*, which is constructed in lengths up to 20 feet and longer, operated by means of an endless chain, and capable of drawing the sheet metal over any shaped wood mould as tightly as if it were cast in one piece. The smaller tools in the shop are similar to those described in the Instruction Papers on Tinsmithing and Sheet Metal Work, Part I.

METHOD EMPLOYED FOR OBTAINING PATTERNS

The principles applied to cylinder developments as explained in the Tinsmithing and Sheet Metal Work courses, under the heading of "Parallel-Line Developments," are also applicable for obtaining the patterns for any moulding where all members run parallel; for it makes no difference what profile is employed, so long as the lines run parallel to one another, the parallel-line method is used. While this method is chiefly employed in cornice work, other problems will arise, in which the "Radial-Line" and Triangulation" methods (explained in previous Papers) will be of service.

The term generally used in the shop for pattern cutting on cornice work is *miter cutting*. To illustrate, suppose two pieces of mouldings are to be joined together at angle of 90°, as shown in Fig. 266. The first step necessary would be to bisect the given angle and obtain the *miter-line* and cut each piece so that they would miter together. If a carpenter had to make a joint of this kind, he would place his moulding in the miter-box, and cut one piece right and one piece left at an angle of 45°, and he would be careful to hold the moulding in its proper position before sawing; or else he may, instead of having a return miter as shown, have a face miter as in a picture frame, shown in Fig. 267. The sheet-metal cornice-maker cannot, after his moulding is formed, place it in the miter-box to cut the miter, but must lay it out—or, in other words, develop it—on a flat surface or sheet of metal. He must also be careful to place the profile in its proper position with the miter-line; or else, instead of having a return miter as shown in Fig. 266, he will have a face miter as shown in Fig. 267. If he lays out his work correctly, he can then cut two pieces, form one right and the other left, when a miter will result between the two pieces of moulding and will look as shown in Fig. 266. If, however, a face miter is desired, as shown in Fig. 267, which is used when miters are desired for panels and other purposes, the method of laying them out will be explained as we proceed. The same principles required for developing Figs. 266 and 267 are used, whether the mouldings are mitered at angles of 90°

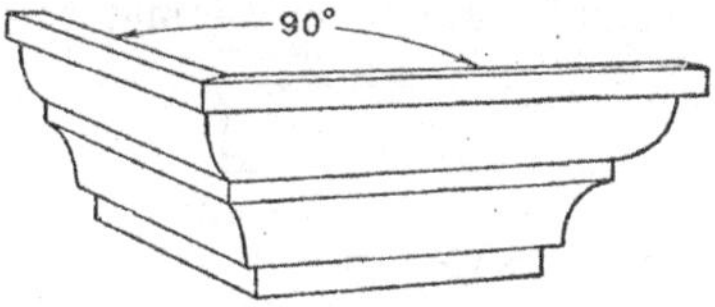

Fig. 266.

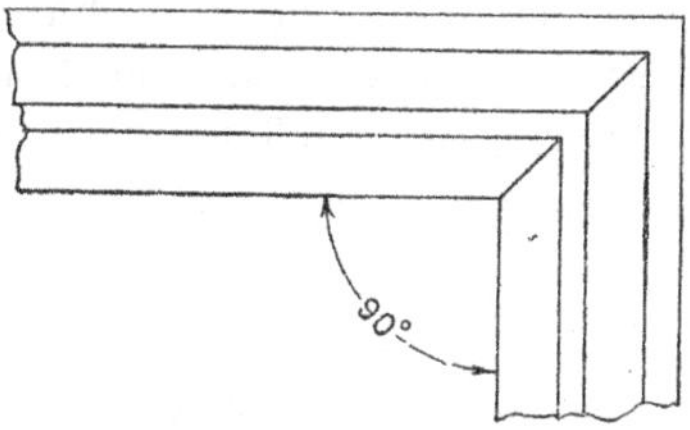

Fig. 267.

or otherwise. The method of *raking* the mouldings—or, in other words, changing their profile to admit the mitering of some other moulding at various angles—will also be thoroughly explained as we proceed.

VARIOUS SHAPES OF MOULDINGS

The style of mouldings arising in the cornice shop are chiefly Roman, and are obtained by using the arcs of a circle. In some cases, Greek mouldings are used, the outlines of which follow the curves of conic sections; but the majority of shapes are arcs of circles. In

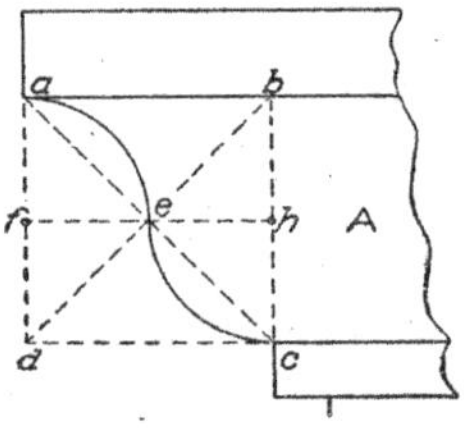

Fig. 268.

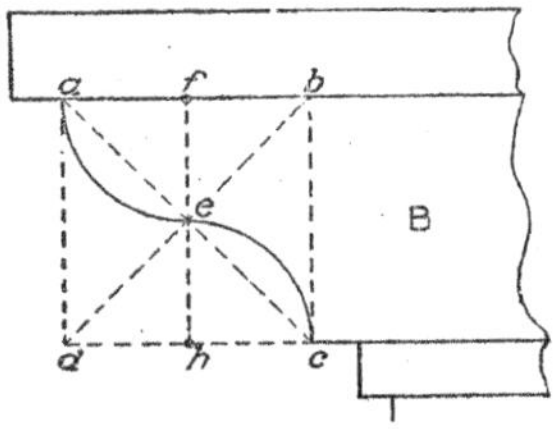

Fig. 269.

Figs. 268 to 272 inclusive, the student is given a few simple lessons on Roman mouldings, which should be carefully followed. As all pattern-cutters are required to draw their full-size details in the shop from small-scale drawings furnished by the architect, it follows that they must understand how to draw the moulds with skill and ease; other-

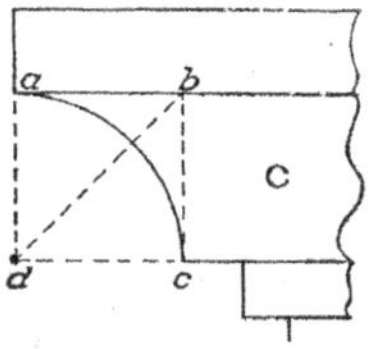

Fig. 270.

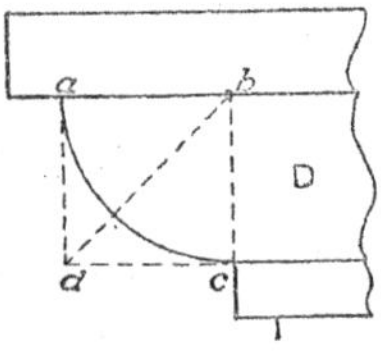

Fig. 271.

wise freehand curves are made, which lack proportion and beauty.

In Fig. 268, A shows the mould known as the *cyma recta*, known in the shop as the *ogee*, which is drawn as follows:

Complete a square *a b c d*; draw the two diagonals *a c* and *b d*, intersecting each other at *e*. Through *e*, draw a horizontal line intersecting *a d* at *f* and *b c* at *h*. Then, with *f* and *h* as centers, draw respectively the two quarter-circles *a e* and *e c*.

In Fig. 269, B shows the *cyma reversa,* known in the shop as the *ogee,* reversed. Complete a square *a b c d,* and draw the two diagonals *b d* and *a c* intersecting at *e*; through *e,* draw a vertical line intersecting *a b* at *f* and *c d* at *h,* which points are the respective centers for the arcs *a e* and *e c.*

C in Fig. 270 shows the *cavetto,* called the *cove* in the shop, which is drawn by completing a square *a b c d.* Draw the diagonal *b d* at 45°, which proves the square; and, using *d* as a center, draw the quarter-circle *a c.*

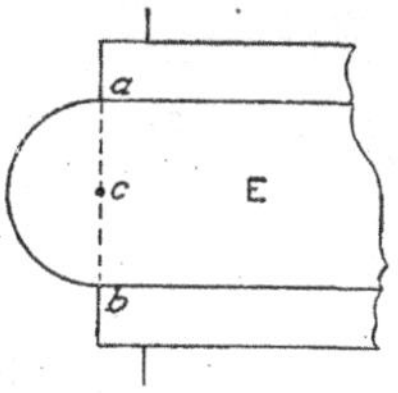

Fig. 272.

In Fig. 271, D represents the *ovolo* or *echinus,* known in the shop as the *quarter-round,* which is constructed similarly to C in Fig. 270, with the exception that *b* in Fig. 271 is used to obtain the curve *a c.*

E in Fig. 272 is known as the *torus,* known in the shop as a *bead-mould.* A given distance *a b* is bisected, thus obtaining *c,* which is the center with which to describe the semicircle *a b.*

All of these profiles should be drawn by the student to any desired scale for practice. In preparing mouldings from sheet metal,

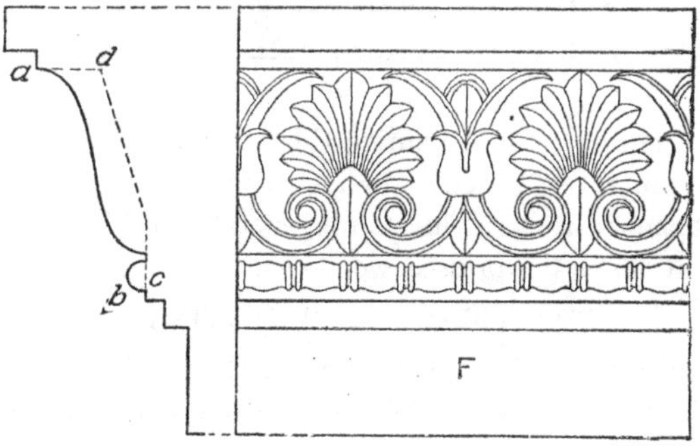

Fig. 273.

it is sometimes required that enrichments are added in the ogee, cove, and bead. In that case the mould must be bent to receive these enrichments, which are usually obtained from dealers in stamped or pressed sheet-metal work. Thus, in Fig. 273, F represents a front view of a crown mould whose ogee is enriched. the section of the enrichment being indicated by *a b* in the section, in which the dotted line *d c* shows the body of the sheet-metal moulding bent to receive the pressed work. In Fig 274, H represents part of a bed-mould in which

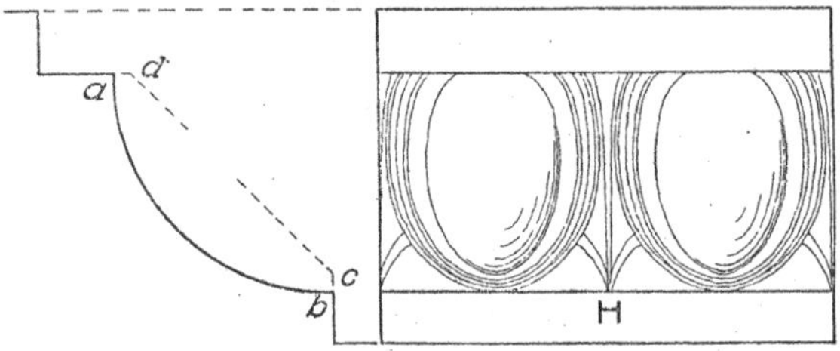

Fig. 274.

egg-and-dart enrichments are placed. In this case the body of the mould is bent as shown by *c d* in the section, after which the egg-and-dart is soldered or riveted in position. J in Fig. 275 represents part

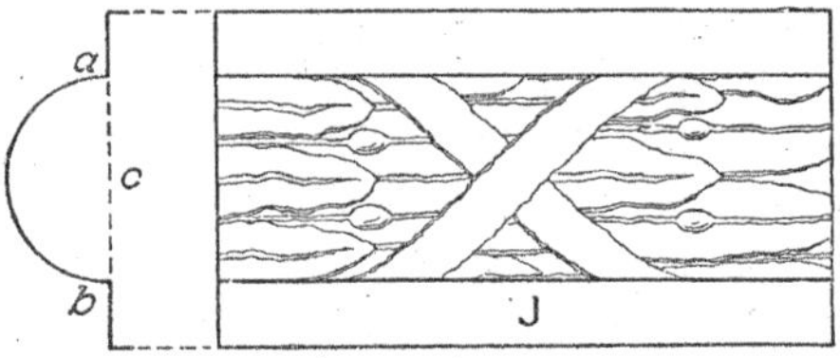

Fig. 275.

of a foot-mould on which an enriched bead is fastened. The body of the mould would be formed as indicated by *c* in the section, and the bead *a b* fastened to it. This same general method is employed, no matter what shape the pressed work has.

PRACTICAL MITER CUTTING

Under this heading come the practical shop problems. The problems which will follow should be drawn to any desired scale by the student, developed, and bent from stiff cardboard to prove the accuracy of the pattern. If the student cannot use the small brake in the shop and test his patterns cut from metal, he can use the dull blade of a table knife, over which the bends can be made, when using cardboard patterns. This at once proves interesting and instructive. Should there be any problem which is not clear, he should write at once for further information; or, should any problem arise on which he desires information, the School will inform him which problem in his text-books contains similar principles, or will prepare such a problem for him.

The first problem will be to obtain the development of a square return miter, such as would occur when a moulding had to return around the corner of a building, as shown in Fig. 276. In Fig. 277 are shown two methods of obtaining the pattern. The first method which will be described is the "long" method, in which are set forth all the principles applicable to obtaining patterns for mouldings, no matter what angle the plan may have. The second method is the "short"

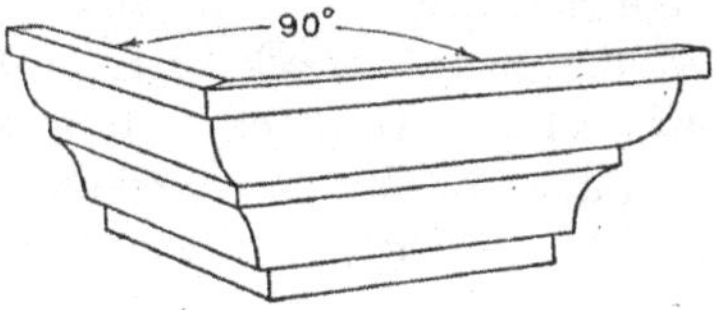

Fig. 276.

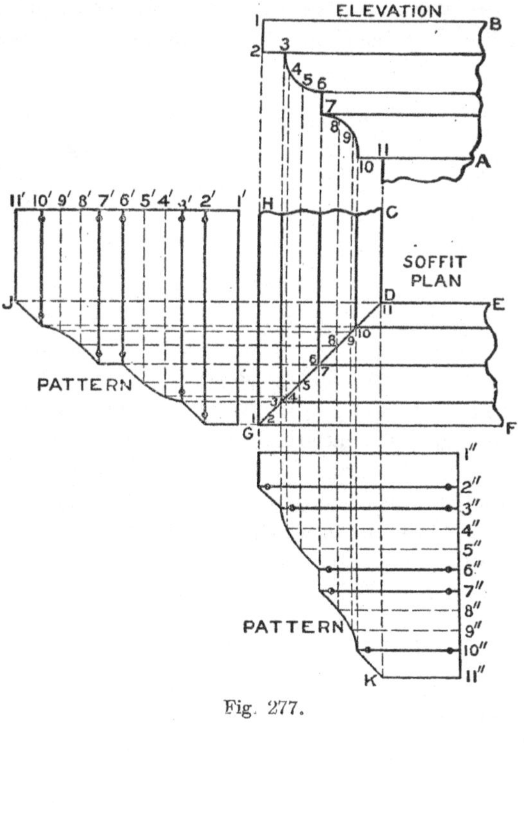

Fig. 277.

rule generally employed in the shop, which, however, can be used only when the angle H G F in plan is 90°, or a right angle.

To obtain the pattern by the first method, proceed as follows: First, draw the elevation of the mould as shown by 1, B, A, 11, drawing the coves by the rule previously given. Divide the curves into equal spaces; and number these, including the corners of the fillets as shown by the small figures 1 to 11. In its proper position below the elevation, draw the soffit plan as shown by C D E F G H. Bisect the angle H G F by the line G D, which is drawn at an angle of 45°. From the various intersections in the elevation, drop lines intersecting the miter-line as shown. At right angles to H G, draw the stretchout line 1′ 11′, upon which place the stretchout of the mould 1 11 in elevation, as shown by similar figures on the line 1′ 11′. At right angles to 1′ 11′, and from the numbered points thereon, draw lines, which intersect by lines drawn at right angles to H G from similarly numbered intersections on the miter-line G D. Trace a line through the intersections

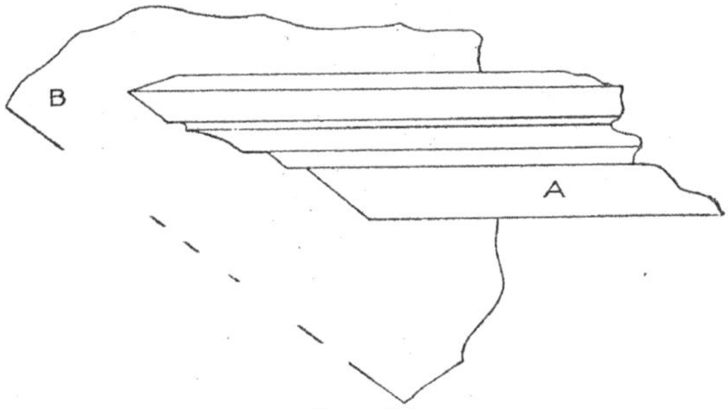

Fig. 278.

thus obtained, as shown by J G. Then will 1′ G J 11′ be the desired pattern. This gives the pattern by using the miter-line in plan.

In developing the pattern by the short method, on the other hand, the plan is not required. At right angles to 1 B in elevation, draw the stretchout line 1″ 11″, upon which place the stretchout of the profile 1 11 in elevation, as shown by similar figures on 1″ 11″, at right angles to which draw lines through the numbered points as shown, which intersect by lines drawn at right angles to 1 B from similarly numbered intersections in the profile in elevation. Trace a line through points thus obtained, as shown by G K. Then will G 1″ 11″ K be similar to J G 1′ 11′ obtained from the plan.

In Fig. 278 is shown a horizontal moulding butting against a plane surface oblique in elevation. A miter cut of this kind would be required when the return moulding of a dormer window would butt against a mansard or other pitched roof. In this case we assume A to be the return butting against the pitched roof B. The method of

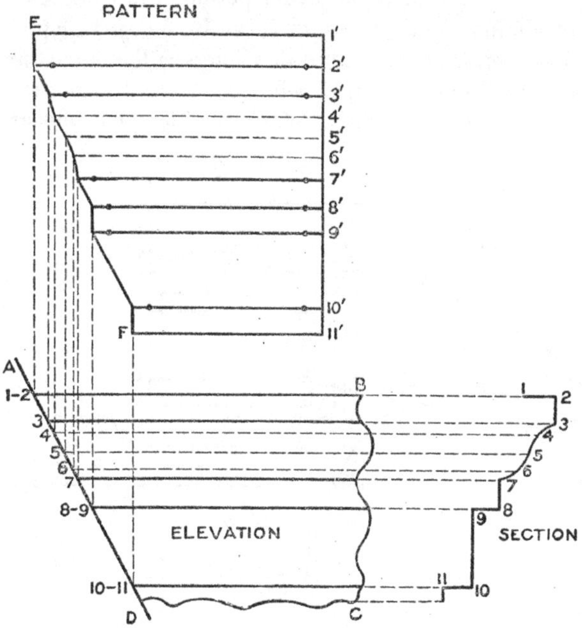

Fig. 279.

obtaining a pattern of this kind is shown in Fig. 279. Let A B C D represent the elevation of the return, A D representing the pitch of the roof. In its proper position as shown, draw the section 1 11, which divide into equal spaces as shown, and from which, parallel to A B, draw lines intersecting the slant line A D from 1 to 11, as shown. At right angles to A B erect the stretchout line 1′ 11′, upon which place the stretchout of the section as shown by similar figures on 1′ 11′. At right angles to 1′ 11′, and through the numbered points thereon, draw lines, which intersect by lines drawn at right angles to A B from similarly numbered intersections on the slant line A D. Through the various intersections thus obtained, draw E F. Then will E F 11′ 1′ be the desired pattern.

It is sometimes the case that the roof against which the moulding butts, has a curved surface either concave or convex, as shown by B C in Fig. 280, which surface is convex. Complete the elevation of the moulding, as D E; and in its proper position draw the section 1 9, which divide into equal spaces as shown by the small figures, from which draw horizontal lines until they intersect the curved line B C, which is struck from the center point A. At right angles to the line of the moulding erect the line 1′ 9′, upon which place the stretchout

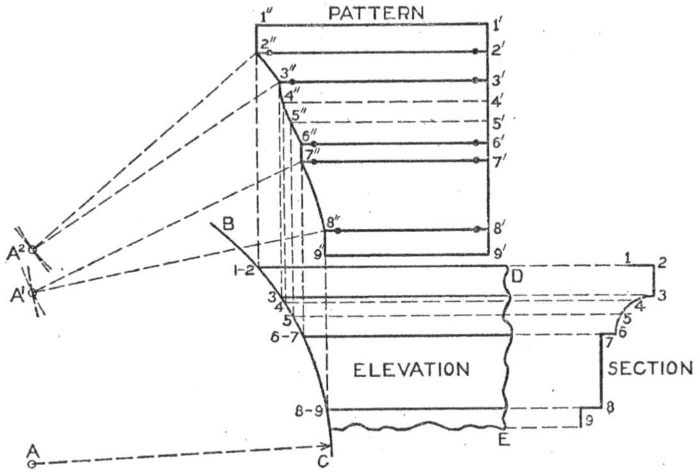

Fig. 280.

of the section, as shown by the figures on the stretchout line. Through the numbered points, at right angles to 1′ 9′, draw lines, which intersect by lines drawn at right angles to 2 D from similarly numbered intersections on the curve B C, thus resulting in the intersections 1″ to 9″ in the pattern, as shown. The arcs 2″ 3″ and 7″ 8″ are simply reproductions of the arcs 2 3 and 7 9 on B C. These arcs can be traced by any convenient method; or, if the radius A C is not too long to make it inconvenient to use, the arcs in the pattern may be obtained as follows: Using A C as radius, and 7″ and 8″ as centers, describe arcs intersecting each other at A^1; in similar manner, using 2″ and 3″ as centers, and with the same radius, describe arcs intersecting each other at A^2. With the same radius, and with A^1 and A^2 as centers, draw the arcs 8″ 7″ and 3″ 2″ respectively. Trace a line through the other various intersections as shown. Then will 1′ 1″ 9″ 9′ be the desired pattern.

In Fig. 281 is shown an elevation of an oblong or rectangular panel for which a miter-cut is desired on the line *a b*—known as a "panel" or "face" miter. The rule to apply in obtaining this pattern is shown in Fig. 282. A shows the part elevation of the panel; *a b* and *c d*, the miter-lines drawn at angles of 45°. In its proper position with the lines of the moulding, draw the profile B, the curve or mould of which divide into equal spaces, as shown by the figures 1 to 7; and from the points thus obtained, parallel to 1 *b*, draw lines intersecting the miter-line *a b* as shown. From these intersections, parallel to *b d*, draw lines intersecting also *c d*. At right angles to *b d* draw the stretchout line 1′ 7′, upon which place the stretchout of the profile B. At right angles to 1′ 7′, and through the numbered points of division, draw lines, which intersect by lines drawn at right angles to *b d* from similarly numbered intersections on the miter-lines *a b* and *c d*. Trace lines through the various points of intersection in the pattern as shown. Then will C D E F be the required cut for the ends of the panel.

Fig. 281.

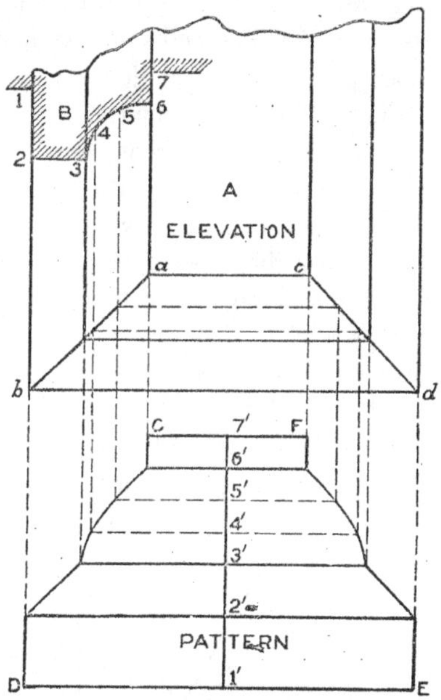

Fig. 282.

The same miter-cuts would be employed for the long side *a c* in

Fig. 281, it being necessary only to make D E in Fig. 282 that length when laying out the patttern on the sheet metal.

Where the miter-cut is required for a panel whose angles are other than right angles, as, for example, a triangular panel as shown in Fig. 283, then proceed as shown in Fig. 284. First draw the elevation of the triangular panel as shown by A B C, the three sides in the case being equal. Bisect each of the angles A, B, and C, thus obtaining the miter-lines A *c*, B *b*, and C *a*. In line with the elevation, place in its proper position the profile E, which divide into equal spaces as shown; and from the numbered division points, parallel to A C, draw lines cutting the miter-line C *a*. From these intersections, parallel to C B, draw lines intersecting the miter-line *b* B. At right angles to C B draw the stretchout line 1′ 7′, upon which place the stretchout of the profile E. Through the numbered points of division and at right angles to 1′ 7′, draw lines as shown, which intersect by lines drawn at right angles to C B from intersections of similar numbers on the miter-lines *a* C and *b* B. Through the points thus obtained, trace the pattern F G H I.

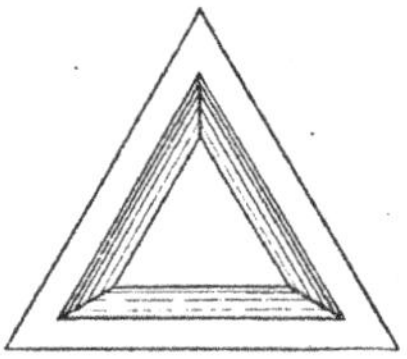

Fig. 283.

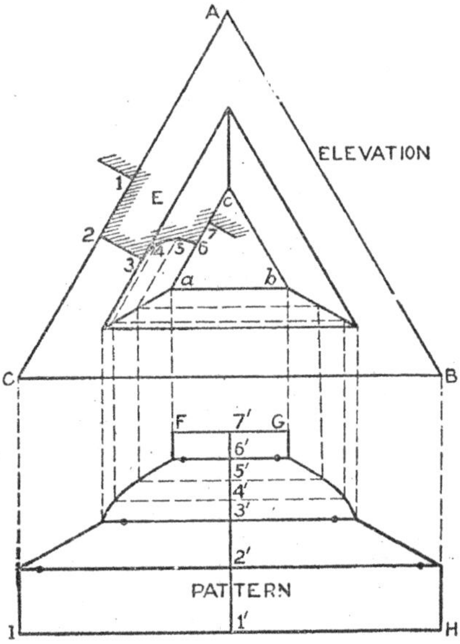

Fig. 284.

It makes no difference what shape or angle the panel may have; the principles above explained are applicable to any case.

In ornamental cornice work, it often happens that tapering moulded panels are used, a plan and elevation of which are shown in Fig. 285.

By referring to the plan, it will be seen that the four parts *b a*, *a b′*, *b′ a′*, and *a′ b* are symmetrical; therefore, in practice, it is necessary only to draw the one-quarter plan, as shown in Fig. 286, and omit the elevation, since the height *d e* (Fig. 285) is known. Thus, in Fig. 286, draw the quarter-plan of the panel, no matter what is its shape, as shown

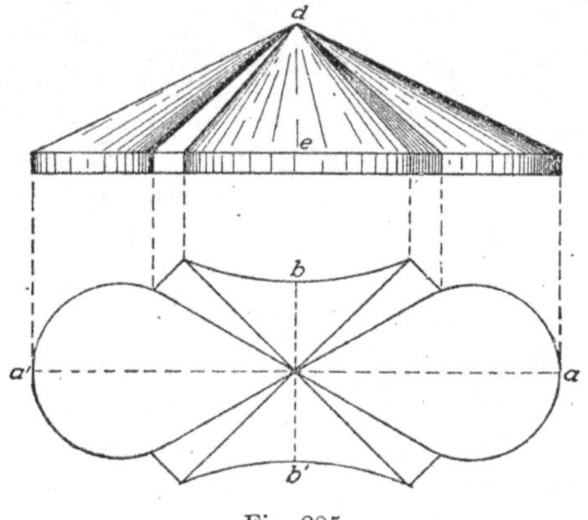

Fig. 285.

by *a* 1 5 6 9. Divide the curves from 1 5 and 6 9 into equa- spaces, indicated respectively by 1, 2, 3, 4, and 5, and 6, 7, 8, and 9. From these points, draw lines to the apex *a*. As the pattern will be developed by triangulation, a set of triangles will be required, as shown in

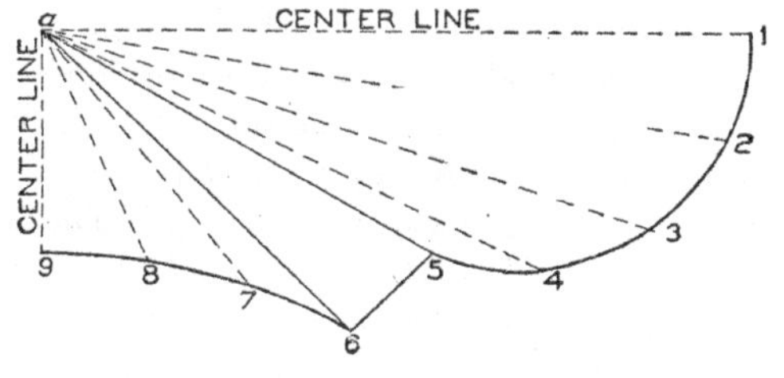

Fig. 286.

Fig. 287, for which proceed as follows: Draw any horizontal line, as *a* 1; and from *a* erect the perpendicular *a a′* equal to the height the panel is to have. Now take the lengths of the various lines in Fig. 286 from *a* to 1, *a* to 2, *a* to 3, etc., to *a* to 9, and place them on the line *a* 1 in Fig. 287, as shown by similar numbers. Then using as radii the various lengths a' 1, a' 2, a' 3, etc., to a' 9, and with any point, as a' in Fig. 288 as center, describe the various arcs shown from 1 to 9. From any point on the arc 1 draw a line to a'. Set the dividers equal to the spaces contained in the curve 1 5 in Fig. 286; and, starting from 1 in Fig. 288 step from one arc to another having similar numbers, as shown from 1 to 5. In similar manner, take the distance from 5 to 6 and the spaces in the curve 6 9 in Fig. 286, and place them on corresponding arcs in Fig. 288, stepping from one arc to the other, resulting in the points 5 to 9. Trace a line through the points thus obtained. Then will a' 1 5 6 9 a' be the quarter-pattern, which can be joined in one-half or whole pattern as desired.

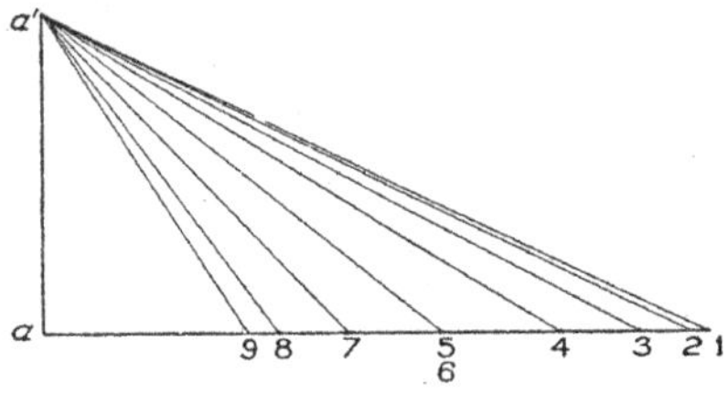

Fig. 287.

In Fig. 289 is shown a perspective of a moulding which miters at an angle other than a right angle. This occurs when a moulding is required for over a bay window or other structure whose angles vary.

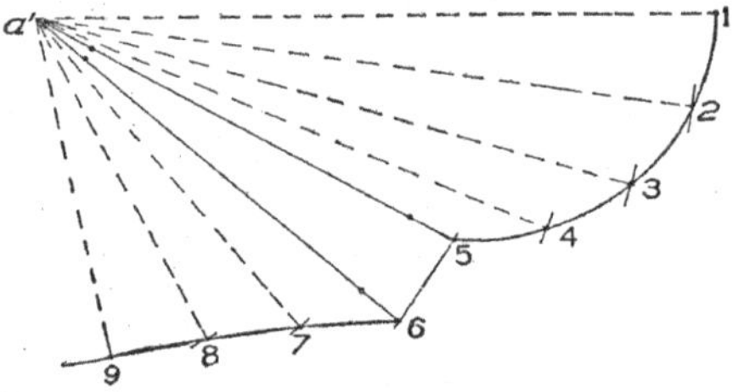

Fig. 288.

The rule given in Fig. 290 is applicable to any angle or profile. First draw a section or an elevation of the moulding as shown by A B 14 1. Directly below the moulding, from its extreme point, as 2 3, draw a plan of the desired angle as shown by C 2 D. Bisect this angle by using 2 as center and, with any radius, describing an arc meeting the sides of the angle at C and E. With the same or any other radius, and with C and E as centers, describe arcs intersecting each other in F. From the corner 2, draw a line through F. Then will 2 H be the

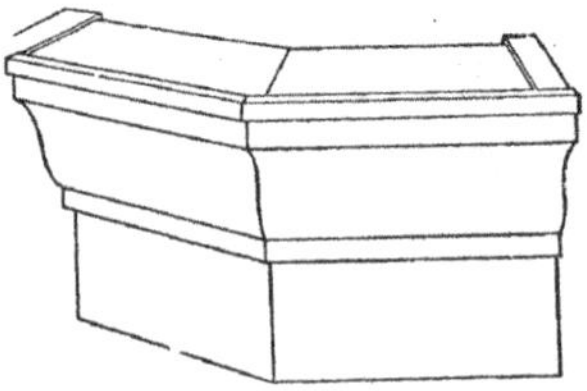
Fig. 289.

miter-line, or the line bisecting the angle C 2 D. Now divide the profile 1 14 into equal spaces as shown by the figures, and from the points thus obtained drop vertical lines intersecting the miter-line 2

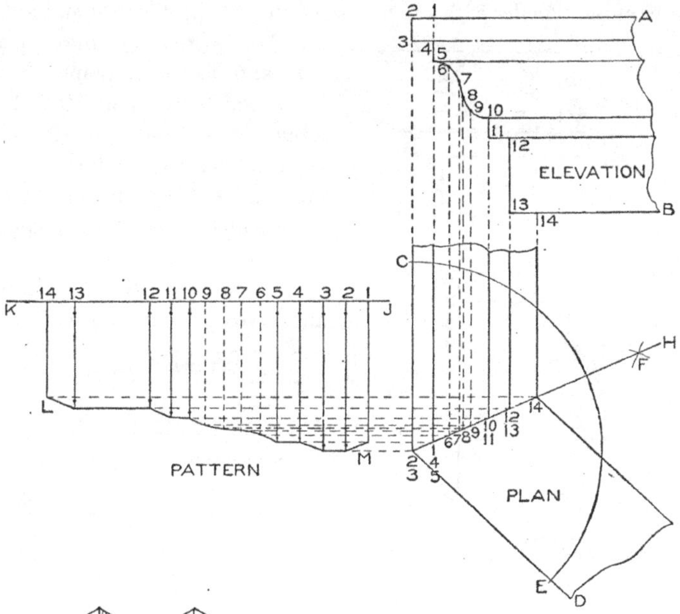

Fig. 290.

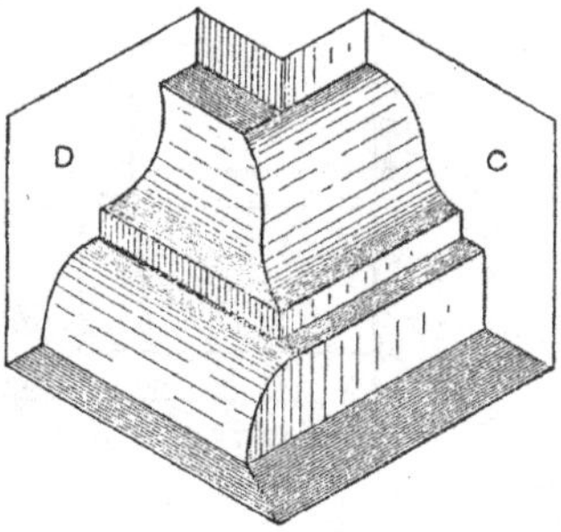

Fig. 291.

H in plan from 1 to 14 as shown. At right angles to C 2, draw the line J K, upon which place the stretchout of the profile in elevation as shown by similar figures on the stretchout line, through which drop lines perpendicular to J K, which intersect with lines drawn parallel to J K from similarly numbered points of intersection on the miter-line 2 H. Trace a line as shown by L M, which is the miter-cut desired.

When two mouldings having different profiles are required to miter together as shown in Fig. 291, where C miters at right angles with D, two distinct operations are necessary, which are clearly shown in Figs. 292 and 293. The first operation is shown in Fig. 292, in which C represents the elevation of an ogee moulding which is to miter at right angles with a moulding of different profile as shown at D. Divide the profile C into equal spaces, from which points draw horizontal lines intersecting the moulding D from 1′ to 10′. At right angles to the line of the moulding C, draw the line A B, upon which place the stretchout of the profile C as shown by similar figures on A B. At right angles to A B, and through the points indicated by the figures, draw lines, which intersect with lines drawn parallel to A B from similarly numbered intersections in the profile D. Trace a line through the points thus obtained, as shown by E H. Then will E F G H be the pattern for C in elevation.

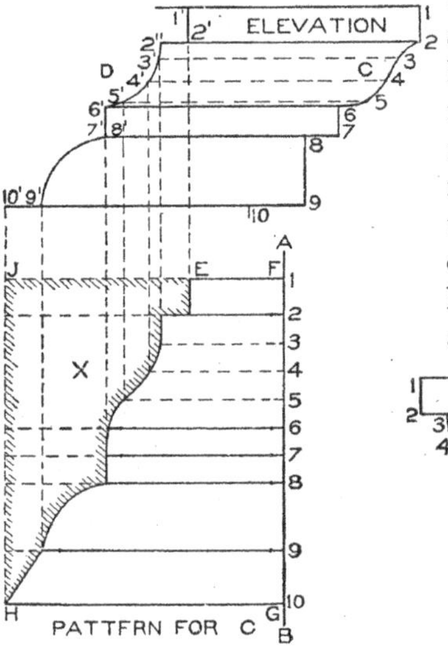

Fig. 292.

To obtain the pattern for D, draw the elevation of D (Fig. 293), which is to miter at right angles with a moulding whose profile is C. Proceed in precisely the same manner as explained in connection with Fig. 292. Divide the profile D in Fig. 293 into equal parts, as shown, from which draw horizontal lines cutting the profile C. At right angles

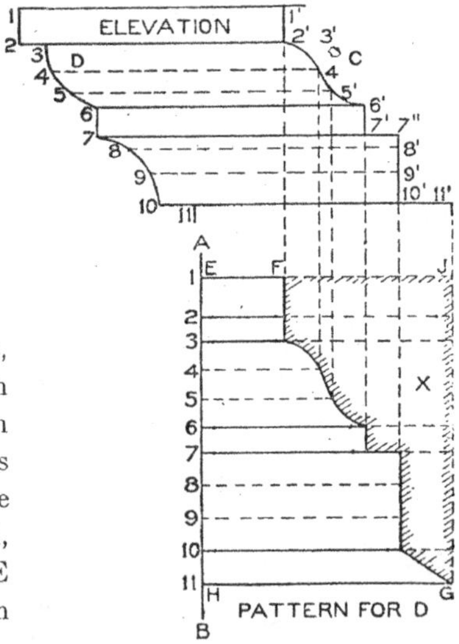

Fig. 293.

to the lines of the moulding D, draw the stretchout line A B, upon which place the stretchout of the profile D. At right angles to A B, and through the numbered points of division, draw lines as shown, which intersect by lines drawn parallel to A B from similarly numbered intersections in the profile C. Through these points of intersection draw F G. Then will E F G H be the desired pattern for D.

It should be understood that when the patterns in Figs. 292 and 293 are formed and joined together, they will form an inside miter, as is shown in Fig. 291. If, however, an outside miter were required, it would be necessary only to use the reverse cuts of the patterns in Figs. 292 and 293, as shown by E J H in Fig. 292 for the mould C, and F J G in Fig. 293 for the mould D.

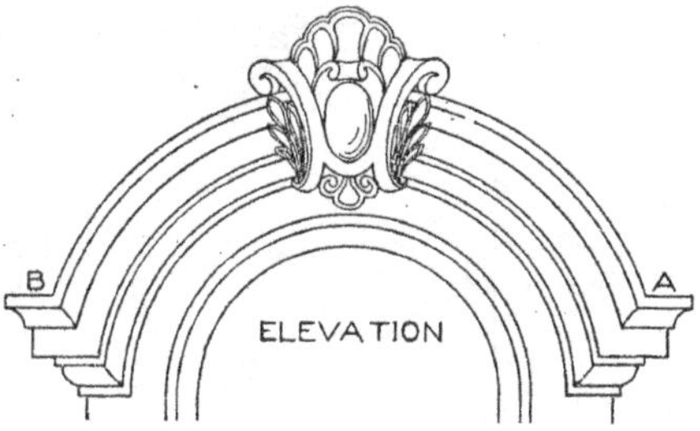

Fig. 294.

When joining a curved moulding with a straight moulding in either plan or elevation even though the curved or straight mouldings each have the same profile, it is necessary to establish the true miter-line before the pattern can be correctly developed, an example being given in Fig. 294, which shows an elevation of a curved moulding which is intersected by the horizontal mouldings A B. The method of obtaining this miter-line, also the pattern for the horizontal pieces, is clearly shown in Fig. 295. First draw the profile which the horizontal moulding is to have, as 1 10. Let the distance 9 B be established. Then, with C on the center line as center, and A C as radius, describe the arc B A. From any point on the line 9 B, as *a*, erect the vertical line *a b*. Through the various divisions in the profile 1 10, draw horizontal lines intersecting the vertical line *a b* from 1 to 10 as shown. From the center C, draw any radial line, as C *d*, cutting the arc B A at *e*. Now take the various divisions on *a b*, and place them from *e* to *d* as shown by points 1′ to 10′. Then, using C as center, with radii determined by the various points on *e d*, draw arcs intersecting horizontal lines of similar numbers drawn through the divisions on *a b*. Through these points of intersection, draw the miter-line shown. The student will note that this line is irregular.

Having obtained the miter-line, the pattern is obtained for the horizontal moulding by drawing the stretchout line E F at right angles to 9 B. On E F lay off the stretchout of the profile 1 10; and through the numbered points and at right angles to E F, draw horizontal lines, which intersect with lines drawn at right angles to 9 B from similarly numbered intersections in the miter-line determined by horizontal lines already drawn through the vertical line *a b*. Trace a line through the points thus obtained, as shown by H I J K, which is the desired pattern.

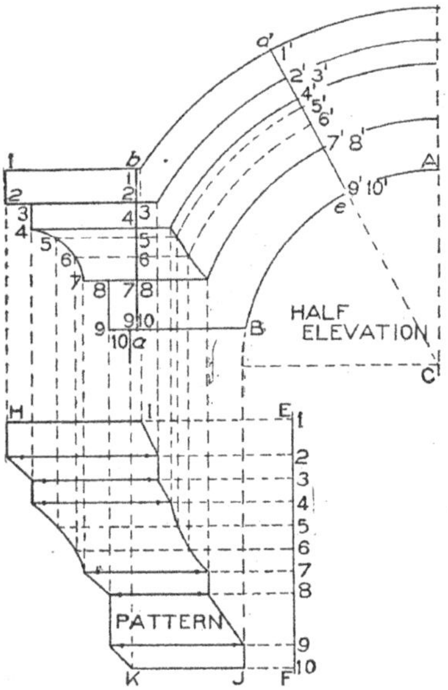

Fig. 295.

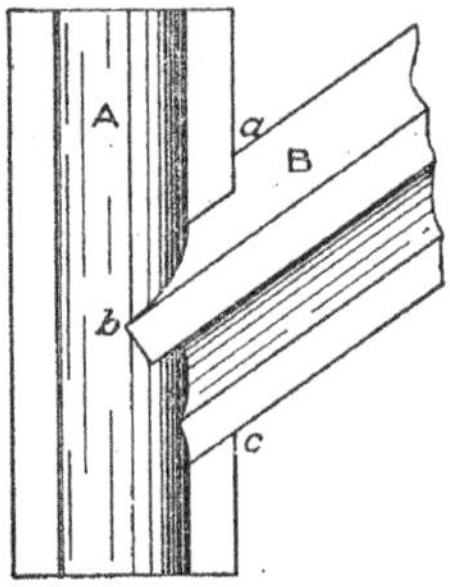

Fig. 296.

In Fig. 296 is shown a shaded view of a gable moulding intersecting a pilaster, the gable moulding B cutting against the vertical pilaster A, the joint-line being represented by *a b c*. To obtain this joint-line, without which the pattern for the gable moulding cannot be developed, an operation in projection is required. This is explained in Fig. 297, in which B C D shows the plan of the pilaster shown in elevation by E. In its proper position in plan, place the profile of the gable moulding, as shown by A, which divide into equal spaces as shown by the figures 1 to 8, through which draw horizontal lines intersecting the plan of the pilaster B C D as shown by similar figures. For convenience in projecting the various points, and to avoid a confusion of lines, number the intersections between the lines drawn from the profile A through the wash B 2, "7°", "4°", and "3°". At the desired point H in elevation, draw the lower line of the gable moulding, as H F. Take a tracing of the profile A in plan, with all of the various intersections on same, and place it in elevation as shown by A¹, placing the line 1 8 at right angles to H F. Through the various intersections 1, 7°, 4°, 3°, 2, 3, 4, 5, 6, 7, and 8 in A¹, and parallel to F H, draw lines indefinitely, which intersect by lines drawn at right angles to C B in plan from similarly numbered intersections in the pilaster C D B, thus obtaining the points of intersection 1ˣ to 8ˣ in elevation.

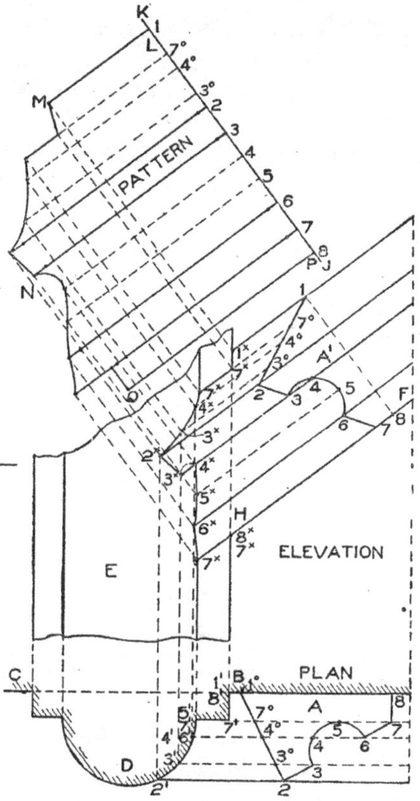

Fig. 297.

For the pattern, proceed as follows: At right angles to H F, draw the stretchout line J K, upon which place the stretchout of the profile A or A¹, with all the points of intersection on the wash 1 2. At right angles to J K, and through the numbered points, draw lines as shown, which intersect by lines drawn at right angles to H F from similarly numbered intersections in the joint-line 1ˣ 8ˣ Through the points thus obtained, trace the miter-cut M N O. Then will L M N O P be the pattern for the gable moulding.

In Fig. 298 are shown gable mouldings mitering upon a wash. The mouldings A A intersect at any desired angle the wash B. In this case, as in the preceding problem, an operation in projection must be gone through, before the pattern can be obtained. This is clearly shown in Fig. 299. Draw the section of the horizontal moulding B¹ with the wash *a b*. From this section project lines, and draw the part elevation D C. Knowing the bevel the gable is to have, draw C B, in this case the top line of the moulding. Draw a section of the gable mould, as A, which divide into equal parts as shown from 1 to 8; and through the point of division draw lines parallel to B C, indefinitely, as shown. Take a tracing of the profile A, and place it in section as shown by A¹. Divide A into the same

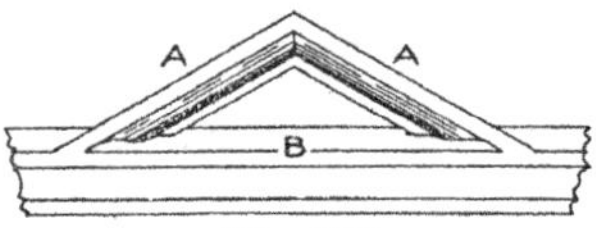

Fig. 298.

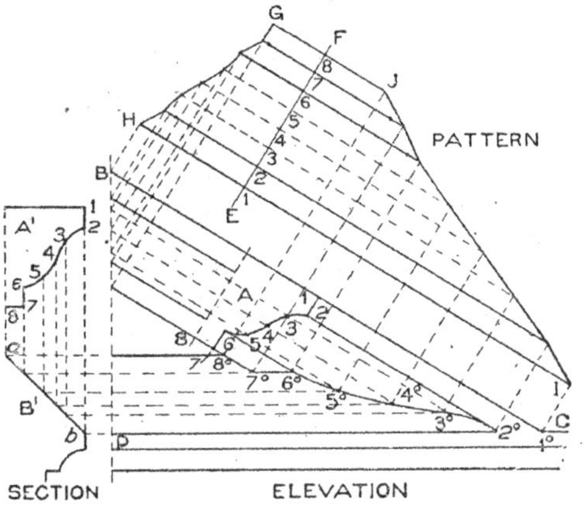

Fig. 299.

number of spaces as A; and from the various divisions in A¹ drop vertical lines intersecting the wash *a b* as shown, from which points draw horizontal lines intersecting lines drawn parallel to B C through similarly numbered points in A, at 1° to 8°. Trace a line through these intersections as shown, which represents the miter-line or line of joint in elevation.

For the pattern, draw any line as E F, at right angles to B C, upon which place the stretchout of the profile A, as shown by similar figures on the stretchout line E F. Through the numbered points of division and at right angles to E F, draw lines as shown, which intersect by lines drawn at right angles to B C from similarly numbered intersections on 1° 8° and on the vertical line B D. A line traced through points thus obtained, as shown by G H I J, will be the desired pattern.

In Fig. 300 is shown a front view of a turret on which four gables are to be placed, as shown by A A; also the roofs over same, as shown by B B. The problem consists in obtaining the developments of the gable mouldings on a square turret. In developing this pattern, the half-elevation only is required, as shown in Fig. 301, in which first draw the center line E F; then establish the half-width of the turret, as C D, and draw the rake B C. At right angles to the line B C, and in its proper position as shown, draw the profile A, which divide into equal spaces as shown by the figures 1 to 6, through which, parallel to B C, draw lines intersecting the center line F E as shown; and extend the lines below C, indefinitely. Now take a tracing of the profile A, and place it in position as shown by A^1, being careful to have it spaced in the same number of divisions, as shown from 1 to 6, through which, parallel to D C, erect lines intersecting similarly numbered lines drawn through the profile A, thus obtaining the intersections 1° to 6°, through which a line is traced, which represents the line of joint at the lower end between the two gables.

Fig. 300.

For the pattern, take a stretchout of A, and place it on the line J K drawn at right angles to B C, as shown by the figures 1 to 6 on J K. At right angles to J K, and through these points of division, draw lines, which intersect by lines drawn from similarly numbered intersections on F B and 1° 6°. Trace a line through the points thus obtained, as shown by F° B° C° 6°, which is the desired pattern, of which eight are required to complete the turret, four formed right and four left.

If the roof shown by B in Fig. 300 is desired to be added to the pattern in Fig. 301, then, at right angles to F° 6°, draw the line F° F^1 equal to F H in the half-elevation, and draw a line from F^1 to 6° in the pattern.

In Fig. 302 is shown front view of an angular pediment with horizontal returns at bottom A and top B. In this problem, as in others which will follow, a change of profile is necessary before the correct pattern for the returns can be developed. In other words, a new profile must be developed from the given or normal profile before the patterns for the required parts can be developed. It should be understood that all given profiles are always divided into equal spaces; therefore the modified profiles will contain unequal spaces, each one of

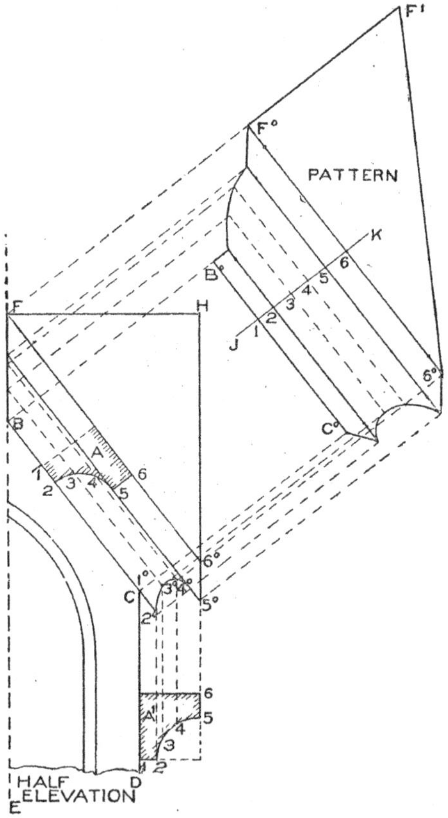

Fig. 301.

which must be carried separately onto the stretchout line. Bearing this in mind, we shall proceed to obtain the modified or changed profiles and patterns for the horizontal returns at top and foot of a gable moulding, as at B and A in Fig. 302, the given profile to be placed in the gable moulding C. In Fig. 303, let C represent the gable moulding placed at its proper angle with the horizontal moulding G H. Assuming that 6^x 6° is the proper angle, place the given profile A at right angles to the rake, as shown; and divide same into equal spaces as shown from 1 to 10, through which points, parallel to 6^x 6°, draw lines towards the top and bottom of the raking moulding. Assuming that the length 6^x 6° is correct, take a tracing of the profile A, and place it in a vertical position below at A^1 and above at A^2, being careful to have the points 6 and 6 in the profiles directly in a vertical position below the points 6^x and 6°, as shown. From the various intersections in the profiles A^1 and A^2 (which must contain the same number of spaces as the given profile A), erect vertical lines intersecting lines drawn through the profile A, as shown at the lower end from 1^x to 10^x, and at the upper end from 1° to 10°. Trace a line through the points thus obtained. Then will 1^x 10^x be the modified profile for the lower horizontal return, and 1° 10° the modified profile for the upper horizontal return.

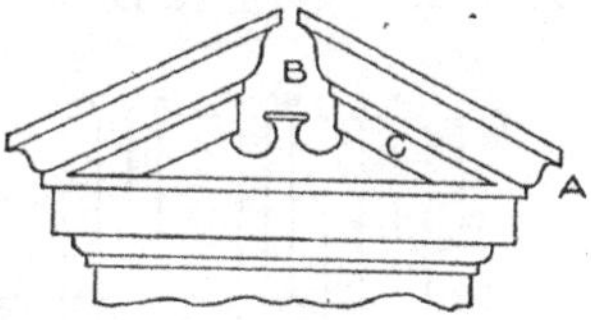

Fig. 303.

Note the difference in the shapes and spaces between these two modified profiles and the given profile A. It will be noticed that a portion of the gable moulding miters on the horizontal moulding G H from 6^x to $10'$.

For the pattern for the gable moulding, proceed as follows: At right angles to E F, draw the stretchout line J K, upon which place the stretchout of the given profile A, as shown by the figures 1 to 10 on J K. Through these figures, at right angles to J K, draw lines as shown, which intersect with lines drawn at right angles to E F from similarly numbered intersections in 1° 10° at the top and 1^x 6^x $10'$ at the lower end. Trace a line through the intersections thus obtained. Then will L M N O be the pattern for C.

For the pattern for the horizontal return at the top, draw a side view as shown at B, making P R the desired projection, and the profile 1 10 on B, with its various intersections, an exact reproduction of 1° 10° in the elevation. Extend the line R T as R S; and, starting from 10, lay off the stretchout of the profile in B as shown by the figures 1 to 10 on R S, being careful to measure each space separately. At right angles to R S draw the usual measuring lines, which intersect by lines drawn parallel to S R from similarly numbered points in the profile in B. Trace a line through points thus obtained. Then will U V 10 1 be the pattern for the return B.

In similar manner, draw the side view of the lower horizontal return as shown at D, making the projection W 10 equal to P R

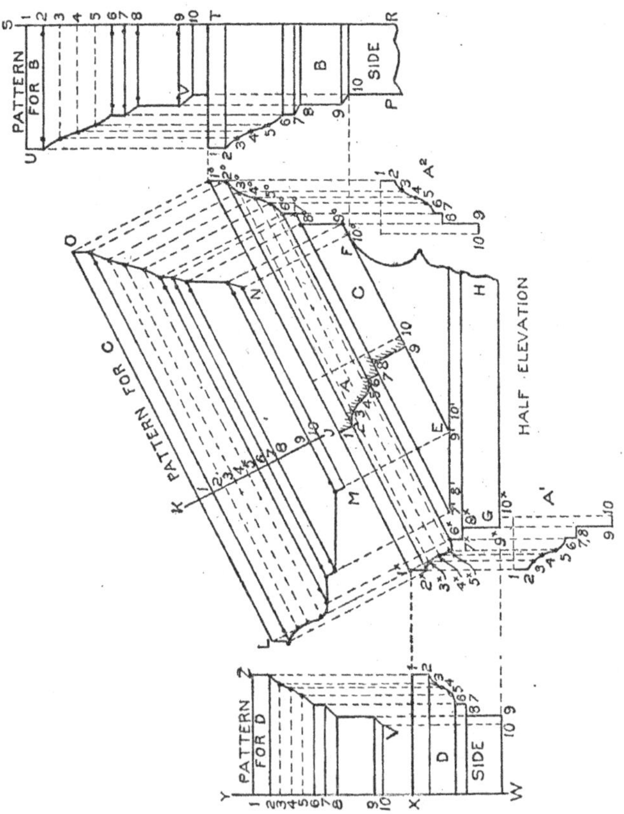

Fig. 303.

in B. The profile shown from 1 to 10 in D, with all its divisions, is to be an exact reproduction of the profile 1^x to 10^x in elevation. Extend the line W X as X Y, upon which lay off the stretchout of the profile 1 10 in D, being careful that each space is measured separately, as they are all unequal. Through the figures on X Y draw lines as shown, which intersect by lines drawn parallel to W Y from the various intersections in the profile in the side D. A line traced through points thus obtained, as shown by Z V, will be the desired cut, and 1 Z V 10 the pattern for the return D.

In Fig. 304 is shown a front view of a segmental pediment with upper and lower horizontal returns. This presents a problem of obtaining the pattern for horizontal returns at top and foot of a segmental pediment, shown respectively at A and B, the given profile to be placed in C. The principles used in obtaining these patterns are similar to those in the preceding problem, the only difference being that the moulding is curved in elevation. In Fig. 305 the true method is clearly given. First draw the center line B D, through which draw the horizon-

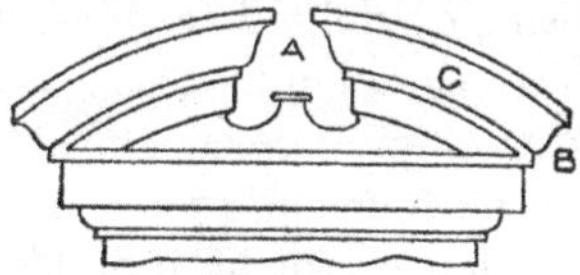

Fig. 304.

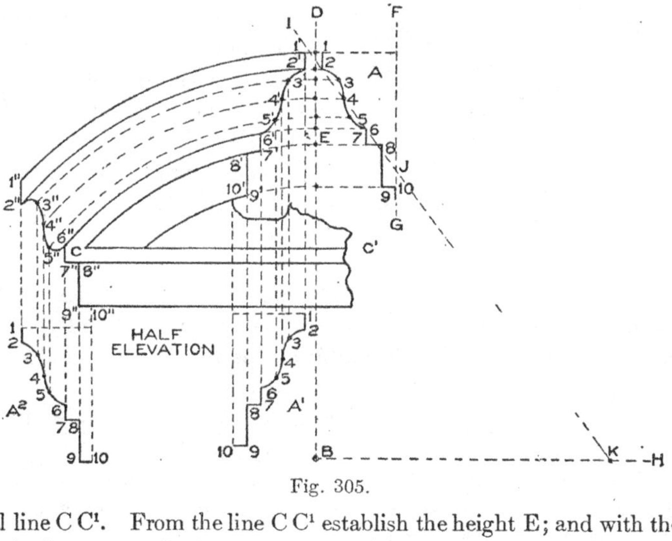

Fig. 305.

tal line C C¹. From the line C C¹ establish the height E; and with the desired center, as B, draw the arc E C intersecting the line C¹ C at C. In its proper position on a vertical line F G, parallel to D B, draw the given profile of the curved moulding as shown by A, which divide into equal spaces as shown from 1 to 10. Through these figures, at right angles to F G, draw lines intersecting the center line D B as shown.

Then, using B as center, with radii of various lengths corresponding to the various distances obtained from A, describe arcs as shown, extending them indefinitely below the foot of the pediment. The point C or 6″ being established, take a tracing of the profile A, with all the various points of intersection in same, and place it as shown by A^2, being careful to have the point 6 in A^2 come directly below the point 6″ in elevation in a vertical position. Then, from the various intersections in A^2 erect vertical lines intersecting similarly numbered arcs drawn from the profile A. Trace a line as shown from 1″ to 10″, which is the modified profile for the foot of the curved moulding.

Establish at pleasure the point 1′ at the top, and take a tracing of the given profile A, placing it in a vertical position below 1′, as shown by A^1. From the various intersections in A^1 erect vertical lines intersecting similarly numbered arcs as before. Through these intersections, shown from 1′ to 10′, trace the profile shown, which is the modified profile for the top return.

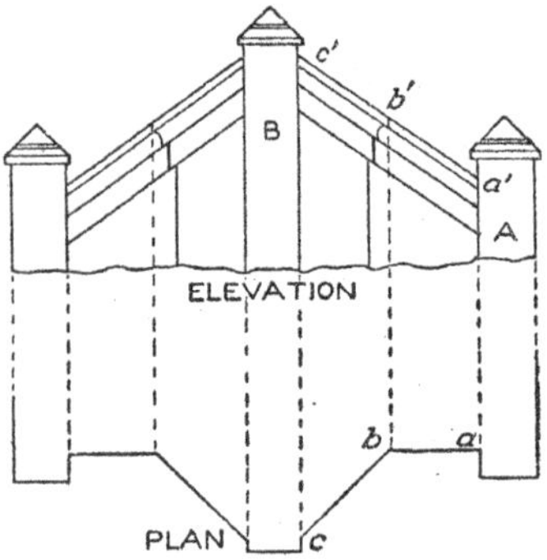

Fig. 306.

The curved moulding shown in elevation can be made either by hand or by machine. The general method of obtaining the blank or pattern for the curved moulding is to average a line through the extreme points of the profile A, as I J, extending it until it intersects a line drawn at right angles to D B from the center B, as B H, at K.

We will not go into any further demonstration about this curved work, as the matter will be taken up at its proper time later on.

To obtain the pattern for the upper and lower return mouldings, proceed in precisely the same manner as explained in connection with returns B and D in Fig. 303.

In Fig. 306 are shown the plan and elevation of a gable moulding in octagon plan. This problem should be carefully followed, as it presents an interesting study in projections; and the principles used in solving this are also applicable to other problems, no matter what angle or pitch the gable has. By referring to the plan, it will be seen that the moulding has an octagon angle in plan *a b c*, while similar points in elevation *a′ b′ c′* run on a rake in one line, the top and foot of the moulding butting against the brick piers B and A.

The method of proceeding with work of this kind is explained in detail in Fig 307, where the principles are thoroughly explained. Let A B C D E represent a plan view of the wall, over which a gable moulding is to be placed, as shown by G H I J, the given profile of the

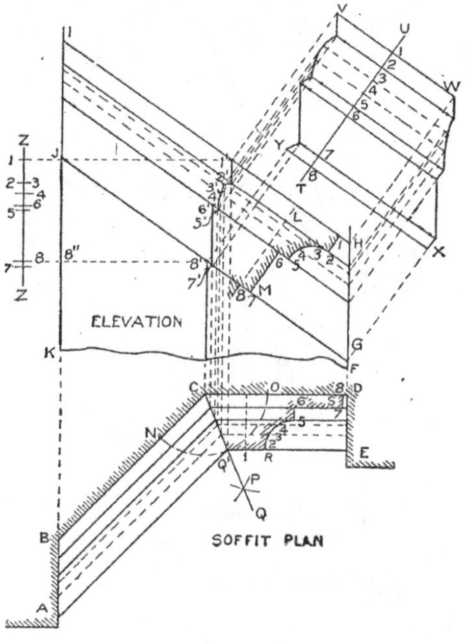

Fig. 307.

moulding being shown by L M. Divide the profile into equal spaces as shown by the figures 1 to 8. Parallel to I H or J G, and through the figures mentioned, draw lines indefinitely as shown. Bisect the angle B C D in plan, and obtain the miter-line as follows: With C as center, and any radius, describe the arc N O. With N and O as centers, and any radius greater than C N or C O, describe arcs intersecting each other at P. From the point C, and through the intersection P, draw the miter-line C Q. Transfer the profile L M in elevation to the position shown by R S in plan, dividing it into the same number of spaces as L M. Through the figures in the profile R S, and parallel to D C, draw lines intersecting the miter-line C Q, as shown. From the intersections on the miter-line, and parallel to C B, draw lines intersecting the surface B A. Now, at right angles to C D in plan, and from the

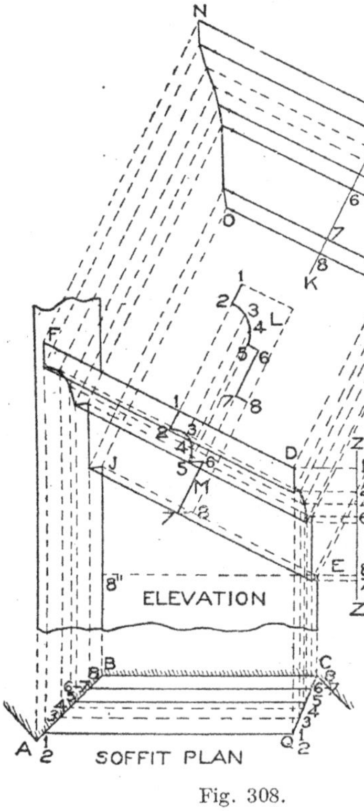

Fig. 308.

intersections on the miter-line C Q, draw vertical lines upward, intersecting lines of similar numbers drawn from points in profile L M in elevation parallel to J G. A line traced through points thus obtained, as shown from 1′ to 8′, will be the miter-line in elevation.

For the pattern for that part of the moulding shown by C D E Q′ in plan, and H G 8′ 1′ in elevation, proceed as follows: At right angles to 1 H in elevation, draw the line T U, upon which place the stretchout of the profile L M, as shown by the figures 1 to 8. At right angles to T U, and through these figures, draw lines, as shown, which intersect with lines of similar numbers drawn at right angles to 1 H from intersections on the miter-line 1′ 8′ and from intersections against the vertical surface H G. Lines traced through points thus obtained, as shown by V W X Y, will be the pattern for that part of the gable shown in plan by C D E Q′ of Fig. 307.

In Fig. 308, on the other hand, the position of the plan is changed, so as to bring the line A Q horizontal. At right angles to B C draw the vertical line C E, on which locate any point, as E. In the same manner, at right angles to C B, draw the vertical line B J indefinitely. From the point E, parallel to B C, draw the line E 8″, intersecting the line J B, as shown. Now take the distance from 8″ to J in elevation, Fig. 307, and set it off from 8″ toward J in Fig. 308. Draw a line from J to E, which will represent the true rake for this portion of the moulding. Now take the various heights shown from 1 to 8 on the line Z Z in elevation in Fig. 307, and place them as shown by Z Z in elevation, Fig 308, being careful to place the point 8 of the line Z Z on the line 8″ E extended. At right angles to Z Z, and from points on same, draw lines, which intersect with lines drawn at right angles to B C from intersections of similar numbers on C Q in plan. A line traced through points thus obtained, as shown by D E in elevation, will be the miter-line on C Q in plan.

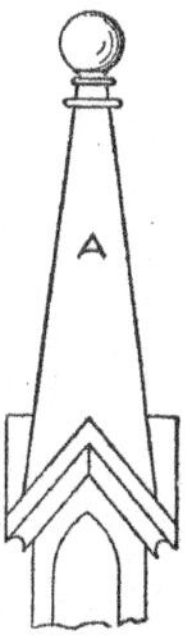

Fig. 309.

From the intersections on the miter-line D E, and parallel to E J, draw lines, which intersect with lines drawn from intersections of similar numbers on A B in plan at right angles to B C. A line traced through points thus obtained, as shown by F J, will be the miter-line or line of joint against the pier shown in plan by B A.

Before obtaining the pattern it will be necessary to obtain a true section or profile at right angles to the moulding F D. To do so, proceed as follows: Transfer the given profile L M in elevation in Fig. 307, with the divisions and figures on same, to a position at right angles to F D of Fig. 308, as shown at L. At right angles to F D, and from the intersections in the profile L, draw lines intersecting those of similar numbers in F D E J. Trace a line through intersections thus obtained, as shown from 1 to 8, thus giving the profile M, or true sections at right angles to F D.

For the pattern, proceed as follows: At right angles to F D, draw the line H K, upon which place the stretchout of the profile M, as shown by the figures. At right angles to H K, and through the figures, draw lines, which intersect with those of similar numbers drawn at

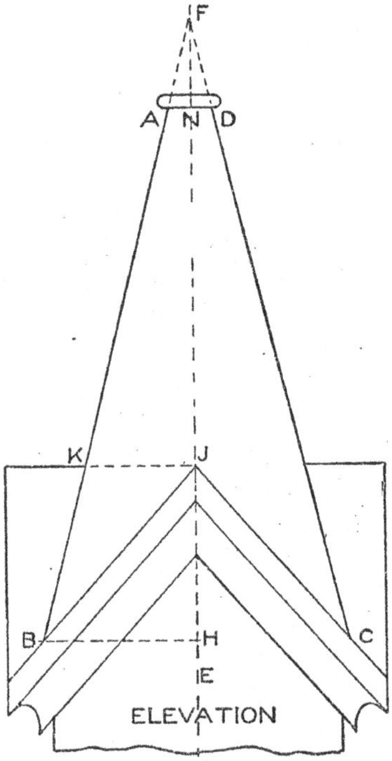

Fig. 310.

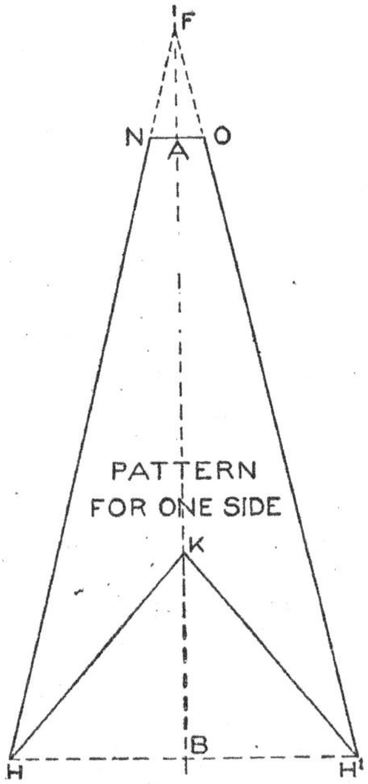

Fig. 311.

right angles to F D from points of intersection in the miter-lines D E and J F, as shown. Lines traced through points thus obtained, as shown by N O P R, will be the pattern for the raking moulding shown in plan, Fig. 307, by A B C Q′.

In Fig. 309 is shown a view of a spire, square in plan, intersecting four gables. In practice, each side A is developed separately in a manner shown in Fig. 310, in which first draw the center line through the center of the gable, as E F. Establish points B and C, from which draw lines to the apex F. At pleasure, establish A D. At right angles to F E, and from B and J, draw the lines B H and J K respectively. For the pattern, take the distances B K, K A, and A F, and place them as shown by similar letters on the vertical line B F in Fig. 311. At right angles to B F, and through points B and A, draw lines as shown, making B H and B H^1 on the one hand, and A N and A O on the other hand, equal respectively to B H and A N in elevation in Fig. 310. Then, in Fig. 311, draw lines from N to H to K to H^1 to O, as shown, which represents the pattern for one side.

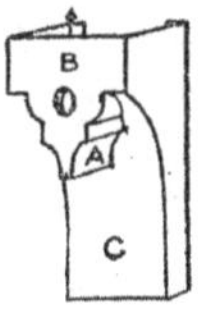

Fig. 312.

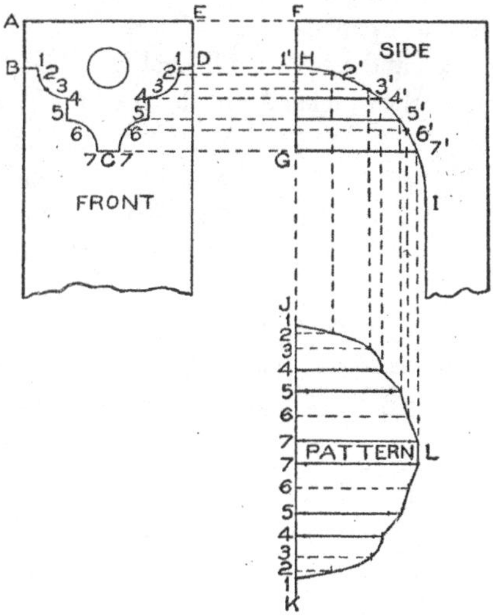

Fig. 313.

In Fig. 312 is shown a perspective view of a drop B mitering against the face of the bracket C as indicated at A. The principles for developing this problem are explained in Fig. 313, and can be applied to similar work no matter what the profiles of the drop or bracket may be. Let A B C D E represent the face or front view of the bracket drop, and F H G I the side of the drop and bracket. Divide one-half of the face, as D C, into equal spaces, as shown by the figures 1 to 7 on either side, from which points draw horizontal lines crossing H G in side view and intersecting the face H I of the bracket at points 1′ to 7′. In line with H G, draw the line J K, upon which place the stretch-out of the profile B C D, as shown by 1 to 7 to 7 to 1 on J K. At right angles to J K, draw the usual measuring lines as shown, which intersect by lines drawn parallel to J K from similarly numbered intersections on H I. Trace a line through the points thus obtained. Then will J K L be the pattern for the return of the drop on the face of the bracket.

In Fig. 314, A shows a raking bracket placed in a gable moulding. When brackets are placed in a vertical position in any raking moulding, they are called "raking" brackets. B represents a raking bracket placed at the center of the gable. The patterns which will be developed for the bracket A are also used for B, the cuts being similar, the only difference being that one-half the width of the bracket in B is formed right and the other half left, the two halves being then joined at the angle as shown.

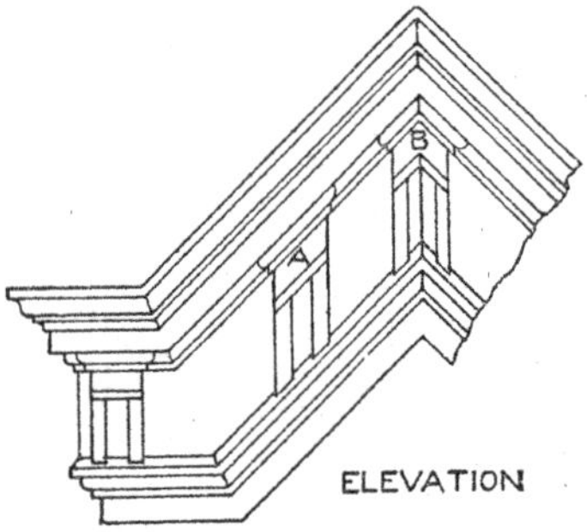

Fig. 314.

In Fig. 315 are shown the principles employed for obtaining the patterns for the side, face, sink strips, cap, and returns for a raking bracket. These principles can be applied to any form or angle in the bracket or gable moulding respectively. Let S U V T represent part of a front elevation of a raking cornice placed at its proper angles with any perpendicular line. In its proper position, draw the outline of the face of the bracket as shown by E G M O. Also, in its proper position as shown, draw the normal profile of the side of the bracket, indicated by 6-Y-Z-15; the normal profile of the cap-mould, as W and X; and the normal profile of the sink strip, as indicated by 10 10′ 15′ 15.

Complete the front elevation of the bracket by drawing lines parallel to E O from points 7 and 9 in the normal profile; and establish at pleasure the width of the sink strip in the face of the bracket, as at J K and L H. To complete the front elevation of the cap-mould of the bracket, proceed as follows: Extend the lines G E and M O of the front of the brackets, as shown by E 6 and O 6, on which, in a vertical position as shown, place duplicates (W^1, W^2) of the normal profiles W and X, divided into equal spaces as shown by the figures 1 to 6 in W^1 and W^2. From these intersections in W^1 and W^2, drop vertical lines, which intersect by lines drawn parallel to E O from similarly numbered intersections in X, and trace lines through the points thus obtained. Then will R E and O P represent respectively the true elevations, also the true profiles, for the returns at top and foot of the cap of the raking bracket.

Now divide the normal profile of the bracket into equal spaces, as shown by the figures 6 to 15, through which, parallel to E O, draw lines intersecting the normal sink profile from 10′ to 15′ and the face lines of the bracket EFG, JH, KL, and ONM, as shown. To obtain the

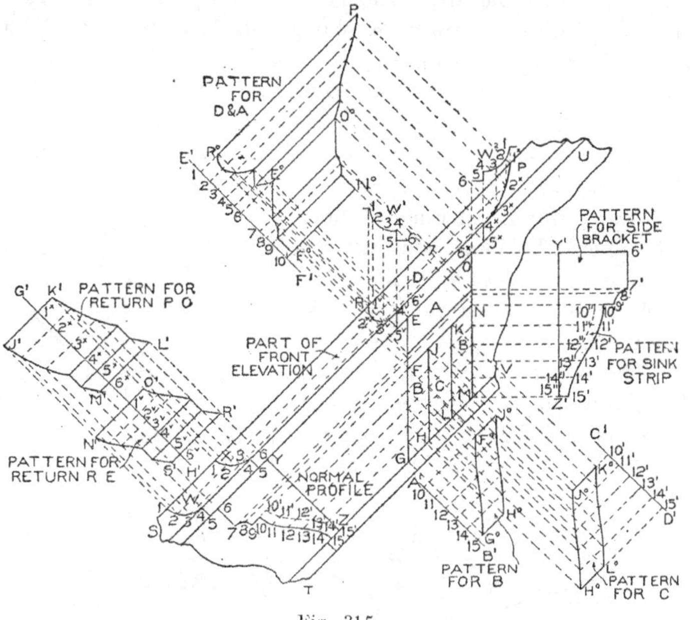

Fig. 315.

true profile for the side of the bracket on the lines OM and GE, proceed as follows: Parallel to OM, draw any line, as $Y^1 Z^1$; and at right angles to OM, and from the various intersections on the same, draw lines indefinitely, crossing to the line $Y^1 Z^1$ as shown. Now, measuring in each instance from the line YZ in the normal profile, take the various distances to points 6 to 15 and 15′ to 10′, and place them on similarly numbered lines measuring in each and every instance from the line $Y^1 Z^1$, thus obtaining the points 6′ to 15′ and 15″ to 10″, as shown. Trace a line through the points thus obtained. Then will Y^1 6′ 7′ 9′ 10′ 15′ Z^1 be the pattern for the side of the raking bracket, and 10′ 10″ 15″ 15′ the pattern for the sink strip shown by the lines K L and H J in the front.

For the pattern for the face strip B, draw any line, as A^1 B^1, at right angles to G M, upon which place the stretchout of 10 15 in the normal profile, as shown from 10 to 15 on A^1 B^1. Through these points, at right angles to A^1 B^1, draw lines as shown, which intersect with lines drawn from similar intersections on the lines F G and H J. Trace a line through points thus obtained as shown by F° G° H° J°, which will be the pattern for the face B, B.

For the pattern for the sink-face C, draw C^1 D^1 at right angles to GM, upon which place the stretchout of 10′ 15′ in the normal profile as shown from 10′ to 15′ on C^1 D^1, through which, at right angles to C^1 D^1, draw lines, which intersect by lines drawn from similar intersections on K L and H J. Trace a line through the points so obtained as J° K° L° H°, which is the pattern for the sink-face C.

The pattern for the cap D and the face A will be developed in one piece, by drawing at right angles to EO the line E^1 F^1. At right angles to E^1 F^1, and through the figures, draw lines, which intersect with lines drawn at right angles to EO from similarly numbered intersections on REF and NOP. A line traced through the points thus obtained, as shown by R° E° F° and N° O° P° will be the pattern for D and A.

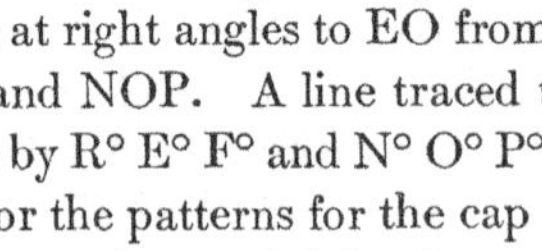

Fig. 316.

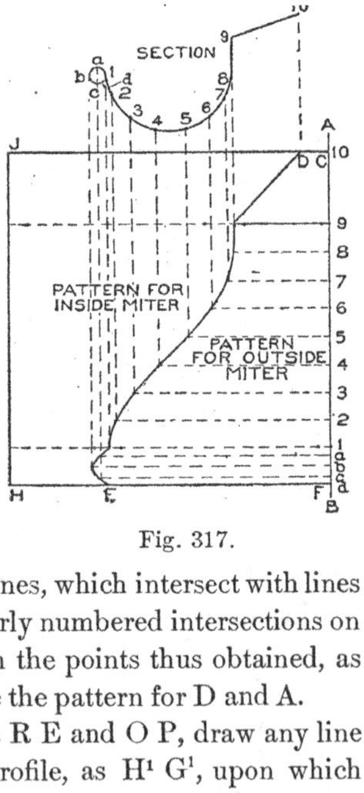

Fig. 317.

For the patterns for the cap returns R E and O P, draw any line at right angles to 1 1 in the normal profile, as H^1 G^1, upon which place the stretchouts of the profiles R E and O P, being careful to carry each space separately onto the line H^1 G^1, as shown respectively by 6^v 1^v and 6^x 1^x. Through these points draw lines at right angles to G^1 H^1, which intersect by lines drawn at right angles to 1 1 from similar numbers in W and X. Trace lines through the points thus obtained. Then will N^1 O^1 R^1 S^1 be the pattern for the lower return of the cap, R E; while J^1 M^1 L^1 K^1 will be the pattern for the upper return, P O.

In Fig. 316 is shown a perspective view of a gutter or eave-trough at an exterior angle, for which an outside miter would be required. It is immaterial what shape the gutter has, the method of obtaining the pattern for the miter is the same. In Fig. 317 let 1 9 10 represent the section of the eave-trough with a bead or wire edge at *a b c*; divide the wire edge, including the gutter and flange, into an equal number of spaces, as shown by the small divisions *d* to 1 to 9 to 10. Draw any vertical line, as A B, upon which place the stretch-out of the gutter as shown by similar letters and numbers on A B, through which, at right angles to A B, draw lines, which intersect by

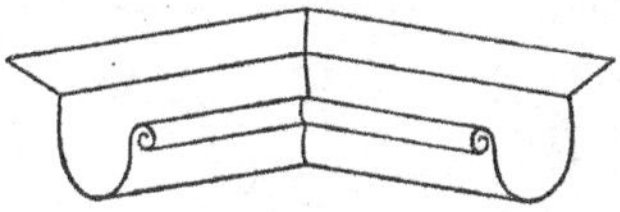

Fig. 318.

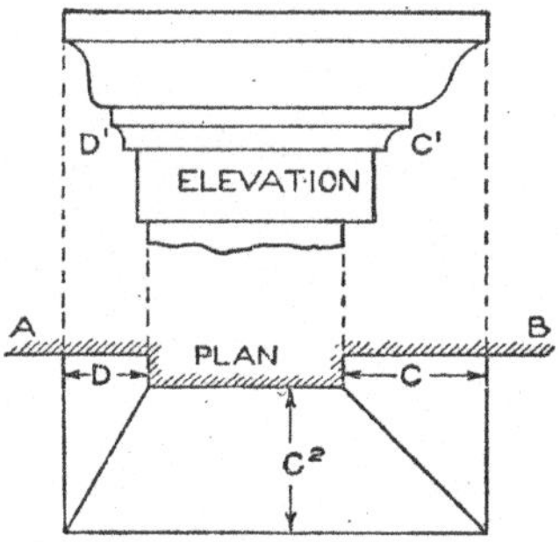

Fig. 319.

lines drawn parallel to A B from similar points in the section. Trace a line through the points thus obtained. Then will C D E F be the pattern for the outside angle shown in Fig. 316.

If a pattern is required for an interior or inside angle, as is shown in Fig. 318, it is necessary only to extend the lines C D and F E in the pattern in Fig. 317, and draw any vertical line, as J H. Then will J D E H be the pattern for the inside angle shown in Fig. 318.

In Fig. 319 are shown a plan and elevation of a moulding which has more projection on the front than on the side. In other words, A B represents the plan of a brick pier, around which a cornice is to be constructed. The projection of the given profile is equal to C, the profile in elevation being shown by C^1. The projection of the front in plan is also equal to C, as shown by C^2. The projection of the left side of the cornice should be only as much as is shown by D in plan. This requires a change of profile through D, as shown by D^1. To obtain this true profile and the various patterns, proceed as shown in Fig. 320, in which A B C D represents the plan view of the wall, against which, in its proper position, the profile E is placed and divided into equal spaces, as shown by the figures 1 to 12. Through 1 2, parallel to C D, draw G F. Locate at pleasure the projection of the re-

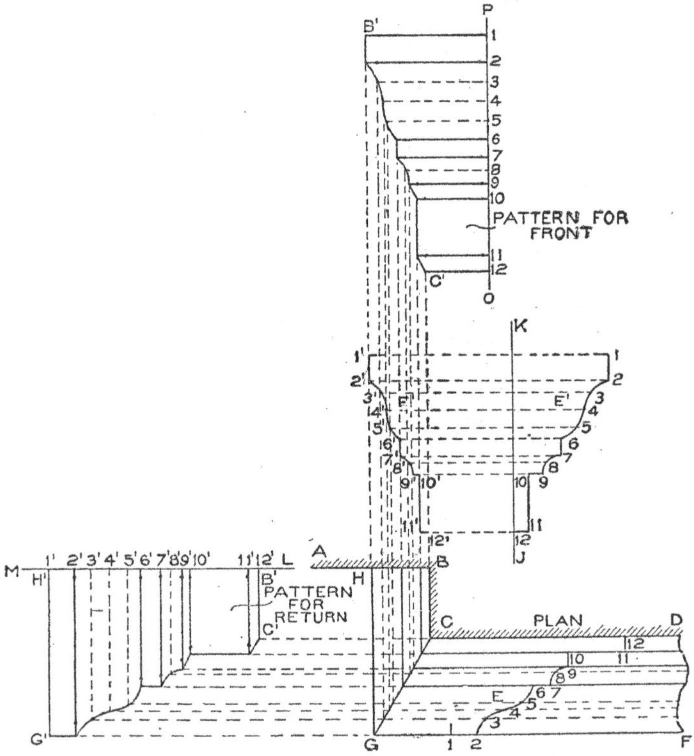

Fig. 320.

turn mould, as B H, and draw H G parallel to B C, intersecting F G at G. Draw the miter-line in plan, G C. From the various divisions in the profile E, draw lines parallel to C D, intersecting the miter-line C G as shown. From these intersections, erect vertical lines indefinitely, as shown. Parallel to these lines erect the line K J, upon which place a duplicate of the profile E, with the various divisions on same, as shown by E¹. Through these divisions draw horizontal lines intersecting the similarly numbered vertical lines, as shown by the intersections 1' to 12'. Trace a line through these points. Then will F^1 be the true section or profile on H B in plan.

For the pattern for the return H G C B in plan, extend the line B A, as B M, upon which place the stretchout of the profile F^1, being careful to measure each space separately (as they are unequal), as shown by figures 1' to 12' on M B.

At right angles to this line and through the figures, draw lines, which intersect by lines drawn at right angles to H G from similar points on C G. Trace a line through the points thus obtained. Then will H^1 G^1 C^1 B^1 be the pattern for the return mould.

The pattern for the face mould G C D F is obtained by taking a stretchout of the profile E and placing it on the vertical line P O, as shown by similar figures, through which, at right angles to P O, draw lines intersecting similarly numbered lines previously extended from C G in plan. Trace a line through these intersections. Then will 1 B^1 C^1 12 be the miter pattern for the face mould.

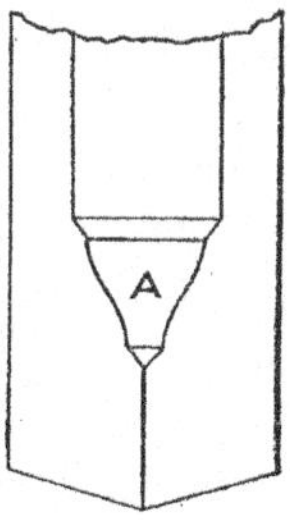

Fig. 321.

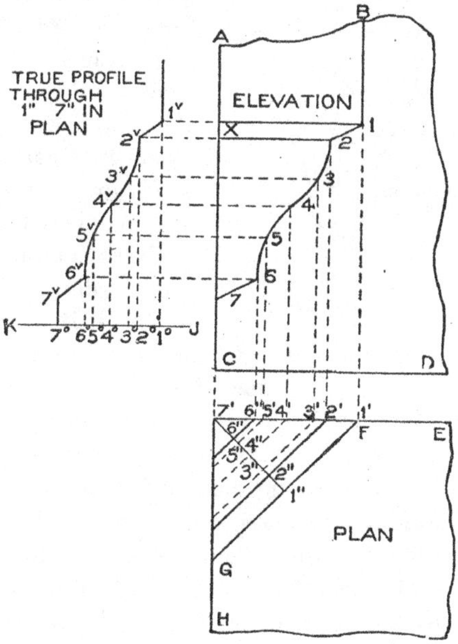

Fig. 322.

In Fig. 321 is shown a perspective view of a gore piece A joined to a chamfer. This presents a problem often arising in ornamental sheet-metal work, the development of which is given in Fig. 322. Let A B C D show the elevation of the corner on which a gore piece is required. H 7′ E in plan is a section through C D, and E F G H is a section through X I, all projected from the elevation as shown. The profile 1 7 can be drawn at pleasure, and at once becomes the pattern for the sides. Now divide the profile 1 7 into an equal number of spaces as shown, from which drop vertical lines onto the side 7′ E in plan, as shown from 1′ to 7′. From these points draw lines parallel to F G, intersecting the opposite side and crossing the line 7′ 1″ (which is drawn at right angles to F G from 7′) at 1″ 2″ 3″ 4″ 5″ 6″. Draw any line parallel to C D, as K J, upon which place all the intersections contained on 7′ 1″ in plan, as shown by 1° to 7° on K J. From these points erect perpendicular lines, which intersect by lines drawn from similarly numbered points in elevation parallel to C D. Through the points thus obtained trace a line. Then will 1^v to 7^v be the true profile on 7′ 1″ in plan.

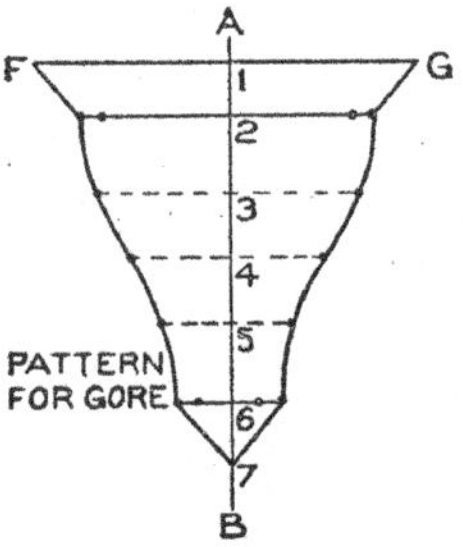

Fig. 323.

For the pattern for the gore, draw any vertical line, as A B in Fig. 323, upon which place the stretchout of the profile 1^v 7^v in Fig. 322, as shown by similar figures on A B in Fig. 323. At right angles to AB, and through the figures, draw lines as shown, Now, measuring in each instance from the line 7′ 1″ in plan in Fig. 322, take the various distances to points 1′ to 7′, and place them in Fig. 323 on similarly numbered lines, measuring in each instance from the line A B, thus locating the points shown. Trace a line through the points thus obtained. Then will F G 7 be the pattern for the gore shown in plan in Fig. 322 by F G 7′.

Fig. 324.

In Fig. 324 is shown a face view of a six-pointed star, which often arises in cornice work. No matter how many points the star has, the principles which are explained for its development are applicable to any size or shape. Triangulation is employed in this problem, as shown in Fig. 325. First draw the half-outline of the star, as shown by A B C D E F G. Above and parallel to the line AG, draw JH of similar length, as shown. Draw the section of the star on A G in plan, as shown by J K H. Project K into plan as shown at I, and draw the miter-lines B I, C I, D I, E I, and F I. As K H is the true length on I G, it is necessary that we find the true length on I F. Using I F as radius and I as center, draw an arc intersecting I G at *a*. From *a* erect a line cutting J H in section at *b*. Draw a line from *b* to K, which is the true length on I F.

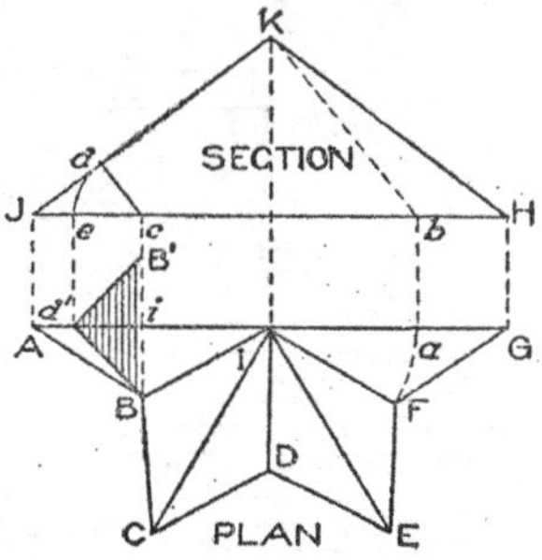

Fig. 325.

For the pattern, proceed as shown in Fig. 326. Draw any line, as K H, equal in length to K H in Fig. 325. Then, using K *b* as radius and K in Fig. 326 as center, describe the arc *b b*, which intersect at *a* and *a* by an arc G G struck from H as center and with F G in plan in Fig. 325 as radius. Draw lines in Fig. 326 from K to *a* to H to *a* to K, which will be the pattern for one of the points of the star of which 6 are required.

When bending the points on the line HK, it is necessary to have a stay or profile so that we may know at what angle the bend should be made. To obtain this stay, erect from the corner B in Fig. 325 a line intersecting the base-line J H at *c*, from which point, at right angles to J K, draw *c d*. Using *c* as center, and *c d* as radius, strike an arc intersecting J H at *e*. From *e* drop a vertical line meeting A G in plan at *d'*. Set off *i* B¹ equal to *i* B, and draw a line from B to *d'* to B¹, which is the true profile after which the pattern in Fig. 326 is to be bent. If the stay in Fig. 325 has been correctly developed, then *d'* B¹ or *d'* B must equal *e a* in Fig. 326 on both sides.

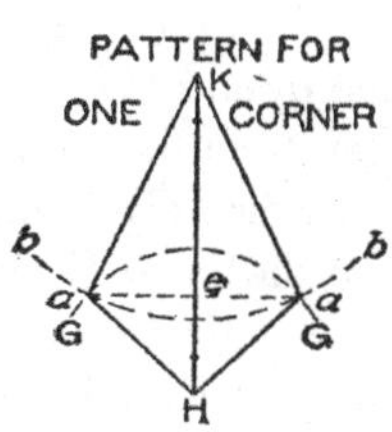

Fig. 326.

In Fig. 327 is shown a finished elevation of a hipped roof, on the four corners of which a hip ridge A A butts against the upper base B and cuts off on a vertical line at the bottom, as C and C. To obtain the true profile of this hip ridge, together with the top and lower cuts and the patterns for the lower heads, proceed as shown in Fig. 328, where the front elevation has been omitted, this not being necessary, as only the part plan and diagonal elevation are required. First draw the part plan as shown by A B C D E F A, placing the hip or diagonal line F C in a horizontal position; and make the distances between the lines F A and C B and between F E and C D equal, because the roof in this case has equal pitch all around. (The same principles, however, would be used if the roofs had unequal pitches.) Above the plan, draw the line G H. From the points F and C in plan, erect the lines F G and C I, extending C I to C^1 so that I C^1 will be the required height of the roof above G I at the point C in plan. Draw a line from G to C^1, and from C^1 draw a horizontal and vertical line indefinitely, as shown. Then will I G C^1 be a true section on the line of the roof on F C in plan.

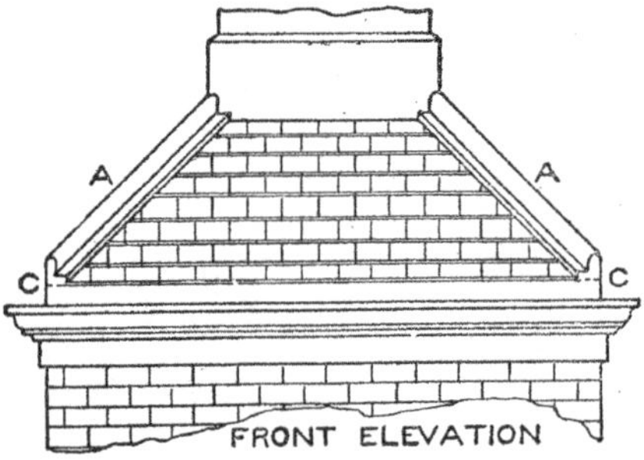

Fig. 327.

The next step is to obtain a true section of the angle of the roof at right angles to the hip line G C^1 in elevation. This is done by drawing at right angles to F C in plan. any line, as *a b*, intersecting the lines F A and F E as shown. Extend *a b* until it cuts the base-line G I in elevation at *c*. From *c*, at right angles to G C^1, draw a line, as *c d*, intersecting G C^1 at *d*. Take the distance *c d*, and place it in plan on the line F C, measuring from *i* to *d'*. Draw a line from *a* to *d'* to *b*, which is the true angle desired. On this angle, construct the desired shape of the hip ridge as shown by J, each half of which divide into equal spaces, as shown by the figures 1 to 6 to 1. As the line G C^1 represents the line of the roof, and as the point *d'* in plan in the true angle also represents that line, then take a tracing of the profile J with the various points of intersection on same, together with the true angle *a d' b*, and place it in the elevation as shown by J^1 and *a' d'' b'*, being careful to place the point *d''* on the line G C^1, making *a' b'* parallel to G C^1. From the various points of intersection in the profile J, draw lines parallel to F C, intersecting B C and A F at points from 1 to 6, as shown. As both sides of the profile J are symmetrical, it is necessary only to draw lines through one-half.

In similar manner, in elevation, parallel to G C^1, draw lines through the various intersections in J^1, which intersect by lines drawn at right angles to F C in plan from similarly numbered points on A F

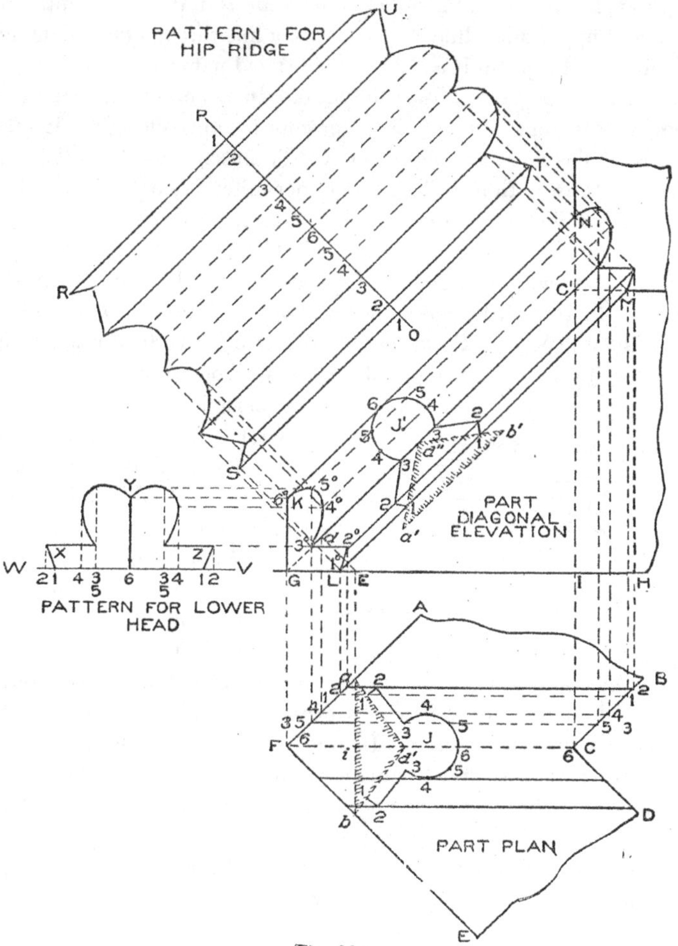

Fig. 328.

and BC. Trace a line through the points thus obtained. Then will K L be the miter-line at the bottom, and M N the miter-line at the top.

For the pattern, draw any line, as O P, at right angles to G C^1, upon which place the stretchout of J in plan or J^1 in elevation, as shown by the figures 1 to 6 to 1 on O P; and through these numbered points, at right angles to O P, draw lines, which intersect by lines drawn at right angles to G C^1 from similar intersections in the lower miter-line K L and upper miter-line N M. Trace a line through the points thus obtained. Then will R S T U be the desired pattern.

In practice it is necessary only to obtain one miter-cut—either the top or the bottom—and use the reverse for the opposite side. In other words, U T is that part falling out of R S, the same as R S is that part which cuts away from U T. The upper miter-cut butts against B in Fig. 327; while the lower cut requires a flat head, as shown at C. To obtain this flat head, extend the line I G in Fig. 328, as I W, upon which place twice the amount of spaces contained on the line A F in plan, as 6, 3–5, 4, 1, 2, as shown by similar figures on either side of 6 on the line V W. From these divisions erect vertical lines, which intersect by lines drawn parallel to V W from similarly numbered intersections in the miter-line K L G. A line traced through the points thus obtained, as shown by X Y Z, will be the pattern for the heads.

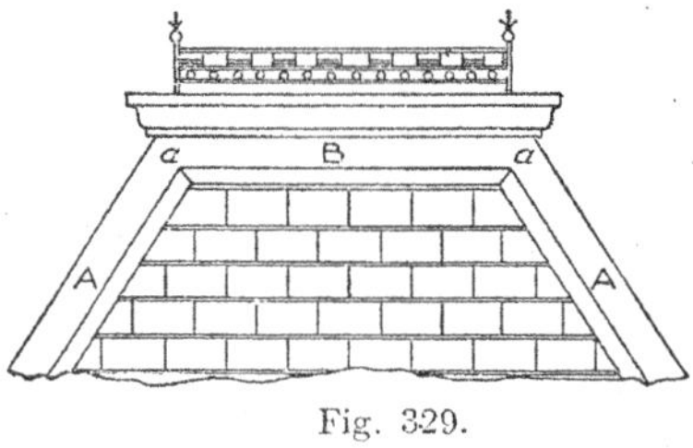

Fig. 329.

Where a hip ridge is required to miter with the apron of a deck moulding, as shown in Fig. 329, in which B represents the apron of the deck cornice, A and A the hip ridges mitering at *a* and *a*, a slightly different process from that described in the preceding problem is used. In this case the part elevation of the mansard roof must first be drawn as shown in Fig. 330. Let A B C K represent the part elevation of the mansard, the section of the deck moulding and apron being shown by D B E. Draw E X parallel to B C. E X then represents the line of the roof. In its proper position, at right angles to B C, draw a half-section of the hip mould, as shown by F G, which is an exact reproduction of B E of the deck mould. Through the corners of the hip mould at Y and G, draw lines parallel to B C, which intersect by lines drawn parallel to B A from V, W, and E in the deck cornice. Draw the miter-line H I, which completes the part elevation of the mansard.

Before the patterns can be obtained, a developed surface of the mansard must be drawn. Therefore, from B (Fig. 330), drop a vertical line, as B J, intersecting the line C K at J. Now take the distance of B C, and place it on a vertical line in Fig. 331, as shown by B C¹. Through these two points draw the horizontal lines B A and C K as shown. Take the projection J to C in Fig. 330, and place it as

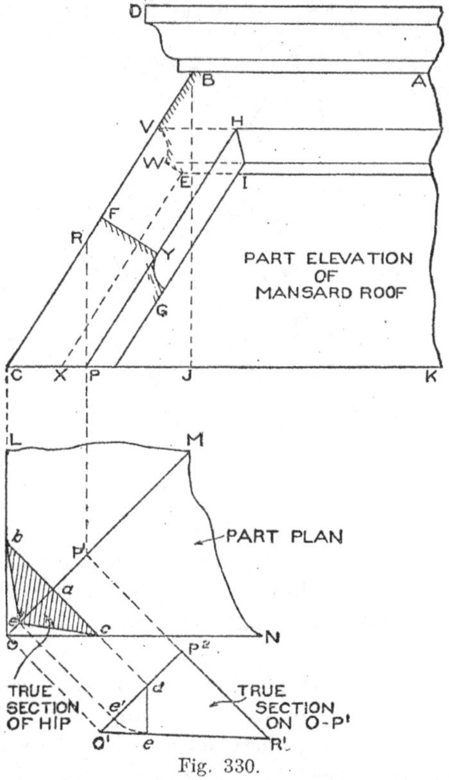

Fig. 330.

shown from C¹ to C in Fig. 331, and draw a line from C to B. Then will A B C K be the developed surface of A B C K in Fig. 330.

As both the profiles B V W E and F Y G are similar, take a tracing of either, and place it as shown by D and D¹ respectively in Fig. 331. Divide both into the same number of equal spaces, as shown. Bisect the angle A B C by establishing *a* and *b*, and, using these as centers, by describing arcs intersecting at c; then draw d B, which represents the miter-line. Through the points in D and D^1, draw lines parallel to their respective moulds, as shown, intersecting the miter-line B d and the base-line C C^1.

For the pattern for the hip, draw any line, as E F, at right angles to B C, upon which place twice the stretchout of D, as shown by the divisions 6 to 1 to 6 on EF. Through these divisions draw lines at

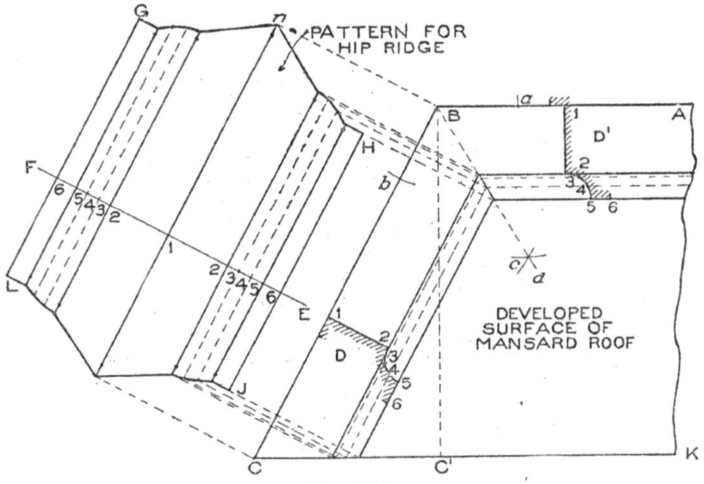

Fig. 331.

right angles to E F, intersecting similarly numbered lines drawn at right angles to B C from the divisions on B d and C C^1. Trace a line through the points thus obtained. Then will G H J L be the pattern for the hip ridge.

When bending this ridge in the machine, it is necessary to know at what angle the line 1 in the pattern will be bent. A true section must be obtained at right angles to the line of hip, for which proceed as shown in Fig. 330. Directly in line with the elevation, construct a part plan L M N O, through which, at an angle of 45 degrees (because the angle L O N is a right angle), draw the hip line O M. Establish at pleasure any point, as P^1 on O M, from which erect the vertical line into the elevation crossing the base-line C K at P and the ridge-line C B at R. Parallel to O M in plan, draw O^1 P^2, equal to O P^1, as shown. Extend P^1 P^2 as P^2 R^1, which make equal to PR in elevation.

Draw a line from R^1 to O^1. Then $O^1 R^1 P^2$ represents a true section on OP^1 in plan. Through any point, as *a*, at right angles to OM, draw *bc*, cutting L O and ON at *b* and *c* respectively. Extend *b c* until it intersects $O^1 P^2$ at *d*. From *d*, at right angles to $O^1 R^1$, draw the line *d e*. With *d* as center, and *de* as radius, draw the arc *e e′*, intersecting $O^1 P^2$ at *e′*, from which point, at right angles to OM in plan, draw a line intersecting OM at *e″*. Draw a line from *b* to *e″* to *c*, which represents the true section of the hip after which the pattern shown in Fig. 331 is formed.

The pattern for the deck mould D B in Fig. 330 is obtained in the same way as the square miter shown in Fig. 277; while the pattern for the apron D^1 in Fig. 331 is the same as the one-half pattern of the hip ridge shown by *n* H 1 6.

In Fig. 332 is shown a front elevation of an eye-brow dormer. In this view A B C represents the front view of the dormer, the arcs being

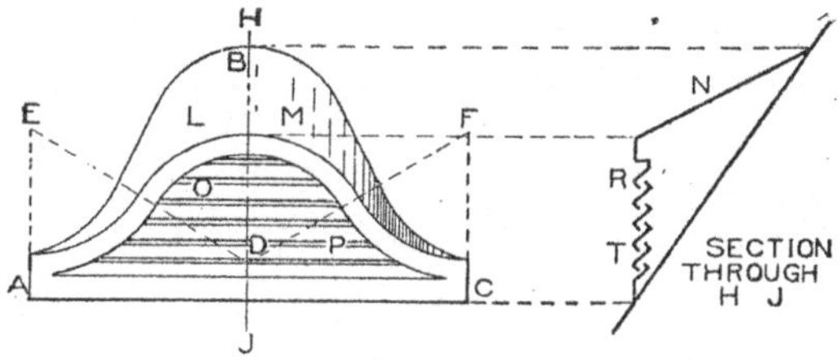

Fig. 332.

struck from the center points D, E, and F. A section taken on the line H J in elevation is shown at the right; L M shows the roof of the dormer, indicated in the section by N; while the louvers are shown in elevation by O P and in section by RT.

In Fig. 333 is shown how to obtain the various patterns for the various parts of the dormer. ABC represents the half-elevation of the dormer, and EFG a side view, of which EG is the line of the dormer, EF that of the roof, and GF the line of the pitched roof against which the dormer is required to miter.

The front and side views being placed in their proper relative positions, the first step is to obtain a true section at right angles to EF. Proceed as follows: Divide the curve A to B into a number of equal spaces, as shown from 1 to 9. At right angles to A C, and from the figures on A B, draw lines intersecting E G in side view as shown.

From these intersections, and parallel to EF, draw lines intersecting the roof-line GF at 1^5, 2^5, 3^5, etc. Parallel to EF, and from the point

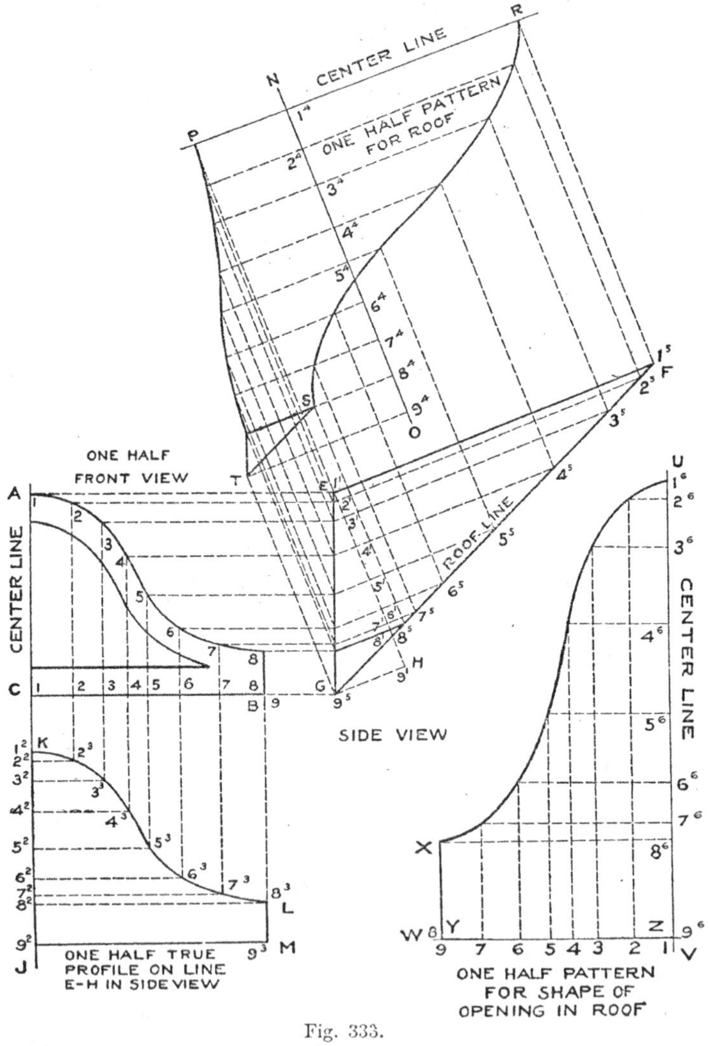

Fig. 333.

G, draw any line indefinitely, as G H. At right angles to EF, and from the point E, draw the line EH, intersecting lines previously drawn, at 1^1, 2^1, 3^1, etc., as shown. Now take a duplicate of the line E H, with the various intersections thereon, and place it on the center line AC extended as KJ. At right angles to KJ, and from the figures 1^2, 2^2, 3^2, etc., draw lines, which intersect with those of similar numbers drawn at right angles to CB, and from similarly numbered points on the curve A B. Trace a line through the points of intersection thus obtained. Then KLMJ will be one-half the true profile on the line E H in side view, from which the stretchout will be obtained in the development of the pattern.

For the pattern for the roof of the dormer, draw at right angles to EF in side view the line N O, upon which place the stretchout of one-half the true profile on the line EH as shown by the small figures 1^4, 2^4, 3^4, etc. Then, at right angles to N O, and through the figures, draw lines, which intersect with those of similar numbers drawn at right angles to EF from intersections on EG and GF. Trace a line through the points thus obtained. Then will PRST represent one-half the pattern for the roof.

To obtain the pattern for the shape of the opening to be cut into the roof, transfer the line GF, with the various intersections thereon, to any vertical line, as UV, as shown by the figures 1^6, 2^6, 3^6, etc. In similar manner, transfer the line CB in front view, with the various intersections on same, to the line ZW, drawn at right angles to UV, as shown by the figures 1, 2, 3, etc. At right angles to UV, and from the figures, draw lines, which intersect with those of similar numbers drawn at right angles to YZ. Through these points, trace a line. Then will UXYZ be the half-pattern for the shape of the opening to be cut into the main roof.

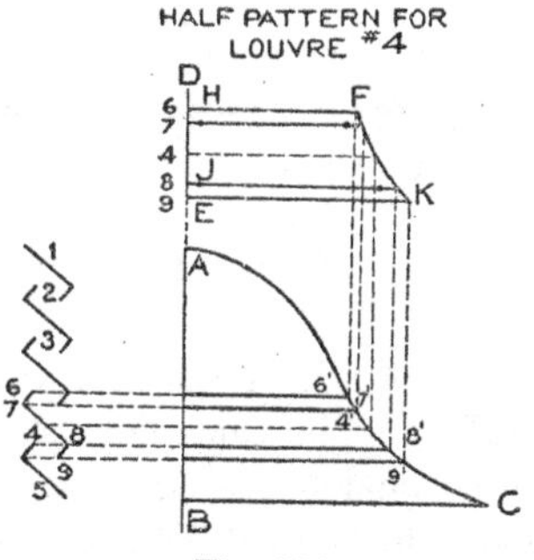

Fig. 334.

For the pattern for the ventilating slats or louvers, should they be required in the dormer, proceed as shown in Fig. 334. In this figure, A B C is a reproduction of the inside opening shown in Fig. 333. Let 1, 2, 3, 4, 5 in Fig. 334 represent the sections of the louvers which will be placed in this opening. As the methods of obtaining the patterns for all louvers are alike, the pattern for louver No. 4 will illustrate the principles employed. Number the various bends of louver No. 4 as shown by points 6, 7, 8, and 9. At right angles to A B, and from these points, draw lines intersecting the curve A C as 6^1, 7^1, 4^1, 8^1, and 9^1. On B A extended as E D, place the stretchout of louver No. 4 as shown by the figures on ED. Since the miter-line AC is a curve, it will be necessary to introduce intermediate points between 7 and 8 of the profile, in order to obtain this curve in the pattern. In this instance the point marked 4 has been added.

Now, at right angles to DE, and through the figures, draw lines, which intersect with those of similar numbers, drawn parallel to AB from intersections 6^1 to 9^1 on the curve AC. A line traced through the points thus obtained, as FKJH, will be the half-pattern for louver No. 4. The pattern for the face of the dormer is pricked onto the metal direct from the front view in Fig. 333, in which A 8 B C is the half-pattern.

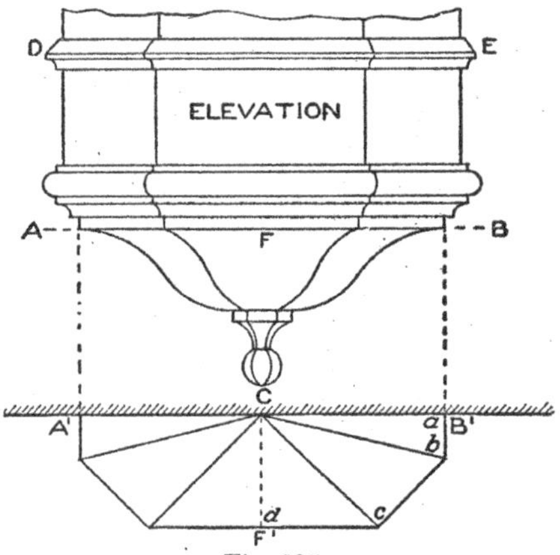

Fig. 335.

In laying out the patterns for bay window work, it often happens that each side of the window has an unequal projection, as is shown in Fig. 335, in which DEF shows an elevation of an octagonal base of a bay window having unequal projections. All that part of the bay above the line AB is obtained by the method shown in Fig. 290, while the finish of the bay shown by ABC in Fig. 335 will be treated here. In some cases the lower ball C is a half-spun ball. A^1 B^1 F^1 is a true section through A B. It will be noticed that the lines C*a*, C*c*, and C*d*, drawn respectively at right angles to *ab*, *bc*, and *cd*, are each of different lengths, thereby making it necessary to obtain a true profile on each of these lines, before the patterns can be obtained. This is clearly explained in connection with Fig. 336, in which only a half-elevation and plan are required as both sides are symmetrical. First draw the center line AB, on which draw the half-elevation of the base of the bay, as shown by CDE. At right angles to AB draw the wall line in plan, as FK; and in its proper position in relation to the line CD in elevation, draw the desired half-plan, as shown by GHIJ. From the corners H and I draw the miter-lines HF and IF, as shown. As DE

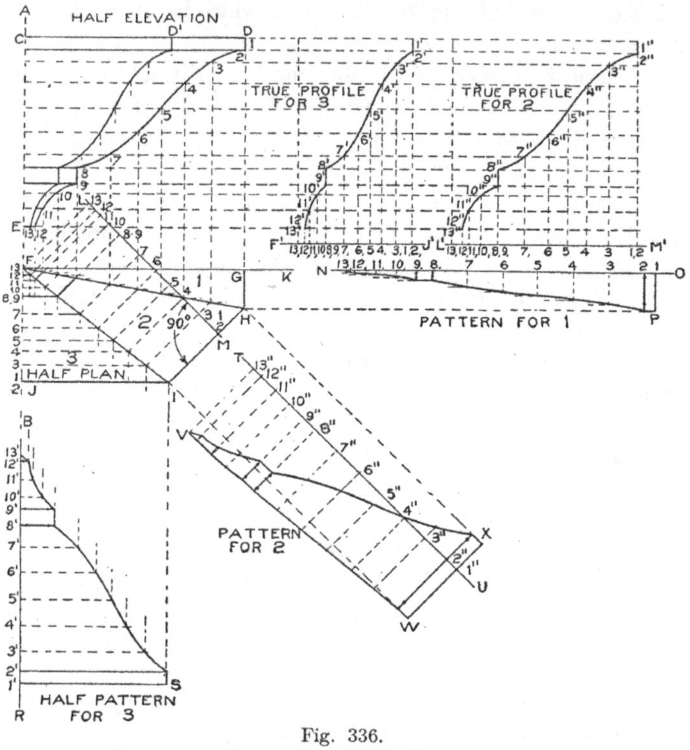

Fig. 336.

represents the given profile through FG in plan, then divide the profile DE into an equal number of spaces as shown by the figures 1 to 13. From these points drop vertical lines intersecting the miter-line FH in plan, as shown. From these intersections, parallel to HI, draw lines intersecting the miter-lines IF, from which points, parallel to IJ, draw lines intersecting the center line FB. Through the various points of intersection in DE, draw horizontal lines indefinitely right and left as shown.

If for any reason it is desired to show the elevation of the miter-line FI in plan (it not being necessary in the development of the pattern), then erect vertical lines from the various intersections on FI, intersecting similar lines in elevation. To avoid a confusion in the drawing, these lines have not been shown. Trace a line through points thus obtained, as shown by D^1 13, which is the desired miter-line in elevation.

The next step is to obtain the true profile at right angles to HI and IJ in plan. To obtain the true profile through No. 3 in plan, take a tracing of J F, with the various intersections thereon, and place it on a line drawn parallel to CD in elevation, as J^1 F^1, with the intersections 1 to 13, as shown. From these intersections, at right angles to J^1 F^1, erect lines intersecting similar lines drawn through the profile DE in elevation. Trace a line through the points thus obtained, as shown by 1′ to 13′, which represents the true profile for part 3 in plan. At right angles to IH in plan, draw any line, as ML, and extend the various lines drawn parallel to IH until they intersect LM at points 1 to 13, as shown.

Take a tracing of LM, with the various points of intersection, and place it on any horizontal line, as L^1 M^1, as shown by the figures 1 to 13, from which, at right angles to L^1 M^1, erect vertical lines intersecting similarly numbered horizontal lines drawn through the profile DE. Trace a line through the points thus obtained. Then will 1″ – 13″ be the true profile through No. 2 in plan at right angles to HI.

For the pattern for No. 1 in plan, extend the line FK, as NO, upon which place the stretchout of the profile DE as shown by the figures 1 to 13 on NO. At right angles to NO, and from the figures, draw lines, which intersect with lines (partly shown) drawn parallel to FG from similar intersections on the miter-line FH. Trace a line through the points thus obtained; then will 1 P 13 be the pattern for part 1 in plan.

At right angles to H I, draw any line, as T U, upon which place the stretchout of profile No. 2, being careful to measure each space separately, as they are all unequal, as shown by the small figures 1″ to 13″ on TU. Through these figures, at right angles to TU, draw lines as shown, which intersect by lines (not shown in the drawing) drawn at right angles to I H from similar points on the miter-lines HF and FI.

Trace a line through the points thus obtained. Then will V W X be the pattern for part 2 in plan.

For the half-pattern for part 3 in plan, extend the center line A B in plan as B R, upon which place the stretchout of the true profile for 3, being careful to measure each space separately, as shown by the figures 1′ to 13′ on BR. At right angles to B R draw lines through the figures, which intersect by lines drawn at right angles to J I from similar points of intersection on the miter-line F I. A line traced through points thus obtained, as 1′ S 13′, will be the half-pattern for part 3.

DEVELOPMENT OF BLANKS FOR CURVED MOULDINGS

Our first attention will be given to the methods of construction, it being necessary that we know the methods of construction before the blank can be laid out. For example, in Fig. 337 is a part elevation of a dormer window, with a semicircular top whose profile has an ogee, fillet, and cove. If this job were undertaken by a firm who had no circular moulding machine, as is the case in many of the smaller shops, the mould would have to be made by hand. The method of construction in this case would then be as shown in Fig. 338, which shows an enlarged section through *a b* in Fig. 337. Thus the strips *a*, *b*, and *c* in Fig. 338 would be cut to the required size, and would be nothing more than straight strips of metal, while *d d′* would be an angle, the lower side *d′* being notched with the shears and turned to the required circle. The face strips *e*, *f*, and *h* would represent arcs of circles to correspond to their various diameters obtained from the full-sized elevation. These face and sink strips would all be soldered together, and form a succession of square angles, as shown, in which the ogee, as shown by *i j*, and the cove, as shown by *m*, would be fitted. In obtaining the patterns for the blanks hammered by hand, the averaged lines would be drawn as shown by *k l* for the ogee and *n o* for the cove. The method or principles of averaging these and other moulds will be explained as we proceed.

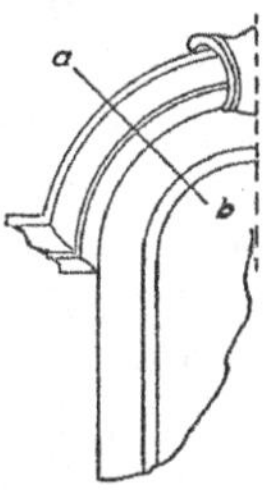

Fig. 337.

In Fig. 339 is shown the same mould as in the previous figure, a different method of construction being employed from the one made by hand and the one hammered up by machine. In machine work this mould can be hammered in one piece, 8 feet long or of the length of the sheets in use, if such length is required, the machine taking in the full

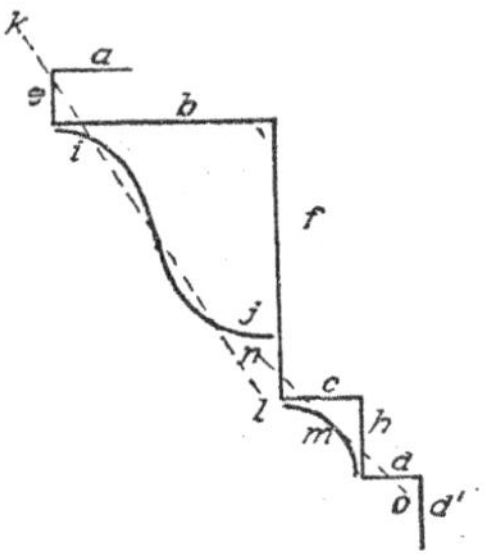

Fig. 338.

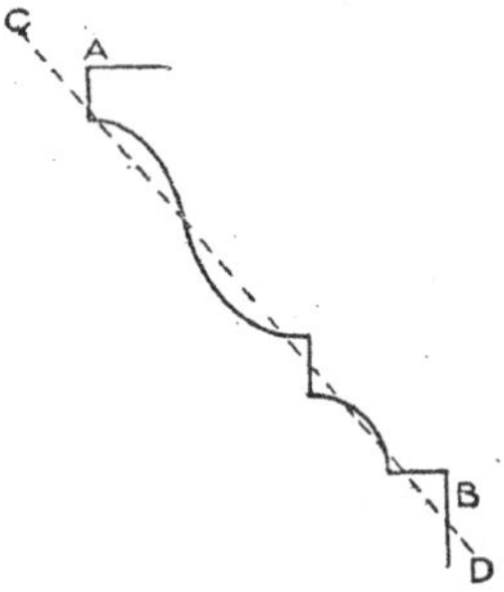

Fig. 339.

mould from A to B. The pattern for work of this kind is averaged by drawing a line as shown by CD. This method will also be explained more fully as we proceed.

SHOP TOOLS EMPLOYED

When working any circular mould by hand, all that is required in the way of tools is various-sized raising and stretching hammers, square stake, blow-horn stake, and mandrel including raising blocks made of wood or lead. A first-rate knowledge must be employed by the mechanic in the handling and working of these small tools. In a thoroughly up-to-date shop will be found what are known as "curved moulding" machines, which can be operated by foot or power, and which have the advantage over hand operation of saving time and labor, and also turning out first-class work, as all seams are avoided.

PRINCIPLES EMPLOYED FOR OBTAINING APPROXIMATE BLANKS FOR CURVED MOULDINGS HAMMERED BY HAND

The governing principles underlying all such operations are the same as every sheet-metal worker uses in the laying out of the simple patterns in flaring ware. In other words, one who understands how to lay out the pattern for a frustum of a cone understands the principles of developing the blanks for curved mouldings. The principles will be described in detail in what follows.

Our first problem is that of obtaining a blank for a plain flare, shown in Fig. 340. First draw the center line A B, and construct the half-elevation of the mould, as C D E F. Extend D E until it intersects the center line A B at G. At right angles to A B from any point, as H, draw H 1 equal to C D, as shown. Using H as center, and with H 1 as radius, describe the quarter-circle 1 7, which is a section on C D. Divide 1 7 into equal spaces, as shown. Now using G as center, with radii equal to G E and G D, describe the arcs D 7′ and E E°. From any point, as 1′, draw the radial line 1′ G, intersecting the inner arc at Eˣ. Take a stretchout of the quarter-section; place it as shown

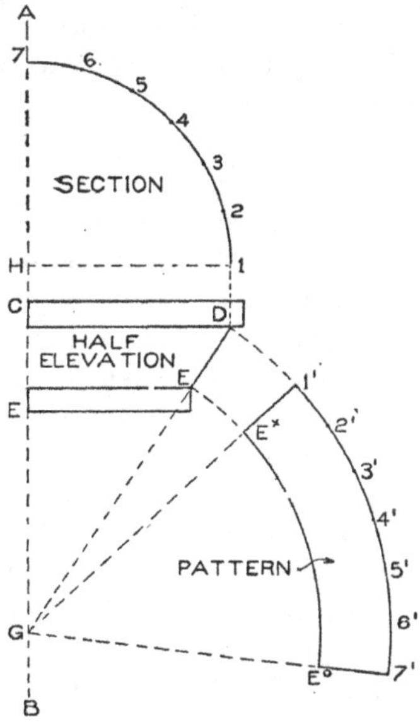

Fig. 340.

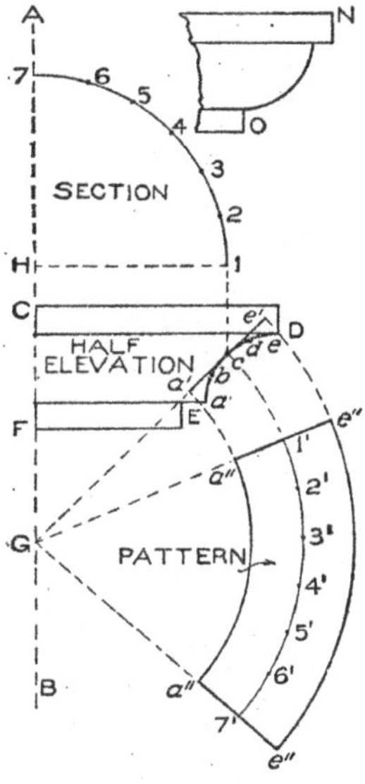

Fig. 341.

from 1′ to 7′; and draw a line from7′ to G, intersecting the inner arc at E°. Then will Eˣ 1′ 7′ E° be the quarter-pattern for the flare D E in elevation. If the pattern is required in two halves, join two pieces; if required in one piece, join four pieces.

In Fig. 341 is shown a curved mould whose profile contains a cove. To work this profile, the blank must be stretched with the stretching hammer. *We mention this here so that the student will pay attention to the rule for obtaining patterns for stretched moulds.* First draw the center line A B; also the half-elevation of the moulding, as C D E F. Divide the cove E D into an equal number of spaces, as shown from

a to e. Through the center of the cove c draw a line parallel to e a, extending it until it meets the center line A B at G, which is the center point from which to strike the pattern. Take the stretchout of the cove c e and c a, and place it as shown by c e' and c a'. When stretching the flare a' e', c remains stationary, e' and a' being hammered towards e and a respectively. Therefore, from c erect a vertical line intersecting H 1, drawn at right angles to A B, at 1. Using H as center and H 1 as radius, describe the arc 1 7, which divide into equal spaces as shown. With G as center, and radii equal to G a', Gc, and G e', describe the arcs e'' e'', 1′ 7′, and a'' a''. Draw a line from e'' to G, intersecting the center and lower arcs at 1′ and a''. Starting from 1′, lay off the stretchout of the quarter-section as shown from 1′ to 7′. Through 7′ draw a line towards G, intersecting the inner arc at a''; and, extending the line upward, intersect the outer arc at e''. Then will a'' e'' e'' a'' be the quarter-pattern for the cove E D in elevation.

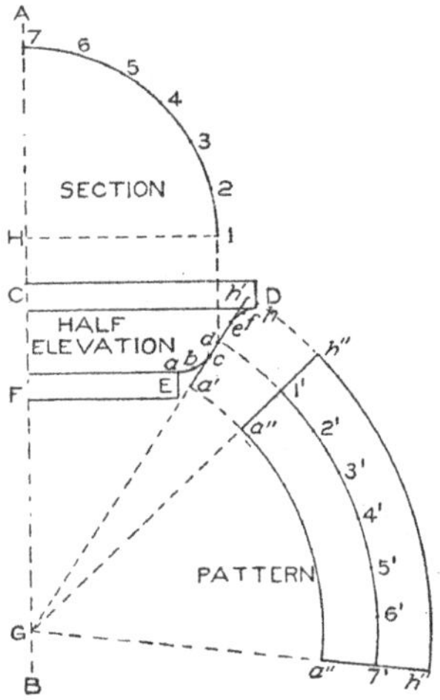

Fig. 342.

If the quarter-round N O were required in place of the cove E D, then, as this quarter-round would require to be raised, the rule given in the former Instruction Paper on Sheet Metal Work would be applied to all cases of raised mouldings.

In Fig. 342 is shown a curved mould whose profile is an ogee. In this case as in the preceding, draw the center line and half-elevation, and divide the ogee into a number of equal parts, as shown from a to h. Through the flaring portion of the ogee, as c e, draw a line, extending it upward and downward until it intersects the center line A B at G. Take the stretchouts from a to c and from e to h and place them respectively from c to a' and from e to h' on the line h' G. Then, in working the ogee, that portion of the flare from c to e remains stationary; the part from e to h' will be stretched to form e h; while that part shown from c to a' will be raised to form c a. From any point in the stationary flare, as d, erect a line meeting the line H 1, drawn at right angles to A B, at 1. Using H as center and H 1 as radius, describe the quarter-section, and divide same into equal spaces, as shown. With G as center and with radii equal to G a', G d, and G h', describe the arcs a'' a'', 1′ 7′, and h'' h''. From h'' draw a line to G. Starting at 1′, lay off the stretchout of the section as shown from 1′ to 7′. Through 7′ draw a line to G, as before described. Then will h'' a'' a'' h'' be the quarter-pattern for the ogee E D.

In Fig. 343 is shown how the blanks are developed when a bead moulding is employed. As before, first draw the center line A^1 B^1 and the half-elevation A B C D. As the bead takes up ¾ of a circle, as shown by *a c e f*, and as the pattern for *f e* will be the same as for *e c*, then will the pattern for *c e* only be shown, which can also be used for *e f*. Bisect *a c* and *c e*, obtaining the points *b* and *d*, which represent the stationary points in the patterns. Take the stretchouts of *b* to *a* and *b* to *c*, and place them as shown from *b* to a' and from *b* to c'; also take the stretchouts of *d* to *c* and *d* to *e*, and place them from *d* to c' and from *d* to e' on lines drawn parallel respectively to *a c* and *c e* from points *b* and *d*. Extend the lines e' c' and c' a' until they intersect the center line A^1 B^1 at E and F respectively. From the points *b* and *d* erect lines intersecting the line G 1, drawn at right angles to A^1

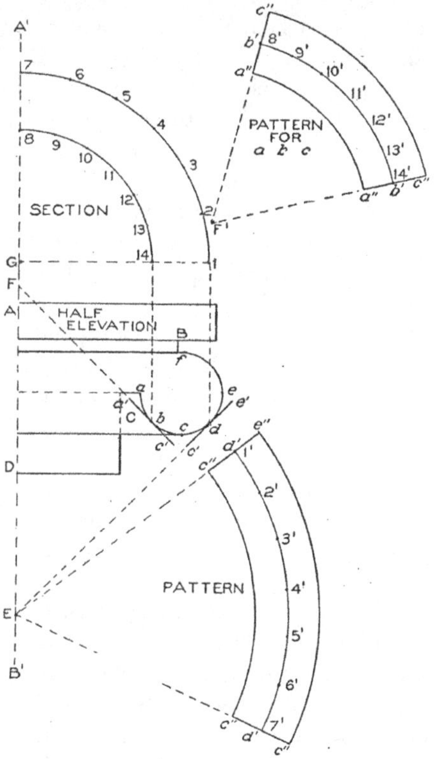

Fig. 343.

B[1], at 14 and 1 respectively. Using G as center, and with radii equal to G 14 and G 1, describe quarter-sections, as shown. Divide both into equal parts, as shown from 1 to 7, and from 8 to 14. With E as center, and with radii equal to E c', E d, and E e', describe the arcs c'' c'', d' d', and e'' e''. From any point on one end, as e'', draw a radial line to E, intersecting the inner arcs at d' and c''. Now take the stretchout of the section from 1 to 7, and, starting at d', lay off the stretchout as shown from 1′ to 7′. Through 7′ draw a line towards E, intersecting the inner arc at c'' and the outer one at e''. Then will c'' e'' e'' c'' be the quarter-pattern for that part of the bead shown by c e, also for e f, in elevation. For the pattern for that part shown by a c, use F[1] as center; and with radii equal to F a', F b, and F c', describe the arcs a'' a'', b' b', and c'' c''. From any point on the arc b' b', as 8′, lay off the stretchout of the quarter-section 8 14, as shown from 8′ to 14′. Through these two points draw lines towards F[1], intersecting the inner arcs at a'' a''; and extend them until they intersect the outer arc at c'' and c''. Then will c'' a'' a'' c'' be the desired pattern.

Fig. 344.

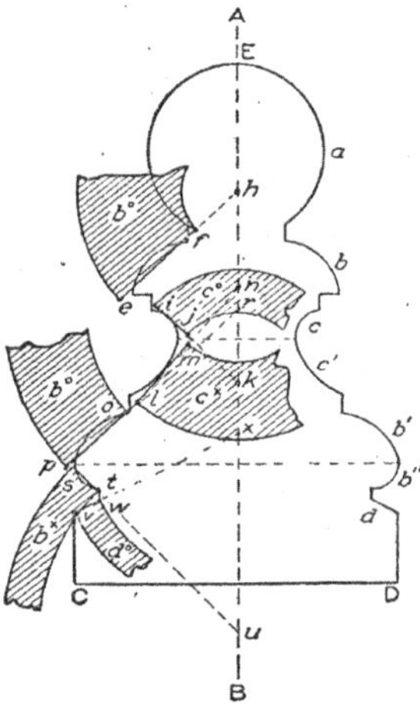

Fig. 345.

In Fig. 344 is shown an illustration of a round finial which contains moulds, the principles of which have already been described in the preceding problems. The ball A is made of either horizontal or vertical sections. In Fig. 345 is shown how the moulds in a finial of this kind are averaged. The method of obtaining the true length of each pattern piece will be omitted, as this was thoroughly covered in the preceding problems. First draw the center line A B, on either side of which draw the section of the finial, as shown by C D E. The blanks for the ball a will be obtained as explained in the Instruction Paper on Sheet Metal Work. The mould b is averaged as shown by the line e f, extending same until it intersects the center line at h, e f representing the stretchout of the mould obtained, as explained in the paper on Sheet Metal Work. Using h as center, with $h\ f$ and $h\ e$ as radii, describe the blank b°.

In the next mould, $c\ c'$, a seam is located in same as shown by the dotted line. Then average C by the line $i\ j$, extending same until it meets the center line at k; also average c' by the line $l\ m$, extending this also until the center line is intersected at n. Then $i\ j$ and $l\ m$ represent respectively the stretchouts of the mould $c\ c'$, the blanks c° and c^{x} being struck respectively from the centers k and n. The mould $b'\ b''$ also has a seam, as shown by the dotted line, the moulds being averaged by the lines $p\ o$ and $s\ t$, which, if extended, intersect the center line at r and u. These points are the centers, respectively, for striking the blanks b° and b^{x}. The flaring piece d is struck from the

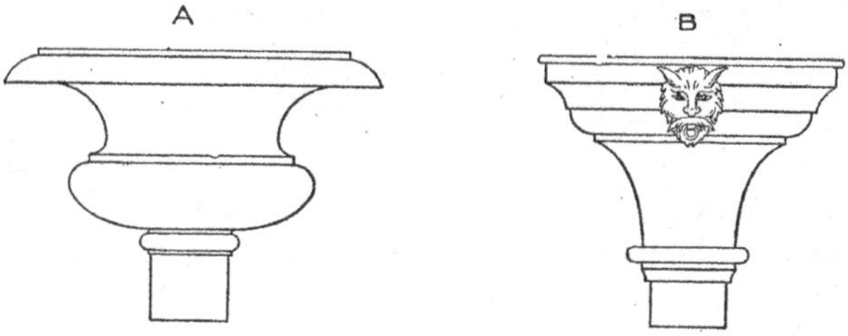

Fig. 346.

center x, with radii equal to $x\ w$ and $x\ v$, thus obtaining the blank d°.

By referring to the various rules given in previous problems, the true length of the blanks can be obtained.

The principles used for blanks hammered by hand can be applied to almost any form that will arise, as, for example, in the case shown in Fig. 346, in which A and B represent circular leader heads; or in that shown in Fig. 347, in which A and B show two styles of balusters, a and b (in both) representing the square tops and bases. Another example is that of a round finial, as in Fig. 348, A showing the hood which slips over the apex of the roof. While these forms can be bought, yet in some cases where a special design is brought out by the architect, it is necessary that they be made by hand, especially when but one is required.

The last problem on handwork is shown in Fig. 349—that of obtaining the blanks for the bottom of a circular bay. The curved moulding A will be hammered by hand or by machine, as will be explained later on, while the bottom B is the problem before us. The plan, it will be seen, is the arc of a circle; and, to obtain the various blanks, proceed as shown in Fig. 350, in which A B C is the elevation of the bottom of the bay, I J K being a plan view on A C, showing the

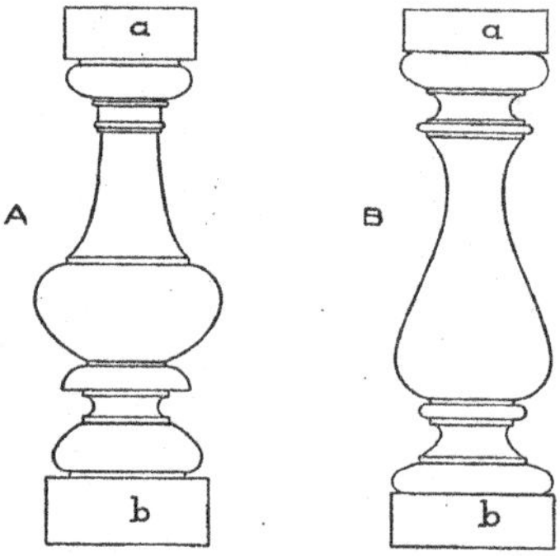

Fig. 347.

curve struck from the center H. In this case the front view of the bottom of the bay is given, and must have the shape indicated by A B C taken on the line I J in plan. It therefore becomes necessary to establish a true section on the center line S K in plan, from which to obtain the radii for the blanks or

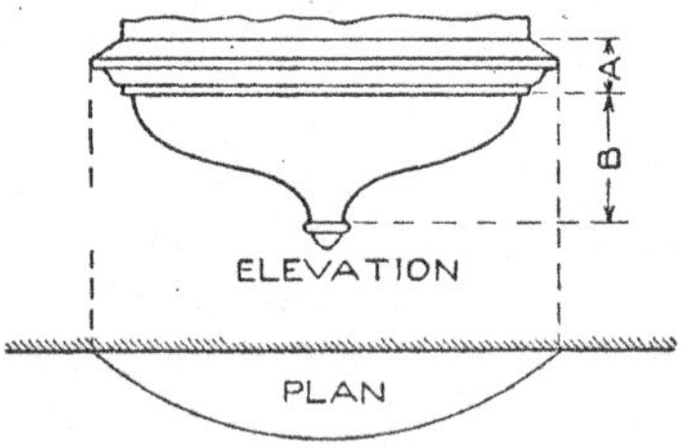

Fig. 349.

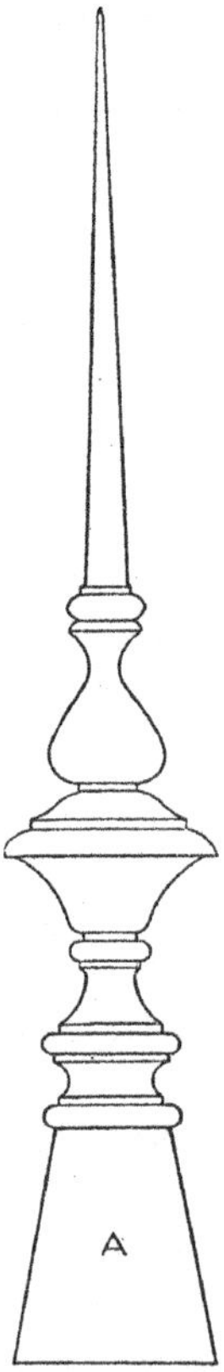

Fig. 348.

patterns. To obtain this true section, divide the curve A B into any number of equal parts, as shown from 1 to 6. From the points of division, at right angles to A C, drop lines as shown, intersecting the wall line I J at points 1′ to 6′. Then, using H as center, and radii equal to H 6′, H 5′, H 4′, H 3′, and H 2′, draw arcs crossing the center line D E shown from 1″ to 6″. At any convenient point opposite the front elevation draw any vertical line, as T U. Extend the lines from the spaces in the profile A B until they intersect the vertical line T U as shown. Now, measuring in every instance from the point S in plan, take the various distances to the num-

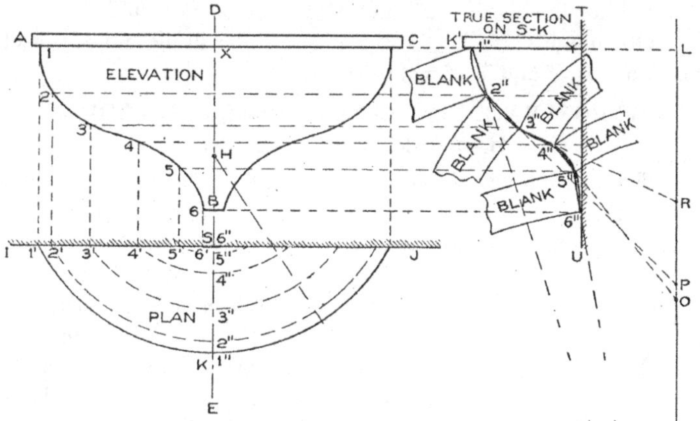

Fig. 350.

bered points in plan and place them upon lines of similar numbers, measuring in every instance from the line T U in section. Thus take the distance S K in plan, and place it as shown from the line T U to K^1; then again, take the distance from S to 2″ in plan, and place it as shown from the line T U to 2″ on line 2 in section. Proceed in this manner until all the points in the true section have been obtained. Trace a line as shown, when 1″ to 6″ to Y will be the true section on the line S K in plan.

It should be understood that the usual method for making the bottom of bays round in plan is to divide the profile of the moulding into such parts as can be best raised or stretched. Assuming that this has been done, take the distance from 1″ in plan to the center point H, and place it as shown from 1″ to L in section. From the point L, draw a vertical line L M, as shown. For the pattern for the mould 1″ 2″, average a line through the extreme points, as shown, and extend the same until it meets L M at N. Then, with N as center, and with radii equal to N 2″ and N 1″, describe the blank shown. The length of this blank is obtained by measuring on the arc 1′ 1″ in plan, and placing this stretchout on the arc 1″ of the blank. The other blanks are obtained in precisely the same manner. Thus P is the center for the blank 2″ 3″; R, for the blank 3″ 4″; O, for the blank 4″ 5″; and M, for the blank 5″ 6″.

The moulds 1″ 2″, 2″ 3″, and 3″ 4″ will be raised; while the blanks 4″ 5″ and 5″ 6″ will be stretched.

APPROXIMATE BLANKS FOR CURVED MOULDINGS HAMMERED BY MACHINE

The principles employed in averaging the profile for a moulding to be rolled or hammered by machine do not differ to any material extent from those used in the case of mouldings hammered by hand. Fig. 351 shows the general method of averaging the profile of a moulding in determining the radius of the blank or pattern. It will be seen that A B is drawn in such a manner, so to speak, as to average the inequalities of the profile D C required to be made. Thus distances *a* and *b* are equal, as are the distances *c* and *d*, and *e* and *f*. It is very difficult to indicate definite rules to be observed in drawing a line of this kind, or, in other words, in averaging the profile. Nothing short of actual experience and intimate knowledge of the material in which the moulding is to be made, will enable the operator

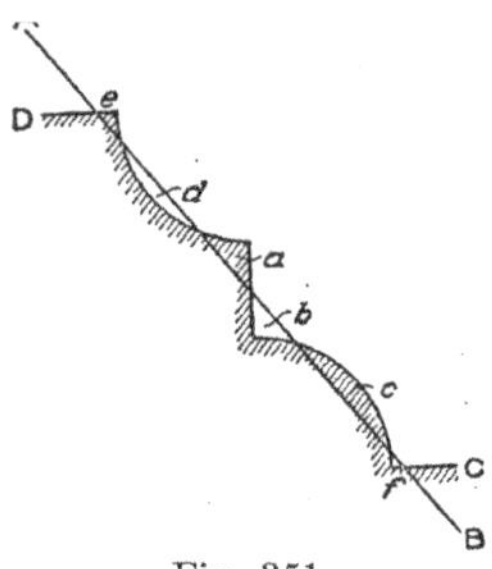

Fig. 351.

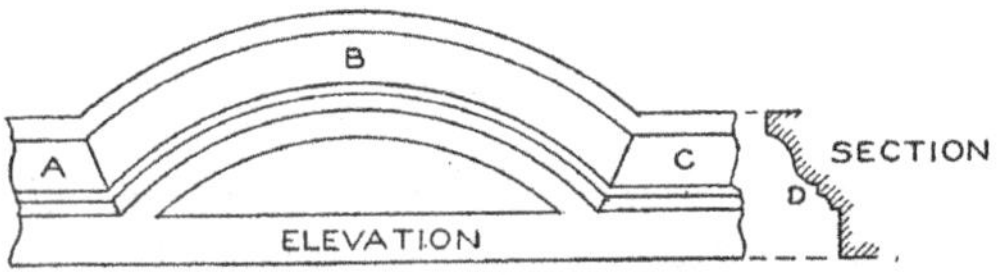

Fig. 352.

to decide correctly in all cases. There is, however, no danger of making very grave errors in this respect, because the capacity of the machines in use is such, that, were the pattern less advantageously planned in this particular than it should be, still, by passing it through the dies or rolls an extra time or two, it would be brought to the required shape.

In Fig. 352 is shown a part elevation of a circular moulding as it would occur in a segmental pediment, window cap, or other structure arising in sheet-metal cornice work. B shows the curved moulding, joining two horizontal pieces A and C, the true section of all the moulds being shown by D.

In this connection it may be proper to remark that in practice, no miters are cut on the circular blanks, the miter-cuts being placed on the horizontal pieces, and the circular moulding trimmed after it has been formed up.

In Fig. 353 is shown the method of obtaining the blanks for mouldings curved in elevation, no matter what their radius or profile

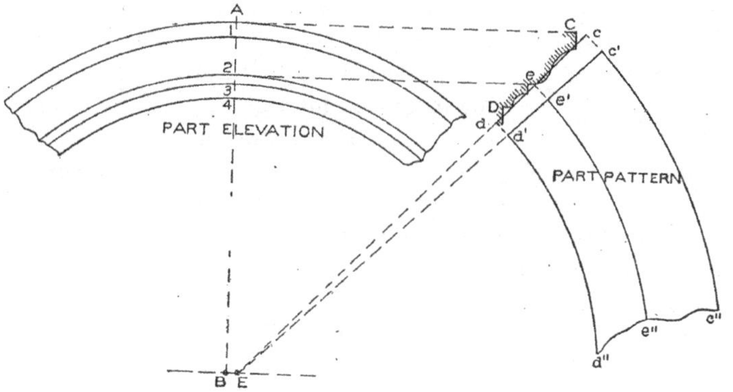

Fig. 353.

may be. First draw the center line A B, and, with the desired center, as B, describe the outer curve A. At right angles to A B, in its proper position, draw a section of the profile as shown by C D. From the various members in this section, project lines to the center line A B, as 1, 2, 3, and 4; and, using B as center, describe the various arcs and complete the elevation as shown by A B C in Fig. 352, only partly shown in Fig. 353. In the manner before described, average the profile C D by the line *c d*, extending it until it intersects the line drawn through the center B at right angles to A B, at E. Then E is the center from which to strike the pattern. Centrally on the section C D, establish *e* on the line *c d*, where it intersects the mould, and take the stretchout from *e* to C and from *e* to D, and place it as shown respectively from *e* to *c* and from *e* to *d* on the line *c d*. Now, using E as center, with radii equal to E d, E e, and E c, describe the arcs $d' \, d''$, $e' \, e''$, and $c' \, c''$. Draw a line from c' to E, intersecting the middle and inner arc at e' and d'. The arc $e' \, e''$ then becomes the measuring line to obtain the length of the pattern, the length being measured on the arc 2 in elevation, which corresponds to the point e in section.

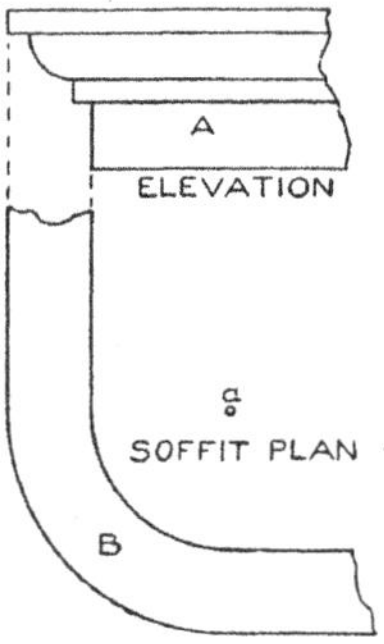

Fig. 354.

In Fig. 354 is shown the elevation of a moulding A curved in plan B, the arc being struck from the given point a. This is apt to occur when the moulding or cornice is placed on a building whose corner is round. To obtain the pattern when the moulding is curved in plan, proceed as shown in Fig. 355. Draw the section of the moulding, as A B, A C being the mould for which the pattern is desired. C B represents a straight strip which is attached to the mould after it is hammered or rolled to shape. In practice the elevation is not required. At pleasure, below the section, draw the horizontal line E D. From the extreme or outside edge of the mould, as b, drop a line intersecting the horizontal line E D at E. Knowing the radius of the arc on b in section, place it on the line E D, thus obtaining the point D. With D as center, describe the arc E F, intersecting a line drawn at right angle to E D from D. Average a line through the section, as G H, intersecting the line D F, drawn vertical from the center D, at J. Establish at pleasure the stationary point a, from which drop a line cutting E D at a'. Using D as center, and with D a' as radius, describe the arc $a' \, a''$, which is the measuring line when laying out the pattern. Now take the stretch-

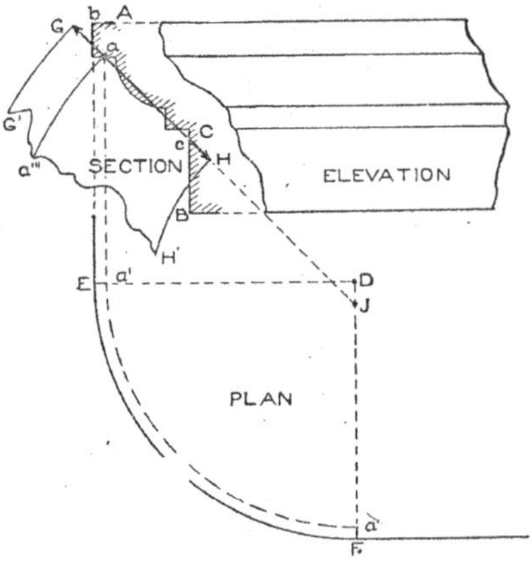

Fig. 355.

outs from a to b and from a to c, and place them on the averaged line from a to G and from a to H respectively. Using J as center, with radii extending to the various points G, a, and H, describe the arcs G G^1, a a''', and H H^1. On the arc a' a''', the pattern is measured to correspond to the arc a' a'' in plan.

In Fig. 356 is shown a front view of an ornamental bull's-eye window, showing the circular mould A B C D, which in this case we desire to lay out in one piece, so that, when hammered or rolled in the machine, it will have the desired diameter. The same principles can be applied to the upper mould E F, as were used in connection with Figs. 352 and 353.

Fig. 356.

To obtain the blank for the bull's-eye window shown in Fig. 356, proceed as shown in Fig. 357. Let A B C D represent the elevation of the bull's-eye struck from the center E. Through E draw the hori-

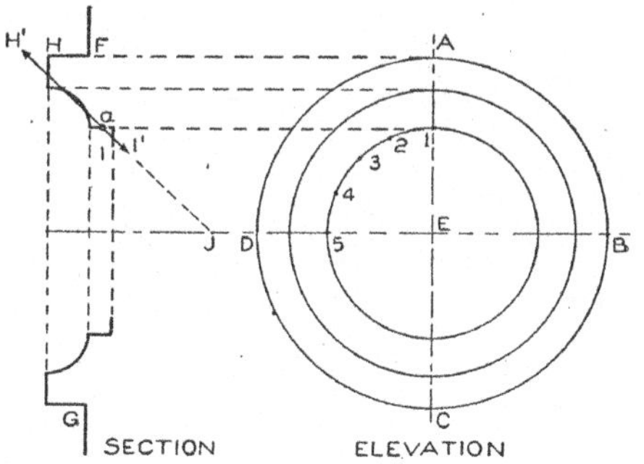

Fig. 357.

zontal and perpendicular lines shown. In its proper position, draw a section of the window as shown by F G. Through the face of the mould, as H I, average the line H^1 I^1, extending it until it intersects the center line B D at J. Where the average line intersects the mould at a, establish this as a stationary point; and take the stretchouts from a to I and from a to H, and lay them off on the line H^1 I^1 from a to I^1 and a to H^1 respectively. As 1 5 in elevation represents the quarter-circle on the point a in section, divide this quarter-circle into equal spaces, as shown. Now, with radii equal to J I^1, J a, and J H^1, and with J in Fig. 358 as center, describe the arcs H H, a a, and I I. From any point, as H, on one side, draw a line to J, intersecting the middle and inner arcs at a and I. Take the stretchout of the quarter-circle from 1 to 5 in elevation in Fig. 357, and place it on the arc a a as shown from 1 to 5. Step this off four times, as shown by 5′, 5″, and 5‴. From J draw a line through 5‴, intersecting the inner and outer arcs at I and H. Then will H a a H be the full pattern.

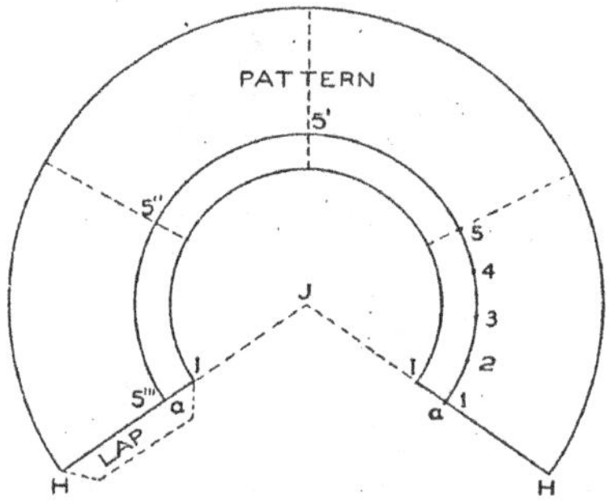

Fig. 358.

PRACTICAL PROBLEMS IN MENSURATION FOR SHEET METAL WORKERS.

A square tank, Fig. 1, is required whose capacity should be 200 gallons, the sides *b a* and *a c* each to be 30 inches; how high must *c d* be, so that the tank will hold the desired quantity?

Suppose the height *c d* is to be $51\frac{1}{3}$ inches, and the tank is to

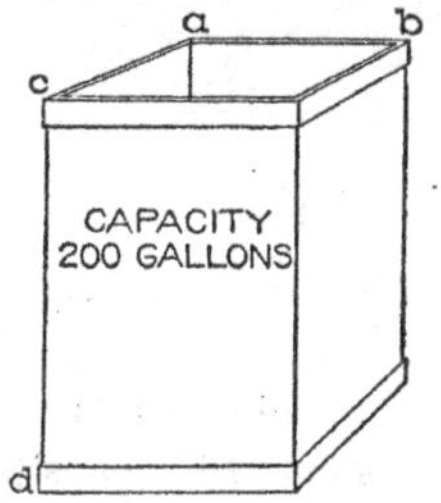

Fig. 1.

Fig. 2.

have similar capacity, and one side *c a* is to be 20 inches wide, how long must the alternate side *a b* be, so that the tank will hold 200 gallons?

A round tank, Fig. 2, is to be constructed whose capacity should equal 510 gallons, and be 5 feet high from *c* to *a*; what must its diameter *a b* be, so as to hold the desired capacity?

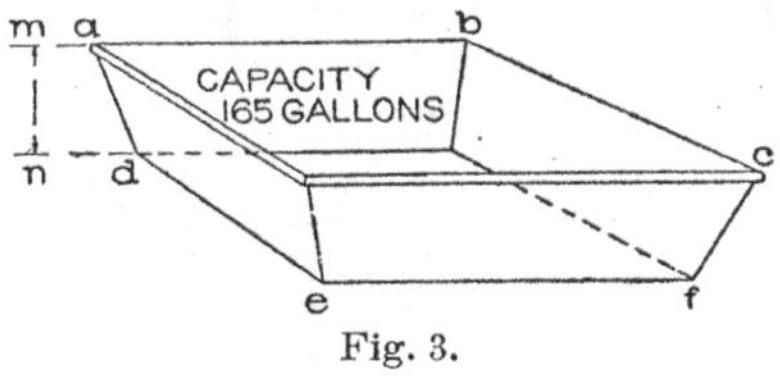

Fig. 3.

Suppose the diameter of the tank is to be 50 inches as *a b*; what must its height *a c* be, so that the tank will hold 510 gallons?

A large drip pan, Fig. 3, is to be constructed whose capacity should be 165 gallons, and whose top measurements *a b* and *b c* are 60 × 40 inches respectively, and bottom measurements *d e* and *e f* 34 × 54 inches respectively; what must its height *m n* be, so as to hold the desired volume?

A round tapering measure, Fig. 4, is to be constructed whose volume will equal 42 quarts; its bottom diameter *a b* is to be 14

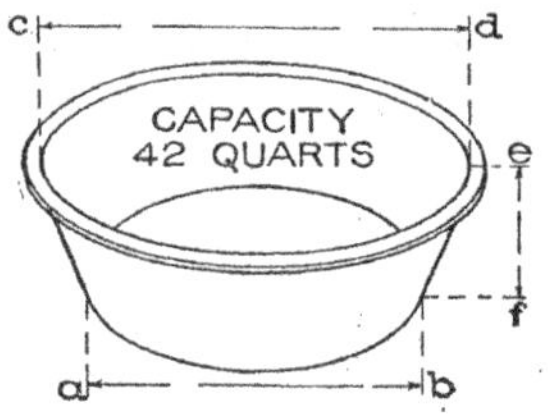

Fig. 4.

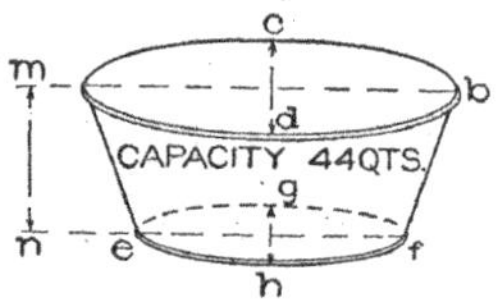

Fig. 5.

inches, its top diameter *c d* 18 inches; what must its height *e f* be to hold the desired quantity?

An elliptical tapering tank, Fig. 5, is to be constructed whose major axis *m b* is 24 inches, and minor axis *c d* 14 inches at the top, while at the bottom the major axis *e f* is 20 inches, and minor axis *g h* 10 inches; the capacity of the tank should equal 44 quarts; what must the height *m n* be, so that the tank will hold the desired amount?

A tank, Fig. 6, is to be constructed with semicircular ends

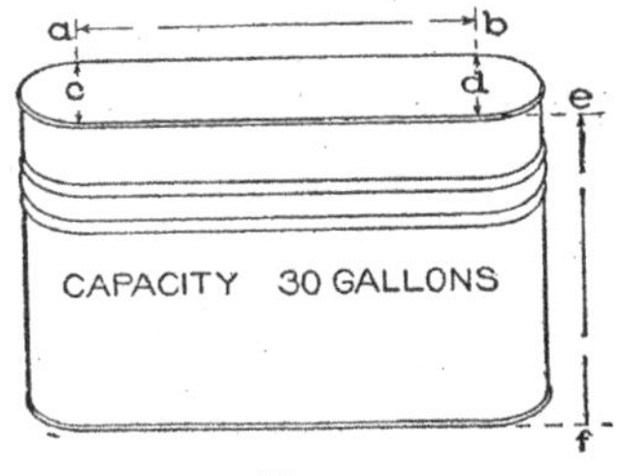

Fig. 6.

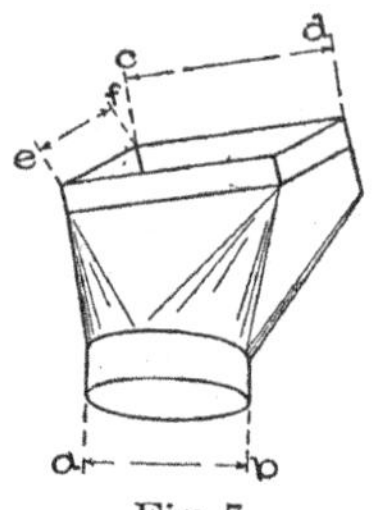

Fig. 7.

whose capacity should equal 30 gallons; the length *a b* to be 20 inches, and the diameters of *c* and *d* to be each 10 inches; what must the height *e f* be, so that the tank will hold the desired quantity?

Suppose the height *e f* is to be 24 inches, the diameters *c* and *d* each 11 inches; what must the length of *a b* be, so that the tank will hold 30 gallons?

In Fig. 7 is shown a fitting used in ventilation piping; the diameter *a b* is $11\frac{1}{4}$ inches and it is desired that the oblong pipe on the opposite end shall have an area similar to the round pipe *a b*; if *e f* must be 5 inches, what must *c d* be so that both areas are alike?

Suppose the pipe is to be square in place of oblong, what must the length of each side be, so that both ends have similar area?

In Fig. 8, *a b* is 40 inches in diameter; and each one of the branches *c*, *d*, and *e* are to have equal diameters, what must the diameter of the branches be, so that the combined area of *c*, *d*, and *e* will equal the area of *a b*?

If *c* is 10 inches in diameter, *d* 12 inches, and *e* 8 inches, what must be the diameter of *a b*, to have the combined area of the branches?

Fig. 9 shows a transition piece from a round pipe *a* to an

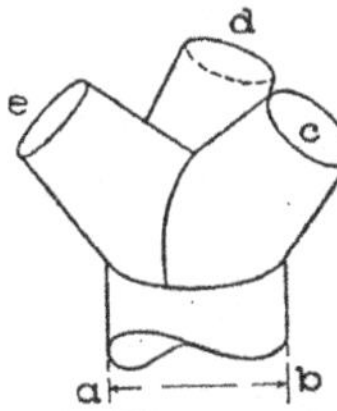

Fig. 8.

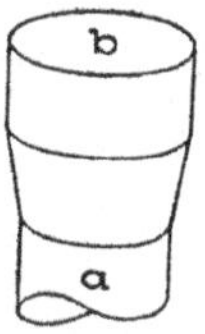

Fig. 9.

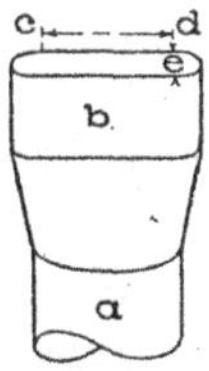

Fig. 10.

elliptical pipe *b*, both sections to have similar area; if the round pipe is 24 inches in diameter, and the major axis of the elliptical pipe must be 32 inches, what must the minor axis of *b* be so that the area at *b* will equal the area of *a*?

If the minor axis of *b* is to be 16 inches and the major axis 35 inches, what must the diameter of *a* be, so that both sections will have similar area?

In Fig. 10, *a* is 20 inches in diameter and forms a transition to an oblong pipe with semicircular end; the semicircular ends are to be 10 inches in diameter; what must the length of *c d* be, so that the area of *b* will be equal to the area of *a*?

If the pipe *b* measured 40×11 inches, having semicircular ends, what must the diameter of *a* be, so that both sections are equal in area?

If *a* is 20 inches in diameter and the upper section was to be rectangular in shape, 8 inches wide, what would the length of the upper section be?

Suppose the upper section *b* was desired to be square, what must the length of each side be, to have an area similar to *a*?

In Fig. 11 is shown the illustration of an ordinary steel square, and the method is given of obtaining accurate diameters of pipes, round or square, without any computation whatever, the rule being based on the geometrical principle that the square of the hypothenuse of a right angle triangle is equal to the sum of the squares of its base and altitude. To illustrate the rule, Fig. 12 has been

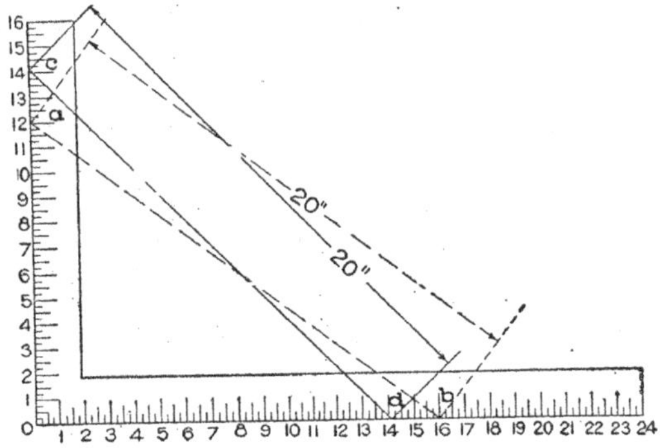

Fig. 11.

prepared. Let A represent a round or square pipe, 20 inches across, and B a round or square pipe 12 inches across; it is desired to take a branch from the main so that the two branches B and C will equal the area of the main A. What must the size of C be?

The size of C is found by simply taking a rule 20 inches long and placing one end on the arm of the square in Fig. 11, on the number 12, when the opposite end of the rule will touch the number 16. Then 16 is the required size of the branch C in Fig. 12. We can prove this by computation which, however, is not necessary in practice. The area of a 20-inch round pipe equals 314.16 in.; area of 12-in. pipe = 113.098 in.; area of 16-in. pipe = 201.062 in.; and 113.098 in. + 201.062 in. = 314.160 in. The area of a 20-in. square pipe = 400 in.; area of 12-in. square pipe = 144 in.; area of 16-in. square pipe = 256 in.; and 256 in. + 144 in. = 400 in.

Suppose any two branches are given as B and C in Fig. 12, what must the size of A be so that its area will have the combined area of the two branches ?

Simply set the rule on the numbers 12 and 16 on the two arms of the square respectively, and the length from *a* to *b* in Fig. 11 will measure 20 inches.

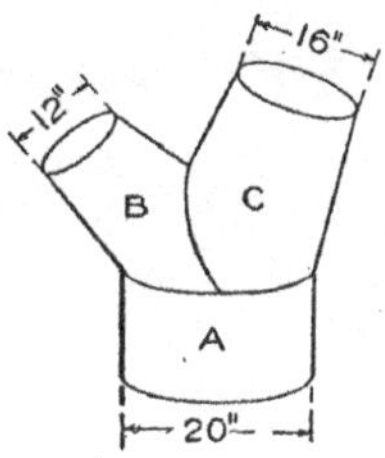

Fig. 12.

If A, Fig. 12, were given, and two branches were required, so that B and C were both of equal size, then simply set the rule 20 inches long, on both arms of the square so that the distance from O to *c* and O to *d* would be equal, as shown in Fig. 11, which would be found to measure $14\frac{1}{7}$ in. plus a least trifle. This rule can be used to advantage for any size round or square pipe in blower, blast, heat, and ventilating piping, saving time and trouble in computation. Where no square is at hand, one can be drawn on paper and used for work of this kind.

INDEX

Spotted something that isn't quite right?

Let us know and we'll fix it for future reprints. And as a thank you for your diligence, we'll send you a free Sothis book of your choice.

publisher@sothis.press

This page intentionally left blank

Except of course for that line above. And this one.
And whilst we have your attention, perhaps you'd
like to follow us on Instagram @sothis.press. Go on.

Made in the USA
Monee, IL
22 October 2025

32622228R10173